WITHDRAWN

**SOLIDIFICATION
PROCESSING**

# McGRAW-HILL SERIES IN MATERIALS SCIENCE AND ENGINEERING

### Editorial Board

MICHAEL B. BEVER
M. E. SHANK
CHARLES A. WERT
ROBERT F. MEHL, Honorary Senior Advisory Editor

AVITZUR: Metal Forming: Processes and Analysis
AZÁROFF: Introduction to Solids
BARRETT AND MASSALSKI: Structure of Metals
BLATT: Physics of Electronic Conduction in Solids
BRICK, GORDON, AND PHILLIPS: Structure and Properties of Alloys
BUERGER: Contemporary Crystallography
BUERGER: Introduction to Crystal Geometry
DE HOFF AND RHINES: Quantitative Microscopy
DRAUGLIS, GRETZ, AND JAFFEE: Molecular Processes on Solid Surfaces
ELLIOTT: Constitution of Binary Alloys, First Supplement
FLEMINGS: Solidification Processing
GILMAN: Micromechanics of Flow in Solids
GORDON: Principles of Phase Diagrams in Materials Systems
GUY: Introduction to Materials Science
HIRTH AND LOTHE: Theory of Dislocations
KANNINEN, ADLER, ROSENFIELD, AND JAFFEE: Inelastic Behavior of Solids
MILLS, ASCHER, AND JAFFEE: Critical Phenomena in Alloys, Magnets, and Super-conductors
MURR: Electron Optical Applications in Materials Science
PAUL AND WARSCHAUER: Solids under Pressure
ROSENFIELD, HAHN, BEMENT, AND JAFFEE: Dislocation Dynamics
ROSENQVIST: Principles of Extractive Metallurgy
RUDMAN, STRINGER, AND JAFFEE: Phase Stability in Metals and Alloys
SHEWMON: Diffusion in Solids
SHEWMON: Transformations in Metals
SHUNK: Constitution of Binary Alloys, Second Supplement
WERT AND THOMSON: Physics of Solids

McGRAW-HILL
BOOK COMPANY
New York
St. Louis
San Francisco
Düsseldorf
Johannesburg
Kuala Lumpur
London
Mexico
Montreal
New Delhi
Panama
Rio de Janeiro
Singapore
Sydney
Toronto

MERTON C. FLEMINGS
*Abex Professor of Metallurgy*
*Massachusetts Institute of Technology*

# Solidification Processing

This book was set in Times Roman.
The editors were B. J. Clark and Michael Gardner;
the production supervisor was Joan M. Oppenheimer.
The drawings were done by John Cordes, J & R Technical Services, Inc.
The printer and binder was The Maple Press Company.

To ELIZABETH

Library of Congress Cataloging in Publication Data

Flemings, Merton C     1929–
   Solidification processing.

  (McGraw-Hill series in materials science and engineering)
   Includes bibliographical references.
   1. Solidification.  2. Alloys.  I. Title.
TN690.F59      669′.9           73-4261
ISBN  0-07-021283-x

**SOLIDIFICATION
PROCESSING**

Copyright © 1974 by McGraw-Hill, Inc. All rights reserved.
Printed in the United States of America. No part of this publication may be reproduced,
stored in a retrieval system, or transmitted, in any form or by any means,
electronic, mechanical, photocopying, recording, or otherwise,
without the prior written permission of the publisher.

1234567890MAMM79876543

# CONTENTS

|   |   |   |
|---|---|---|
|   | **Preface** | ix |
| 1 | **Heat Flow in Solidification** | 1 |
|   | Growth of Single Crystals | 1 |
|   | Solidification of Castings and Ingots | 5 |
|   | Casting Processes Employing Insulating Molds | 6 |
|   | Casting Processes in which Interface Resistance is Dominant | 12 |
|   | Analytic Solutions for Ingot Casting | 17 |
|   | Solidification of Alloys | 21 |
|   | Problems in Multidimensional Heat Flow | 24 |
| 2 | **Plane Front Solidification of Single-phase Alloys** | 31 |
|   | Introduction | 31 |
|   | Equilibrium Solidification | 33 |
|   | No Solid Diffusion | 34 |
|   | Limited Liquid Diffusion, No Convection | 36 |
|   | Effect of Convection | 41 |
|   | Czochralski Growth (Crystal Pulling) | 44 |

|  |  |  |
|---|---|---|
|  | Zone Melting | 46 |
|  | Crystal Growth with Volatile Constituents | 49 |
|  | Diffusion in the Solid | 51 |
|  | The Facet Effect | 51 |
|  | Growth of Single Crystals of High Perfection | 53 |
| **3** | **Cellular Solidification** | **58** |
|  | Constitutional Supercooling and Cell Formation | 58 |
|  | Interface Stability Theory | 64 |
|  | Cell Structure | 66 |
|  | Formation of Dendrites | 73 |
|  | Cellular-Dendritic Transition | 75 |
|  | Solute Redistribution in Cellular Solidification | 77 |
|  | Cell Spacing | 83 |
|  | Solid-state Diffusion | 85 |
|  | Ternary Alloys | 87 |
| **4** | **Plane Front Solidification of Polyphase Alloys** | **93** |
|  | Introduction | 93 |
|  | Lamellar Eutectic Growth | 94 |
|  | Rod Eutectic Growth | 104 |
|  | Faceted—Nonfaceted Eutectic Growth | 105 |
|  | Interface Stability | 107 |
|  | Perturbation Analysis for Stability | 112 |
|  | Crystallographic Features of Directionally Solidified Eutectics | 114 |
|  | Nonsteady State Growth | 115 |
|  | Other Two-phase Structures | 117 |
|  | Ternary Alloys | 120 |
|  | Effect of Convection | 127 |
| **5** | **Solidification of Castings and Ingots** | **134** |
|  | Grain Structure | 134 |
|  | Columnar Structures | 135 |
|  | Microsegregation in Columnar Structures | 141 |
|  | Dendrite Arm Spacing | 146 |
|  | Equiaxed Grains | 154 |
|  | Faceted Dendrites and Preferred Growth Directions | 157 |
|  | Temperature and Fraction Solid | 160 |
|  | Solidification of Undercooled Melts | 167 |
|  | Structure of Undercooled Melts | 172 |

| | | |
|---|---|---|
| **6** | **Solidification of Polyphase Alloys: Castings and Ingots** | 177 |
| | Peritectic Solidification | 177 |
| | Eutectic Solidification | 180 |
| | Cast Irons | 183 |
| | Eutectic Solidification in Cast Irons | 187 |
| | Solute Redistribution in Ternary Alloys | 188 |
| | Monotectic Solidification | 191 |
| | Primary Inclusions | 193 |
| | Secondary Inclusions | 200 |
| | Dissolved Gas | 203 |
| | Gas Removed by Bubble Formation | 207 |
| | Formation of Gas Porosity on Solidification | 208 |
| **7** | **Fluid Flow** | 214 |
| | Introduction | 214 |
| | Gating | 215 |
| | Fluidity | 219 |
| | Convection in the Bulk Liquid | 224 |
| | Risering Pure Materials | 229 |
| | Interdendritic Fluid Flow | 234 |
| | Risering Alloys | 239 |
| | Macrosegregation | 244 |
| | Quantitative Aspects of Macrosegregation | 246 |
| | Movement of Liquid Plus Solid | 252 |
| **8** | **Thermodynamics of Solidification** | 263 |
| | Introduction | 263 |
| | Pure Materials | 264 |
| | Binary Alloys; Stable-phase Equilibrium | 267 |
| | The Partition Ratio | 272 |
| | Effect of Curvature | 273 |
| | Effect of Pressure | 274 |
| | Binary Alloys; Metastable-phase Equilibrium | 275 |
| | Composition at the Liquid-Solid Interface | 279 |
| | The Equilibrium Shapes of Phases | 284 |
| | Anisotropy of Liquid-Solid Surface Energy | 286 |
| **9** | **Nucleation and Interface Kinetics** | 290 |
| | Homogeneous Nucleation | 290 |
| | Heterogeneous Nucleation and Grain Refinement | 295 |

|    |                                                      |     |
|----|------------------------------------------------------|-----|
|    | Growth                                               | 301 |
|    | Continuous Growth                                    | 305 |
|    | Lateral Growth                                       | 308 |
|    | Growth by Two-dimensional Nucleation                 | 309 |
|    | Growth by Screw Dislocations                         | 312 |
|    | Growth by Propagation of Twin Planes                 | 318 |
|    | Growth Morphologies                                  | 319 |
| 10 | **Processing and Properties**                        | 328 |
|    | Homogenization                                       | 328 |
|    | Solution Treatment                                   | 331 |
|    | Malleable Iron                                       | 335 |
|    | Ferritic, Pearlitic, and Martensitic Irons           | 338 |
|    | Mechanical Properties of Equiaxed Cast Structures    | 341 |
|    | Properties of Columnar Structures                    | 344 |
|    | Aligned Composites                                   | 347 |
|    | Effects of Working                                   | 349 |
|    | **Appendixes**                                       | 356 |
| A  | Tabulation of Error Functions                        | 356 |
| B  | Tables of Approximate Thermal Data                   | 357 |
|    | **Index**                                            | 359 |

# PREFACE

This book has grown largely out of a lecture course given to senior-level and graduate students at Massachusetts Institute of Technology. It is intended for use in courses of this type, and also for the practicing engineer and research worker. The essential aim of the book is to treat the fundamentals of solidification processing and to relate these fundamentals to practice. Processes considered include crystal growing, shape casting, ingot casting, growth of composites, and splat-cooling.

The book builds on the fundamentals of heat flow, mass transport, and interface kinetics. Starting from these fundamentals, the basic similarities of the widely different solidification processes become evident. Problems at the end of each chapter relate principles to practice, illustrating important differences, as well as similarities between processes. Two years of college-level mathematics provides ample background for solving the problems given and for adequate comprehension of the text. In addition, it is desirable, though not necessary, that the student have a previous course in structure of materials. Emphasis of the book is on metallic alloys, but other materials are also considered.

An essential element of all solidification processes is heat flow. This subject is treated in the first chapter, primarily to lend cohesiveness to the material to follow. It provides an excellent basis for description and comparison of solidification processes, and it can be treated with rather simple assumptions regarding the solidification mechanism. Chapter 2 deals with mass transport ("solute redistribution") in single-crystal growth. A quantitative description of transport in this type of solidification is greatly simplified by the fact that the liquid-solid interface is single phase and planar. Equations derived in this chapter are also useful in describing dendritic solidification, except that they must be applied to tiny regions on the order of the dendrite arm spacing.

Chapter 3 deals with the important question of how to maintain a plane front in crystal growth, and of how solute redistribution occurs when the plane front breaks down to form "cells." Plane-front solidification is considered again in Chap. 4, this time for polyphase alloys, such as eutectics and off-eutectic "composites" solidified with an essentially planar liquid-solid interface. This chapter is the first to utilize the concept that the equilibrium melting point of a solid depends on its radius of curvature.

Solidification as it occurs in usual castings and ingots is considered in Chaps. 5 and 6. More specifically, these chapters consider the microscopic aspects of such solidification, including dendritic growth, microsegregation, inclusion formation, and gas-pore formation. They draw heavily on the heat- and mass-transport concepts presented in earlier chapters.

Fluid flow plays a larger role in solidification processes than is generally recognized. Flow is caused by introducing the metal to a mold, by density differences due to thermal or solute effects, or by solidification contractions. Fluid flow, treated in Chap. 7, has important effects on structure and segregation in solidification processes; many of these have been only recently recognized. An important part of this chapter deals with interdendritic fluid flow and its relation to porosity and segregation in castings and ingots.

The major portions of the first seven chapters, and all quantitative treatments in these chapters, assume equilibrium at the liquid-solid interface. That is, they assume that the kinetic driving force necessary to advance a solidifying interface, is negligibly small. This assumption is not valid when facets form, but it appears to be an excellent approximation for the many alloys that solidify without facets. Implications of this assumption are considered in Chap. 8, which deals with the thermodynamics of liquid-solid equilibria. A portion of this chapter also deals with what is possible (and impossible) at the liquid-solid interface when conditions are such that equilibrium is not maintained.

Kinetic effects at the liquid-solid interface, including nucleation, are discussed in Chap. 9. An understanding of growth kinetics, however qualitative, provides a basis for understanding the faceted growth morphologies observed in many real systems, and for understanding such solidification processes as growth of "ribbon crystals" by a twin-plane, reentrant growth mechanism. The final chapter deals with relations between the structure and properties of cast materials and with properties of wrought material produced from cast structures. An essential aim of many solidification processes is to obtain optimum properties in the resultant material. This chapter gives examples showing how the principles presented in earlier chapters can be utilized to produce structures with improved mechanical or physical properties.

The book draws heavily on research conducted over the last decade at Massachusetts Institute of Technology by students and associates of the author. A special note of thanks is due them. Critical comments and suggestions of John Cahn have been received and acted on with pleasure. The bulk of the book was written while the author was on sabbatical leave as Overseas Fellow at Churchill College, Cambridge University, England. He is grateful for the unique combination of stimulation and relaxation provided by that environment, and by his colleagues there.

MERTON C. FLEMINGS

# 1
# HEAT FLOW IN SOLIDIFICATION

## GROWTH OF SINGLE CRYSTALS

A variety of different techniques are employed to produce single crystals from melts. These can be grouped in three categories as those in which the entire charge is melted and then solidified from one end, a large charge is melted and a small crystal withdrawn slowly from it, and only a small *zone* of the crystal is melted at any one time. Figure 1-1 shows the methods schematically.

The first category of crystal-growing techniques is termed *normal freezing*. A commonly used normal freezing method for low-melting-point metals is growth in a horizontal *boat*. Here, a charge of metal is contained within a long crucible of small cross section open at the top. A *seed* crystal may be placed at one end of the boat to obtain a crystal of predetermined orientation. The charge and part of the seed are first melted in a suitable furnace. Next, the furnace is withdrawn slowly from the boat so that growth proceeds from the seed; alternatively, the boat is withdrawn slowly from the furnace and the solid-liquid interface moves until the whole charge is solid. In a similar crystal-growing method, the crucible is vertical and open at the top; this is often termed the *Bridgeman method*. In a minor modification of these techniques, neither the furnace nor the crucible moves. The charge is melted and

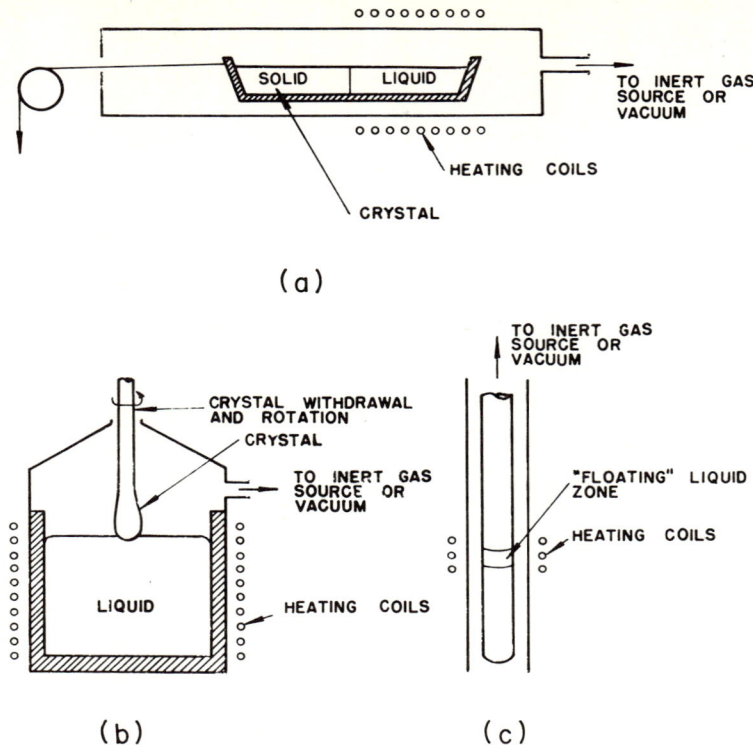

FIGURE 1-1
Examples of crystal-growing methods. (*a*) Boat method; (*b*) crystal pulling; (*c*) floating zone.

equilibrated in a furnace constructed so that one end of the furnace is substantially colder than the other end (*temperature-gradient* furnace). The temperature gradient is maintained constant in the furnace, and crystal growth is obtained by slowly lowering overall furnace temperature.

In growing single crystals, it is not necessary that the entire charge be molten. For some purposes, it is desirable to melt initially only a portion of the charge and move this molten *zone* slowly through the charge (*zone melting* and *zone freezing*). Many types of heat sources are used for zone melting, including induction, resistance, electron beam, and laser beam. The zone is moved either by mechanically moving the power source with respect to the crystal or vice versa. Zone melting is done either with or without crucible. The latter type, crucibleless zone melting, or *floating* zone melting, is widely used for reactive and high-melting-point materials. The molten zone is held in place by surface tension forces sometimes aided by a magnetic field.

Another single-crystal-growing technique, used widely for growing single crystals of silicon, germanium, and nonmetals, is the *crystal-pulling*, or *Czochralski*, technique. In this case, the charge material is placed in a crucible and melted. A seed crystal is attached to a vertical *pull* rod, lowered until it touches the melt, allowed to come to thermal equilibrium, and then raised slowly so that crystallization proceeds from the seed crystal. The crystal is rotated slowly as it is pulled, and crystal diameter is controlled by adjusting pull rate and/or heat input to the melt.

Many variations of these crystal-growing techniques are described in the literature. Crystals are generally grown in vacuum but may be grown in air or inert atmosphere. Highly volatile materials are encapsulated and grown under pressure. Crystals with volatile species as alloy elements are also encapsulated, or grown under flux. In one rather old process, the liquid is carried by a plasma arc as small droplets from an electrode (*Verneuil method*). These and other techniques for growth of specific materials have been described.[1]

The basic heat-flow objectives of all crystal-growing techniques are to (1) obtain a thermal gradient across a liquid-solid interface which can be held at equilibrium (e.g., stable with no interface movement) and (2) subsequently to alter or move this gradient in such a way that the liquid-solid interface moves at a controlled rate. A heat balance at a planar liquid-solid interface in crystal growth from the melt is written

$$K_s G_s - K_L G_L = \rho_s H R \qquad (1\text{-}1)$$

where  $K_s$ = thermal conductivity of solid metal, cal/(cm)(°C)(s)

$K_L$ = thermal conductivity of liquid metal, cal/(cm)(°C)(s)

$G_s$ = temperature gradient in solid at the liquid-solid interface, °C/cm

$G_L$ = temperature gradient in liquid at the liquid-solid interface, °C/cm

$R$ = growth velocity, cm/s

$\rho_s$ = density of solid metal, g/cm$^3$

$H$ = heat of fusion, cal/g

Note from Eq. (1-1) that growth velocity $R$ is dependent, not on absolute thermal gradient, but on the difference between $K_s G_s$ and $K_L G_L$. Hence, thermal gradients can be controlled independently of growth velocity. This is an important attribute of single-crystal-growing furnaces since growing good crystals of alloys requires that the temperature gradients be high and growth rate be low. $K_s$, $K_L$, $H$, and $\rho_s$ are constants of the materials being solidified; $G_L$ is directly proportional to the heat flux in the liquid at the liquid-solid interface.

**4** HEAT FLOW IN SOLIDIFICATION

Growth velocity would be at a maximum when $G_L$ becomes negative (undercooled melt); however, good crystals cannot be grown in undercooled liquids, and so the practical maximum growth velocity occurs when $G_L \to 0$, or from Eq. (1-1)

$$R_{max} = \frac{K_s G_s}{\rho_s H} \quad (1\text{-}2)$$

$G_s$, thermal gradient in the solid at the interface, is evaluated by experiment or heat-flow calculations. As a simple illustrative example of calculation of solid gradient $G_s$, consider the case of floating-zone (crucibleless) crystal growth in which (1) crystal is of circular cross section, (2) heat transfer from the crystal to surroundings is by convection, (3) growth is at steady state, and (4) temperature gradients within the crystal transverse to the growth direction are low. Consider a cylindrical element in the solid crystal $dx'$ in thickness, moving at the velocity $R$ of the liquid-solid interface, Fig. 1-2. Then, for steady state, the temperature of the moving element remains constant and a heat balance is written (for unit time)

$$\begin{array}{c}\text{Net heat change}\\ \text{from conduction}\end{array} + \begin{array}{c}\text{net heat change from}\\ \text{moving boundary}\end{array} + \begin{array}{c}\text{net heat change from}\\ \text{loss to surroundings}\end{array} = 0$$

$$\alpha_s \frac{d^2 T}{dx'^2}(\rho_s c_s \pi a^2 \, dx') - R \frac{dT}{dx'}(\rho_s c_s \pi a^2 \, dx') - h(T - T_0)(2\pi a \, dx') = 0 \quad (1\text{-}3)$$

where $x'$ = distance from liquid-solid interface (negative in solid), cm

$c_s$ = specific heat of solid metal, cal/(g)(°C)

$a$ = radius of crystal, cm

$h$ = heat transfer coefficient for heat loss to surrounding, cal/(cm²)(°C)(s)

$T$ = temperature at $x'$, °C

$T_0$ = ambient temperature, °C

$\rho_s$ = density of the solid crystal, g/cm³

$\alpha_s$ = thermal diffusivity of the solid crystal ($K_s/\rho_s c_s$), cm²/s

Now, integrating Eq. (1-3) with the boundary conditions that at $x' = 0$, $T = T_m$ ($T_m$ = melting point of metal), at $x' = -\infty$, $T = T_0$, the temperature in the solidifying metal is given by

$$\frac{T - T_0}{T_M - T_0} = \exp\left\{-\left[\frac{R}{2\alpha_s} - \sqrt{\left(\frac{R}{2\alpha_s}\right)^2 + \frac{2h}{aK_s}}\right]x'\right\} \quad (1\text{-}4)$$

The thermal gradient in the solid at the liquid-solid interface $G_s = (dT/dx')_{x'=0}$ is then

$$G_s = -(T_M - T_0)\left[\frac{R}{2\alpha_s} - \sqrt{\left(\frac{R}{2\alpha_s}\right)^2 + \frac{2h}{aK_s}}\right] \quad (1\text{-}5)$$

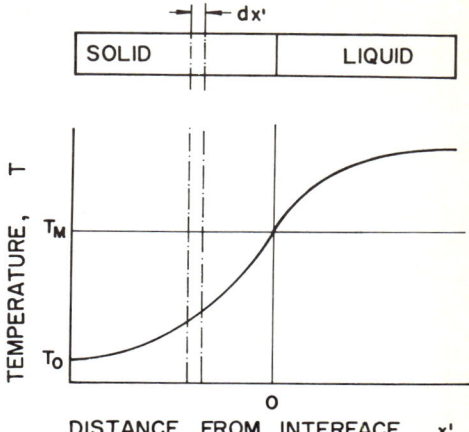

**FIGURE 1-2**
Temperature distribution in crystal growth (schematic).

and when $R/2\alpha_s \ll 1$

$$G_s \approx \left(\frac{2h}{aK_s}\right)^{1/2}(T_M - T_0) \qquad (1\text{-}6)$$

For crystals of high melting point, where $(T_M - T_0)$ is large and the coefficient of heat transfer $h$ is increased by radiation heat transfer, thermal gradients attainable are quite high, 100°C/cm or more. For lower-melting-point crystals, other cooling is necessary to attain steep gradients. As an example, Mollard achieved gradients of the order of 500°C/cm in $\frac{1}{8}$-in-diameter tin crystals by using a thin steel crucible, resistance-heating the crucible at just above the liquid-solid interface and water-cooling it just below the interface.[2] Equation (1-6) is equally applicable to this arrangement, with $h$ now representing the total resistance to radial heat flow from the metal crystal to the cooling water. For the arrangement employed, this resistance was primarily at the metal-crucible interface and was about 0.04 cal/(cm²)(°C)(s). In Prob. 1-1 at the end of this chapter we illustrate that calculated thermal gradients obtained using Eq. (1-6) are about those attained experimentally, 500°C/cm.

## SOLIDIFICATION OF CASTINGS AND INGOTS

In most casting and ingot-making processes, heat flow is not at steady state as in the above examples. Hot liquid is poured into a cold mold; specific heat and heat of fusion of the solidifying metal pass through a series of thermal resistances to the cold mold until solidification is complete. Figure 1-3 shows this process schematically for solidification of a pure metal. Thermal resistances which, in general, must be considered are those across the liquid, the solidifying metal, and the metal-mold interface

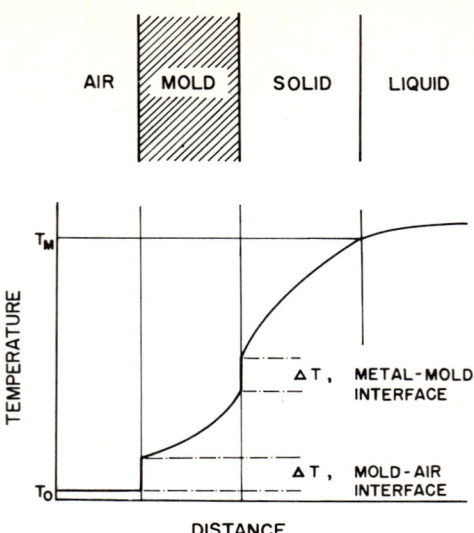

FIGURE 1-3
Temperature profile in solidification of a pure metal.

and those in the mold itself. The problem is mathematically and physically complex and becomes even more so when anything other than simple geometries are considered, when thermal properties are allowed to vary with temperature, or when alloys are considered. Problems such as these are now usually handled by computer methods, and some examples will be considered later. There are, however, certain simplifying approximations that can be made for a number of cases of engineering interest. Some of these will be examined before considering the general problem in further detail.

## CASTING PROCESSES EMPLOYING INSULATING MOLDS

*Sand casting* and *investment casting* are two processes for making shaped castings which employ relatively insulating molds. Both are very old processes, and both are important commercially today.[3]

Figure 1-4 illustrates the sand-casting process used to make a segment of a household radiator. Three sand-mold segments are made separately and assembled to produce a mold cavity the shape of the final casting described. Patterns of wood or metal are employed to make the proper impression in the upper and lower mold halves; the sand is *rammed* in place over the pattern. The outer mold segments (*cope* and *drag*) retain the shape of the impression because the sand used contains a few percent of water and clay and sometimes other binding agents. The internal segment

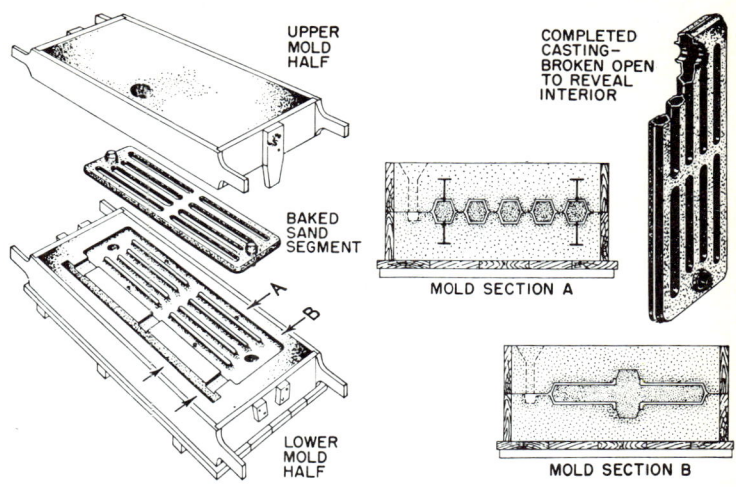

FIGURE 1-4
Sketch of sand-casting process as used in manufacture of a household radiator (*From Taylor, Flemings, and Wulff.*[4])

(*core*) is generally made of baked oil or resin-bonded sand to achieve greater strength and to reduce the amount of volatile components.

In *cope-and-drag* investment casting, mold pieces are made, not by ramming a dry or nearly dry sand mixture, but by pouring a slurry of investment material. One such material widely used for nonferrous alloys is plaster. For ferrous materials, a suitable material is mullite bonded with ethyl silicate.

In *lost-wax* casting, the pattern is originally made of wax and then *invested* with a suitable slurry which is subsequently baked at high temperature. During the high-temperature baking, the wax melts and drips out or volatilizes along with moisture in the mold. Two types of lost-wax casting are common. In the older process, the wax is placed in a box, or *can*, and a slurry poured to fill the box (Fig. 1-5). In the new *shell*-investment-casting process, the pattern is dipped successively in a slurry and then in a fluidized bed of fine particulate material until a shell of desired thickness is built up. The basic advantage of both types of lost-wax casting is that the process allows intricate parts to be made without regard to the problem of pattern removal from the mold. The major advantages of investment-casting processes as compared with sand castings are the greater complexity, thinner sections, and better dimensional accuracy and surface finish that can be obtained. The major disadvantage is their usually greater cost and size limitation.

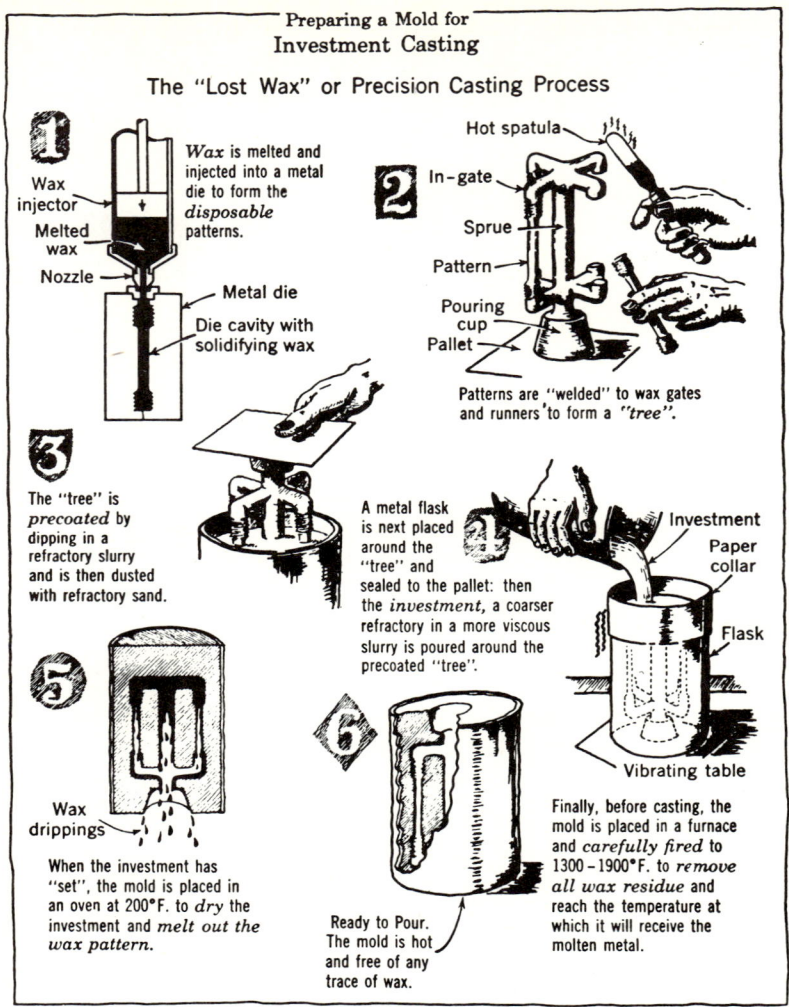

FIGURE 1-5
Preparing a mold for investment casting. (*From Taylor, Flemings, and Wulff.*[4])

From a heat-flow standpoint, the important characteristic of solidification of a metal in processes such as those discussed above is that the metal is a much better conductor of heat than the mold. Thus, solidification rate depends primarily on thermal properties of the mold. The thermal conductivity of the metal has practically no influence. Also, except in relatively heavy-section shell investment castings, the mold can be considered to be *semi-infinite* in extent; i.e., the outside of the mold does not heat up during solidification. The heat-flow problem sketched in Fig. 1-3 is now very much simplified, especially if we assume further that the metal is poured with no *superheat*, that is, exactly at its melting point $T_M$, as shown in Fig. 1-6.

Consider first the problem of unidirectional heat flow. Metal is poured exactly at its melting point against a thick, flat mold wall initially at room temperature $T_0$. Thus, the mold surface is heated suddenly to $T_M$ at time $t = 0$. This is a transient, one-dimensional heat-flow problem, and the solution must conform with the partial differential equation

$$\frac{\partial T}{\partial t} = \alpha_m \frac{\partial^2 T}{\partial x^2} \tag{1-7}$$

where
$\alpha_m$ = thermal diffusivity of mold, cm²/s
$K_m$ = thermal conductivity of mold, cal/(cm)(°C)(s)
$\rho_m$ = density of mold, g/cm³
$t$ = time, s
$x$ = distance from mold wall, cm (negative into the mold)

The solution to this equation for the boundary conditions stated above gives the temperature $T$ in the mold as a function of time $t$ at distance from the mold surface $x$:

$$\frac{T - T_M}{T_0 - T_M} = \operatorname{erf} \frac{-x}{2\sqrt{\alpha_m t}} \tag{1-8}$$

where erf denotes the error function. The error function of zero is zero, and the error function of infinity is unity. A list of tabulated error functions is given in Appendix A.

The rate of heat flow into the mold at the mold-metal interface is given by

$$\left(\frac{q}{A}\right)_{x=0} = -K_m \left(\frac{\partial T}{\partial x}\right)_{x=0} \tag{1-9}$$

where $x$ increases positively from left to right in Fig. 1-6, $q$ is rate of heat flow, and $A$ is area of the mold-metal interface. By partial differentiation of Eq. (1-8) with respect to $x$, letting $x = 0$ and combining the results with Eq. (1-9), the rate of heat flow across the mold-metal interface is seen to be

$$\left(\frac{q}{A}\right)_{x=0} = -\sqrt{\frac{K_m \rho_m c_m}{\pi t}} (T_M - T_0) \tag{1-10}$$

where $c_m$ is specific heat of the mold material. Now, the heat entering the mold comes only from heat of fusion of the solidifying metal since the solid as well as the liquid metal is exactly $T_M$ (Fig. 1-6). Thus,

$$\left(\frac{q}{A}\right)_{x=0} = -\rho_s H \frac{\partial S}{\partial t} \tag{1-11}$$

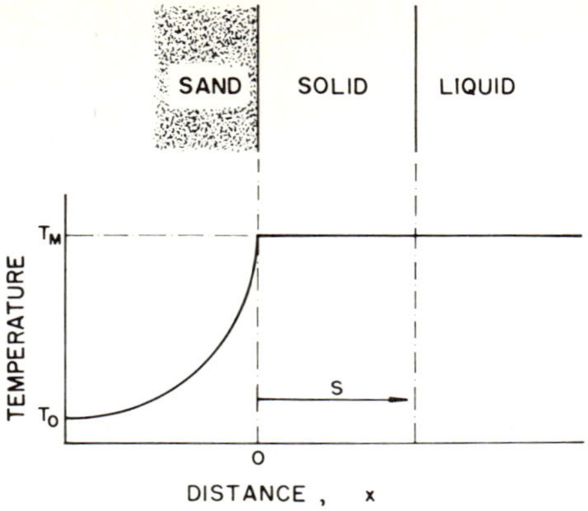

**FIGURE 1-6**
Approximate temperature profile in solidification of a pure metal poured at its melting point against a flat, smooth mold wall.

where $S$ = thickness solidified. Combining (10) and (11) and integrating from $S = 0$ at $t = 0$,

$$S = \frac{2}{\sqrt{\pi}} \underbrace{\left(\frac{T_M - T_0}{\rho_s H}\right)}_{\text{Metal}} \underbrace{\sqrt{K_m \rho_m c_m}}_{\text{Mold}} \sqrt{t} \qquad (1\text{-}12)$$

Equation (1-12) predicts the manner in which thermal properties of the metal and the mold combine to determine the freezing rate of a metal cast into a relatively insulating mold. The relationship is more accurate for sand castings of high conductivity such as nonferrous metals (Cu-, Al-, and Mg-base alloys) than for iron and steel. Note that high melting temperature and low heat of fusion (on a volume basis) favor rapid solidification. The product $K_m \rho_m c_m$ is a measure of the rate at which the mold can absorb heat and is sometimes called the *heat diffusivity*. The thickness of solid metal is a parabolic function of time, which means the solidification rate is initially very rapid and decreases as the mold becomes heated. Figure 1-7 shows the wide range of solidification rates attained in practice, depending on metal and mold and mold temperature. Data used in calculation of this figure are given in Appendix B.

The one-dimensional freezing problem serves to illustrate many of the important thermal aspects of solidification, but for some purposes it is important to evaluate freezing times and rates of complex shapes. Consider again the question of heat flow into a mold wall; the contour of the mold wall has some influence on its ability to absorb heat. For example, the geometry of heat flow into a concave or convex contoured mold wall may be compared with that into a plane mold wall. Heat flow into the concave surface will be divergent and therefore slightly more rapid, and into the convex surface less rapid than into a plane wall. For simple shapes, however, the

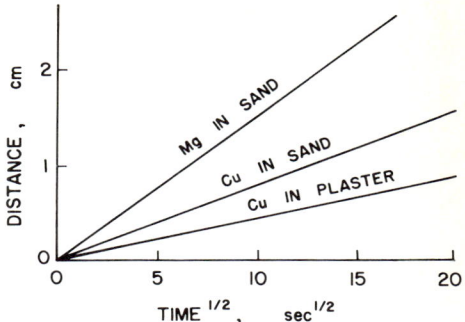

FIGURE 1-7
Distance solidified versus square root of time for several pure metals in insulating molds.

differences will not be large and a useful approximation is to assume that a given square centimeter of mold surface has a fixed ability to absorb heat regardless of its contour or location on the casting. With this approximation, we may now replace $S$ in Eq. (1-12) with $V_s/A$, where $V_s$ is volume solidified at time $t$ and $A$ is area of the mold-metal interface. Or, letting $t = t_f$, where $t_f$ is the total solidification time of a casting of volume $V$,

$$\frac{V}{A} = \frac{2}{\sqrt{\pi}} \left( \frac{T_M - T_0}{\rho_s H} \right) \sqrt{K_m \rho_m c_m} \sqrt{t_f} \qquad (1\text{-}13)$$

and

$$t_f = C \left( \frac{V}{A} \right)^2 \qquad (1\text{-}14)$$

where $C$ is a constant for a given metal-mold material and mold temperature.

Equation (1-14) is the well-known *Chvorinov's rule* used to compare solidification times of simple-shaped castings. It states that the total time of solidification of such castings is proportional to the square of the volume-to-area ratio of the castings. Experimental confirmation of this result is seen in original experiments of Chvorinov on steel castings of different sizes and shapes, varying from 10 mm thick to a 65-ton casting (Fig. 1-8). Chvorinov's times of complete solidification were obtained from thermocouple measurement at the thermal center of the casting.[5]

For shapes such as spheres or cylinders, it is possible to derive a more exact expression than Eq. (1-13) relating $t_f$ to $V/A$ without retaining the assumption of nondivergency of heat flow. In these cases, the applicable partial differential equation for heat flow in the mold is

$$\frac{\partial T}{\partial t} = \alpha_m \left( \frac{\partial^2 T}{\partial r^2} + \frac{n}{r} \frac{\partial T}{\partial r} \right) \qquad (1\text{-}15)$$

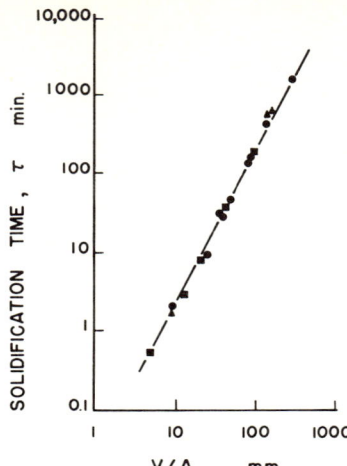

FIGURE 1-8
Chvorinov's experimental results on solidification time of castings versus their volume-to-area ratio.[5]

where $r$ = casting radius and $n = 1$ for cylinder, 2 for sphere. Following a procedure identical to that used to derive Eq. (1-13), the resulting equivalent expression is

$$\frac{V}{A} = \frac{T_M - T_0}{\rho_s H}\left(\frac{2}{\sqrt{\pi}}\sqrt{K_m \rho_m c_m}\sqrt{t_f} + \frac{nK_m t_f}{2r}\right) \qquad (1\text{-}16)$$

By comparison of (1-16) with (1-13) it may be seen that the simple Chvorinov approximation becomes increasingly valid as thermal diffusivity $K_m/\rho_m c_m$ decreases. It is also more nearly valid for cylinders than for spheres. For a given volume-to-surface-area ratio, a sphere freezes more rapidly than a cylinder and a cylinder more rapidly than a plate.

## CASTING PROCESSES IN WHICH INTERFACE RESISTANCE IS DOMINANT

In a large number of important casting processes, heat flow is controlled to significant extent by resistance at the mold-metal interface. These processes include the permanent-mold-casting, die-casting, splat-cooling, and powder manufacturing processes to be discussed below. When the mold-metal interface resistance is of overriding importance, temperature distribution across the solidifying metal and mold is as in Fig. 1-9. All temperature drop is across the interface. The mold, being assumed infinite in extent, remains at its original temperature $T_0$. Rate of heat flow across this interface

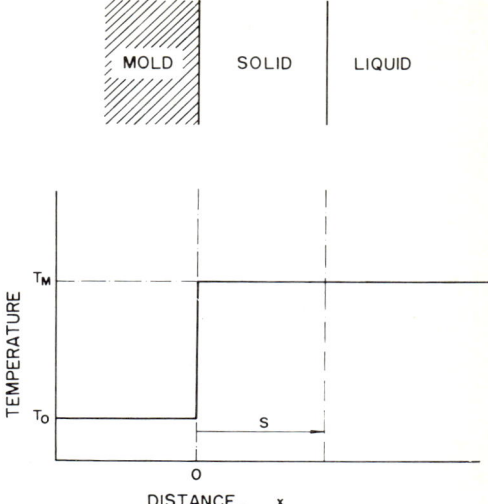

FIGURE 1-9
Temperature profile during solidification against a large flat mold wall with mold-metal interface resistance controlling.

for metal poured at its melting point $T_M$ is

$$\left(\frac{q}{A}\right)_{x=0} = -h(T_M - T_0) \qquad (1\text{-}17)$$

Combining (1-17) with (1-11) for the case of a large flat mold wall and integrating from $S = 0$ at $t = 0$ yields

$$S = h\frac{T_M - T_0}{\rho_s H} t \qquad (1\text{-}18)$$

Furthermore, since shape in no way alters the heat transfer across the interface, Eq. (1-18) may be generalized for simple-shaped castings to calculate the solidification time $t_f$ in terms of the volume-to-area ratio of the casting:

$$t_f = \frac{\rho_s H}{h(T_M - T_0)} \frac{V}{A} \qquad (1\text{-}19)$$

Equations (1-18) and (1-19) are valid when the resistance to heat flow across the mold-metal interface is large compared with other resistances in the metal and mold. Except when the mold is relatively insulating, this condition pertains when

$$h \ll \frac{K_s}{S} \qquad (1\text{-}20)$$

When the mold is relatively insulating, there is the added necessary condition that

$$h^2 \ll \frac{K_m \rho_m C_m}{t} \qquad (1\text{-}21)$$

*Die casting* is the most economical of all casting processes for production of large quantities of simple nonferrous parts. It is a high-first-cost–high-production-rate process, with production rates between 100 and 400 shots per hour and up to 40 castings in a single shot (40-impression die). Figure 1-10 is a schematic diagram of one type of die-casting machine used for die-casting aluminum alloys. In highly mechanized foundries, the hand-ladling method shown is replaced by a metal pump or automatic metal ladle. A typical mold-metal interface resistance for die casting is about 0.1 cal/(cm²)(°C)(s) or less, and it is readily seen from Eq. (1-20) that interface resistance controls overall heat-flow rate for the common metals over the range of thicknesses usually cast (generally 0.3 cm thick or less).

Today, large quantities of aluminum- and zinc-base alloys are die-cast, with smaller amounts of magnesium- and copper-base alloys being made. It is thought by some that magnesium die casting will become more competitive with aluminum primarily because it has a lower volumetric heat of fusion; and therefore, from Eq. (1-19), it should be possible to achieve higher production rates with properly designed machines. There is a small but growing production of copper-base alloys by die casting. This production will expand rapidly if mold life and shot-chamber life can be increased. With equipment and materials now used, the high temperature of copper-base alloys causes relatively rapid failure of these components. There has been much interest in die casting of ferrous alloys because of the potential economies of producing large quantities of simple parts. It seems likely that a commercial process will eventually result, but the thermal and materials problems here are even greater than in the case of copper-base alloys.

In a newer, quite different type of die casting termed *low-pressure die casting*, metal is forced directly from a crucible through a refractory tube and into a mold cavity by low-pressure air or inert gas. In this case, the ambient pressure is raised around the crucible itself. Low-pressure die casting is used for moderately rapid production of relatively large aluminum castings that might otherwise be sand-cast or permanent-mold cast. An essentially similar process is used for casting steel in graphite molds for manufacture, for example, of steel wheels and ingots.

*Permanent-mold casting* is generally similar to die casting except that pressure is not employed to fill the mold. Metal is fed directly into the mold under gravity, Fig. 1-11. Generally, the process is employed for larger, heavier section castings where large quantities are desired. Examples include outdoor steel light reflectors and fire-alarm boxes. *Washes* (ceramic slurry coatings) are often used in permanent-mold castings to intentionally increase the heat-transfer coefficient to make it easier to fill

CASTING PROCESSES IN WHICH INTERFACE RESISTANCE IS DOMINANT  15

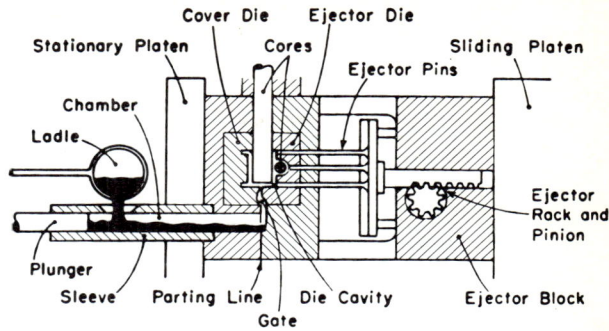

(a) METAL LADLED INTO CHAMBER.

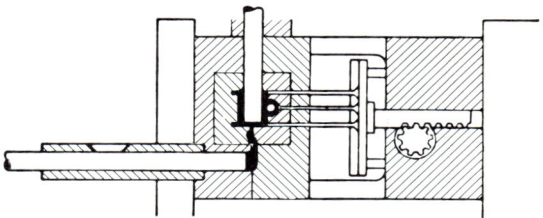

(b) METAL FORCED INTO DIE CAVITY.

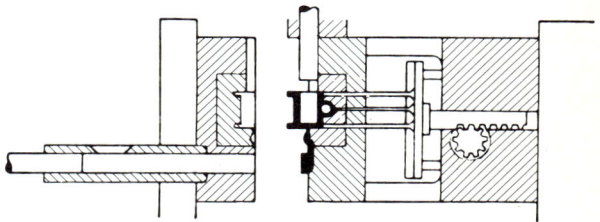

(c) DIE OPENED, CORES WITHDRAWN.

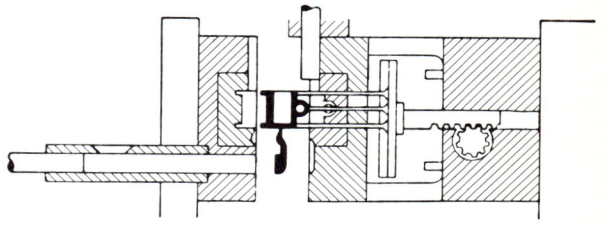

(d) CASTING AND EXCESS METAL EJECTED.

FIGURE 1-10
Die casting, cold chamber machine. (*From Taylor, Flemings, and Wulff.*[4])

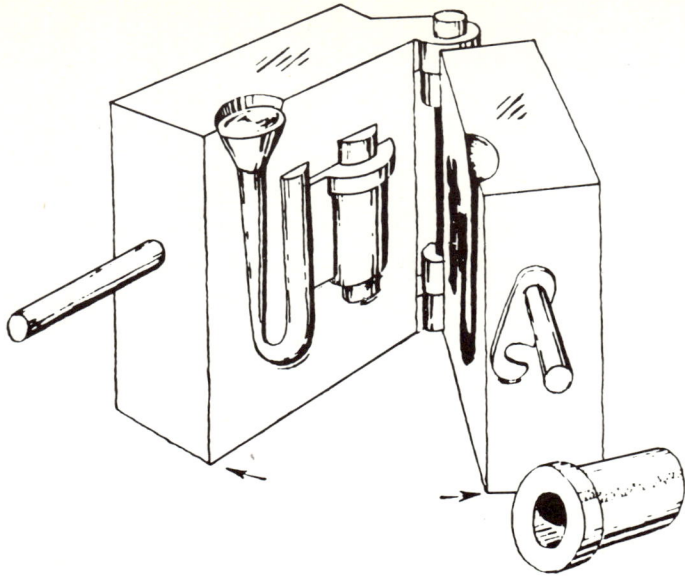

FIGURE 1-11
Permanent-mold casting; a solidified casting is shown in the left half of the mold and another in the foreground. (*From Taylor, Flemings, and Wulff.*[4])

the mold or to help reduce solidification-shrinkage cavities in the casting. Heat-transfer coefficients are in the range from 0.04 cal/(cm$^2$)(°C)(s) (for simple thin graphite-base washes) to much lower values for insulating washes.

*Splat cooling* is another example of a casting process in which heat flow is generally interface-limited. This process was developed by Duwez in 1959 as a method of obtaining very high cooling rates during solidification.[6] In his original apparatus, a small droplet of liquid metal is ejected by a shock wave onto a metallic *ski slide*; samples as thin as 1 μm are obtained. In later *hammer-and-anvil* devices, somewhat thicker specimens are usually obtained (20 to 50 μm). Metallurgical aspects of this process are discussed in Chap. 8. From the heat-flow standpoint, it is of interest to note here that the thinness of the sample cast means [from Eq. (1-20)] that heat flow may be interface-limited even though solidification rate is very high. Strachan[7] has found $h$ coefficients in one hammer-and-anvil apparatus in excess of unity, and higher coefficients are probably obtained in the shock-tube apparatus.

As a final example of $h$-controlled solidification, consider the manufacture of metal-alloy powders by atomization of the liquid alloys and cooling in air, inert atmosphere, hydrogen, or water. Here, $h$ coefficients are the order of $10^{-3}$ for air or inert atmosphere and $10^{-2}$ for water. Droplet size is generally in the range of a few microns to as much as a centimeter. From Eq. (1-20) heat flow is again clearly interface-limited, and Eq. (1-19) again applies. There is much current interest in obtaining atomized metal-alloy powders that are cooled at maximum rate and thus are solidified with maximum $h$. For gas quenching, $h$ is influenced by extent of convection and by type of gas employed: it is also somewhat dependent on particle size. For higher-melting-point metals such as steel, heat transfer from the droplet occurs by radiation as well as by convection; the radiation can be viewed as contributing to an "effective" interface heat-transfer coefficient $h_e$, which is then strongly dependent on temperature.[8]

## ANALYTIC SOLUTIONS FOR INGOT CASTING

For many important types of ingot-casting processes, the thermal profile sketched in Fig. 1-3 applies and thermal resistance of the metal, mold-metal interface, mold, and mold surroundings must be considered for complete solution. Exact analytic solutions are available for the limiting cases of engineering interest where heat flow is one-dimensional, mold-metal interface resistance is negligible, and the mold either is held at constant temperature (as by water cooling) or is very thick. Temperature distributions for these cases are sketched in Fig. 1-12.

Consider, first, the case of the water-cooled chill. The solution to this problem must conform with the applicable partial differential equation [Eq. (1-7)], with the boundary conditions that at $x = 0$, $T = T_0$, at $x = S$, $T = T_M$, and at the liquid-solid interface

$$K_s \left( \frac{\partial T}{\partial x} \right)_{x=S} = H\rho_s \frac{\partial S}{\partial T} \quad (1\text{-}22)$$

The solution, given by Carslaw and Jaeger,[9] is

$$S = 2\gamma \sqrt{\alpha_s t} \quad (1\text{-}23)$$

$\gamma$ is determined from

$$\gamma e^{\gamma^2} \operatorname{erf} \gamma = (T_M - T_0) \frac{C_s}{H\sqrt{\pi}} \quad (1\text{-}24)$$

The temperature distribution in the solidifying metal is given by

$$\frac{T - T_0}{T_0' - T_0} = \operatorname{erf} \frac{x}{2\sqrt{\alpha_s t}} \quad (1\text{-}25)$$

where $T_0'$ = integration constant.

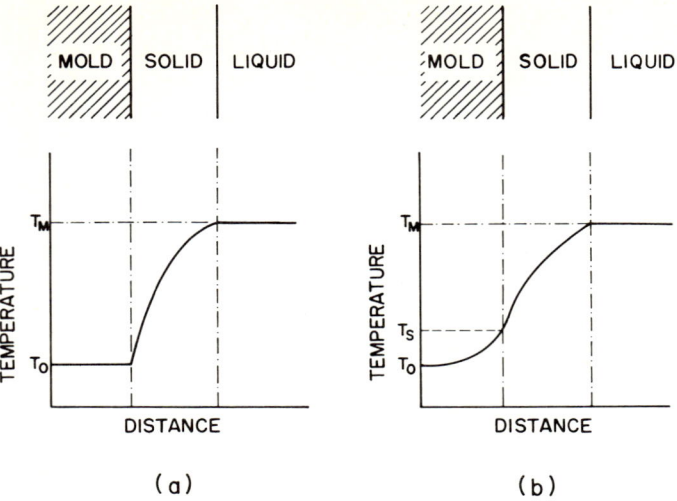

FIGURE 1-12
Temperature profile: solidification against a flat mold wall when (a) resistance of solidifying metal is controlling and when (b) combined resistances of metal and mold are controlling.

The solution when the mold is not water-cooled (but is semi-infinite) is only a little more complicated. It is

$$\gamma e^{\gamma^2} \left( \sqrt{\frac{K_s \rho_s C_s}{K_m \rho_m C_m}} + \text{erf } \gamma \right) = \frac{C_s}{H\sqrt{\pi}} (T_M - T_0) \quad (1\text{-}26)$$

Equation (1-25) must now be rewritten as, replacing $T_0$ by the temperature of the mold-metal interface $T_S$,

$$\frac{T - T_S}{T'_0 - T_S} = \text{erf } \frac{x}{2\sqrt{\alpha_s t}} \quad (1\text{-}27)$$

and Eq. (1-23) applies for this case as before. Thus, Eqs. (1-23), (1-26), and (1-27) constitute the solution to the heat-flow problem for unidirectional solidification in a non-water-cooled metal mold. These equations, as well as Eq. (1-25) for the water-cooled chill and Eq. (1-12) for the insulating mold, represent special cases of Schwartz' solution[10] (for zero superheat). Note that for both the water-cooled and non-water-cooled chill, solidification is parabolic with time. This result has been observed experimentally by a large number of investigators studying rate of solidification in steel ingots. Some typical calculated results, employing thermal data of Appendix B, are shown in Fig. 1-13. These results agree well with experimental measurements

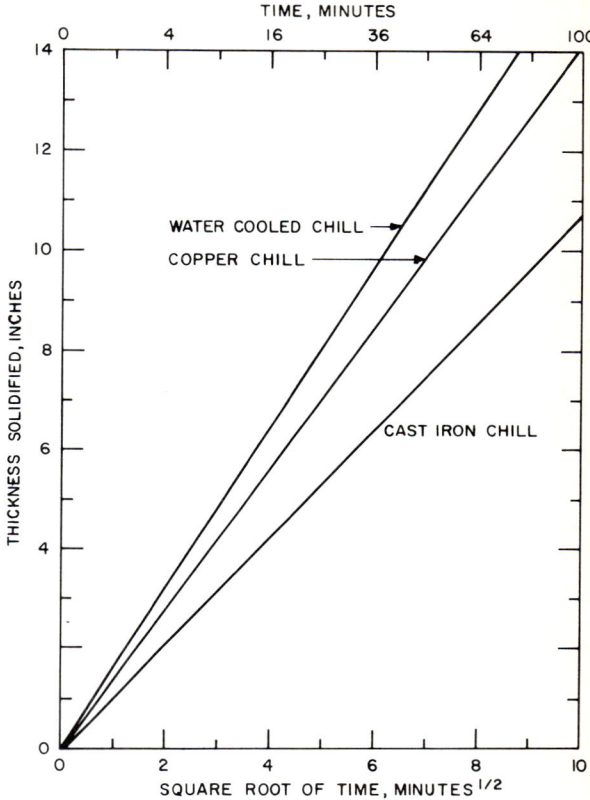

**FIGURE 1-13**
Unidirectional solidification of pure iron against three types of chills. No mold-metal interface resistance.[26]

made of unidirectional solidification against metal chill walls except that the experimental curves obtained are generally displaced slightly on the time axis.

As an example, many investigators[11-13] have obtained experimental results for steel solidification against cast-iron-mold walls which are of the form

$$S = a\sqrt{t} - b \quad (1\text{-}28)$$

where the constants $a$ and $b$ are approximately unity and 0.5, respectively, when the units are inches and minutes (as in Fig. 1-13). The apparent delay in beginning of solidification that is found experimentally results from two causes. First, convection during and just after pouring results in rapid removal of superheat from the liquid, thus slowing the start of solidification and its initial rate. The second effect is finite

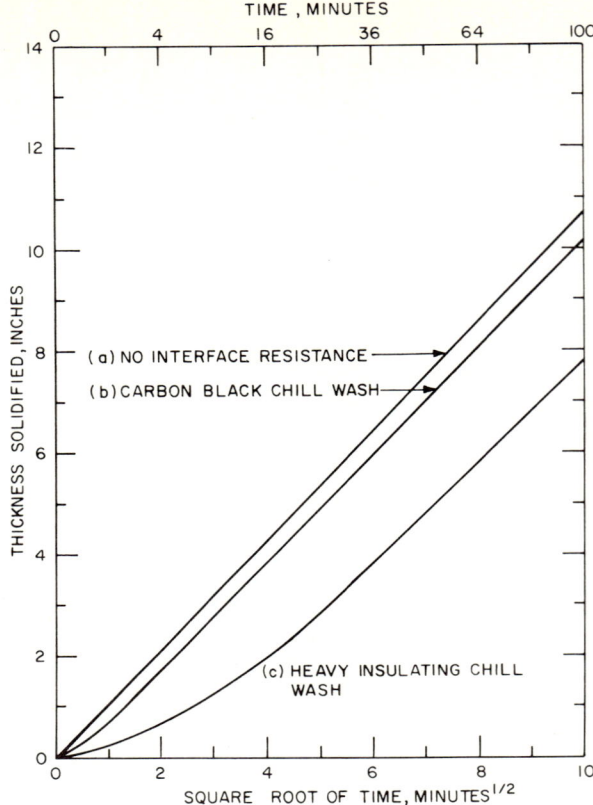

**FIGURE 1-14**
Unidirectional solidification of pure iron against a cast-iron chill. (*a*) No interface resistance; (*b*) carbon-black chill wash [$h$ assumed to be 0.04 cal/(cm$^2$)(°C)(s)]; (*c*) thick insulating wash [$h$ assumed to be 0.004 cal/(cm$^2$)(°C)(s)].[26]

mold-metal resistance to heat transfer. In this case, solidification rate is actually finite at zero time but is interface-controlled and so proceeds at a slower rate than in absence of this resistance. Initially, the amount solidified increases linearly with time, as given by Eq. (1-18). Later, as resistance of the solidifying metal becomes large compared with mold-metal resistance, the amount solidified increases linearly with the square root of time as shown in Fig. 1-14.

Approximate analytical solutions for treating this problem have been given by several investigators and compared by Jones.[14] Figure 1-14 shows results for pure iron with no superheat using Adams' method[15] and values of the mold-metal heat-transfer coefficient that are in the range of those encountered in practice. Note that in the case of iron the interface resistances have relatively little effect on solidification behavior except very near the chill. Similar interface resistances have a relatively large effect with high conductivity metals such as aluminum since the distance from the chill at which they will remain of importance depends on the ratio of the interface resistance to the metal thermal conductivity.

In all the solutions given in the foregoing paragraphs, metal superheat has been neglected. Exact analytical solutions are available to account for this superheat in simple one-dimensional solidification (in absence of interface resistance) providing thermal properties are independent of temperature and thermal convection is absent. However, it is clear that even very small horizontal temperature gradients result in vigorous convection in liquid metals and, therefore, that the superheat is dissipated early in solidification. Experimentally also it is well established that superheat is nearly or completely eliminated in usual ingot- and shaped-casting processes very soon after pouring. In numerical solutions to be discussed below, the effect of convection is sometimes approximately accounted for by assuming an artificially high liquid thermal conductivity (10 to 100 times that of the true thermal conductivity). In sand castings or in cases of interface-controlled solidification the effect of superheat on the time to complete solidification can be approximately accounted for by adding the heat content due to superheat to that of the heat of fusion.

## SOLIDIFICATION OF ALLOYS

Most engineering alloys solidify over a range of temperatures rather than at a discrete melting point $T_M$, and the temperature profile of such an alloy solidifying unidirectionally is as shown in Fig. 1-15. One analytic solution for the case is available, assuming semi-infinite metal and mold, no interface resistance, constant thermal properties, and heat of fusion distributed evenly over the solidification range.[9] Other methods which may be used and which place less restrictions on the problems that can be solved are approximate analytic techniques, integral profile methods, and the simple but powerful technique of numerical solution of the basic differential equations with appropriate boundary conditions. These methods have been described.[15-18]

Figure 1-16 gives results (from Campagna[19]) of the last-mentioned technique as applied to unidirectional solidification of an Al–4.5% Cu alloy. In the absence of interface resistance and superheat, a fully solid "skin" forms immediately, and both the liquidus and solidus isotherms then move linearly with the square root of time. The velocity of the liquidus isotherm is not affected by approaching the upper end of the ingot, but the velocity of the solidus isotherm is significantly altered by this end effect. The velocity begins to increase shortly after the liquidus isotherm reaches the end.

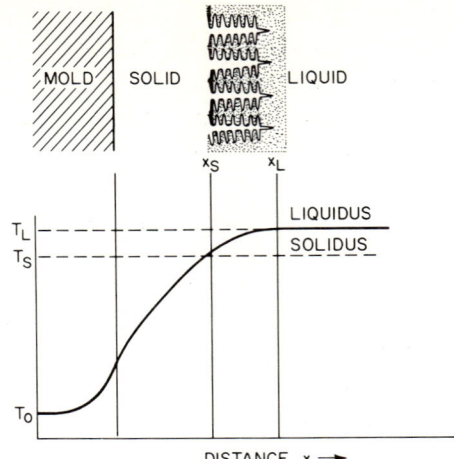

**FIGURE 1-15**
Unidirectional solidification of an alloy against a flat mold wall.

The presence of interface resistance (bottom of Fig. 1-16) alters behavior of the isotherms only in the vicinity of the chill face. Here, initial rate of isotherm movement is slowed, and there is a finite delay time before movement of the solidus isotherm begins at all. Some metallurgically important aspects of curves such as those of Fig. 1-16 are the following:

1. The vertical distance between the liquidus and solidus curves represents the region of the ingot over which solid and liquid coexist at a given time during solidification. This distance is the width of the *mushy zone*, and casting characteristics such as feeding, hot tearing, and macrosegregation are strongly influenced by the width of this zone.
2. The horizontal distance between these curves is a measure of the *local solidification time* $t_f$ (that is, the time required for a given fixed location to go from the liquidus temperature to the solidus temperature). Local solidification time is inversely proportional to average cooling rate at a given location during solidification. Important aspects of the solidification structure (including dendrite arm spacing and inclusion size) depend strongly on this local solidification time.
3. In the case of finite $h$ coefficient, liquid remains in contact with the mold surface a finite time during solidification. This is a necessary condition for formation of certain types of macrosegregation including inverse segregation and exudation.

When alloys are cast in sand or other relatively insulating molds with no chilling, temperature differences in the solidifying metal are always small and may be very

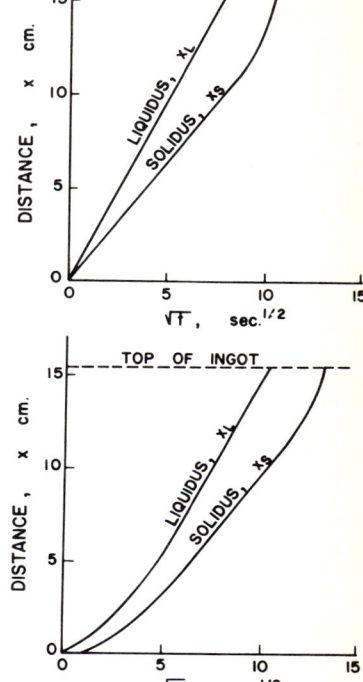

FIGURE 1-16
Progress of liquidus and solidus isotherms in unidirectional solidification of Al–4.5% Cu alloy; no superheat, water-cooled chill. *Top*: $h = \infty$; *bottom*: $h = 0.04$ cal/(cm$^2$)(°C)(s).[7]

much smaller than the solidification range itself. As example, it has been shown by a number of investigators, as summarized by Ruddle,[20] that solidification of commercial aluminum alloys in sand molds is such that liquid and solid coexist throughout the entire casting during most of the solidification process. Moreover, there is only a very small difference in fraction solid between the center and surface of the casting at any time during solidification. This type of solidification is discussed further in a later chapter. From a heat-flow standpoint, the major differences between this case and that of a pure metal solidifying in a sand mold are that mold-interface temperature is not constant during solidification but decreases over the temperature range of solidification, and that the specific heat of the liquid and solid in the solidification range must be extracted to complete solidification. When the range of solidification is small, solidification times of such alloys are closely given by the equations developed earlier for pure metals.

**24** HEAT FLOW IN SOLIDIFICATION

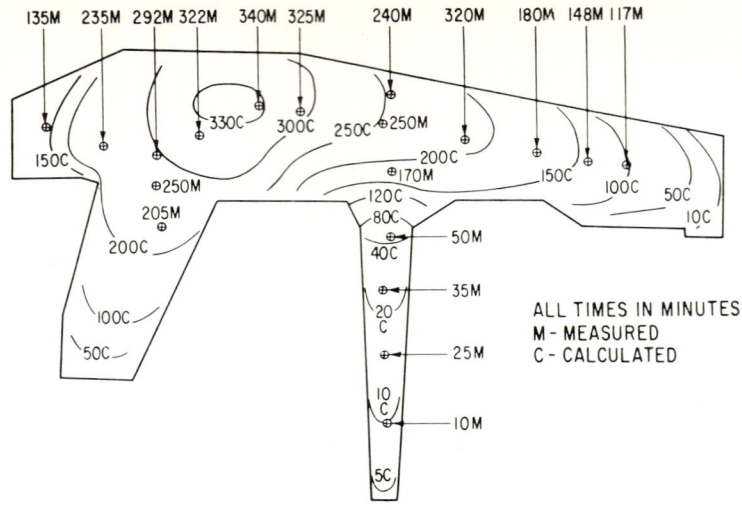

**FIGURE 1-17**
Comparison of calculated end of freeze waves with measured data. (*From Henzel and Keverian.*[21])

## PROBLEMS IN MULTIDIMENSIONAL HEAT FLOW

Except for several simple approximate treatments, all analyses given above have been of one-dimensional heat flow, or of problems with radial symmetry. Two examples are given here of application of heat-flow theory to solution of problems of engineering interest of more complex heat-flow geometry. In each case, the solution is numerical, by computer. The first example is from Henzel and Keverian of movement of isotherms in a 20,000-lb steel-turbine inner-shell sand casting.[21] In calculations, the mold is assumed sufficiently thick that its surface does not heat up significantly and temperature drop at the mold-metal interface is small. Heat of fusion is accounted for by assuming an artificially high specific heat over the range of solidification (i.e., within the liquid-solid "mushy" zone). Now, the only equation to be solved (in both mold and metal) is

$$\frac{\partial T}{\partial t} = \alpha \nabla^2 T \qquad (1\text{-}29)$$

where $\alpha$ is thermal diffusivity which is allowed to vary as function of position and time. Figure 1-17 shows a cross section through one wall of the turbine casting with calculated isotherms and measured times of completion of solidification at various

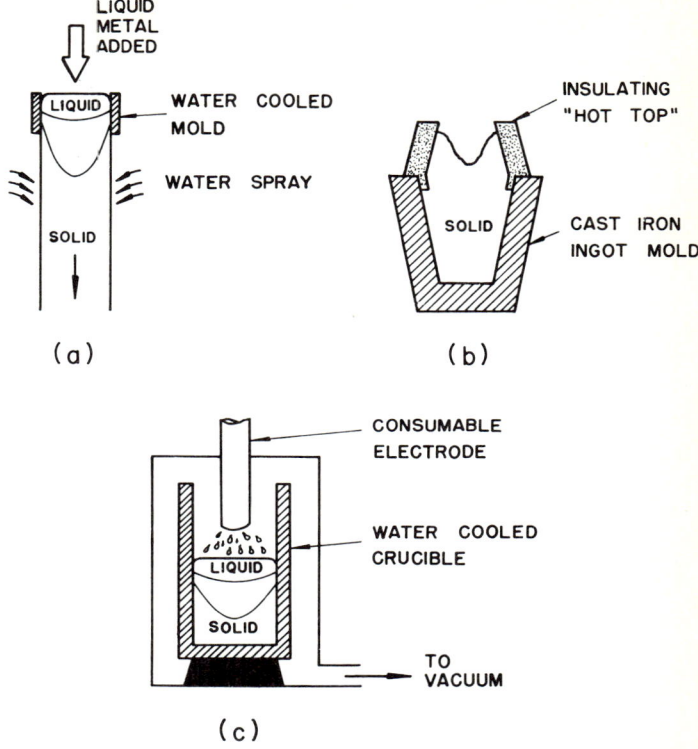

FIGURE 1-18
Three ingot casting processes: (*a*) Continuous casting; (*b*) static casting; (*c*) consumable-electrode casting.

locations. Note that the last point to solidify is within the wall section, and so some centerline solidification shrinkage is expected at this point.

Figure 1-18 is a sketch of the three most widely used ingot-casting processes. At the upper right is the traditional static cast ingot with the *riser*, or *hot top*, to maintain a reservoir of liquid metal to feed solidification shrinkage. At the upper left is the continuous-casting process (when it is run continuously) or *direct-chill* process (when it is run semicontinuously). There are many modifications of this process for making continuous castings ranging from less than an inch in diameter to several feet, for making plates and slabs, and even for making hollow shapes. Figure 1-18 also shows the *consumable-electrode* ingot-making process in which metal is melted from an electrode into a copper water-cooled crucible by electric arc. Another ingot-making technique of growing importance is the electroslag remelting process. This process is schematically similar to the consumable-electrode process except that it is run in air with a conducting slag layer between the ingot and the electrode. Extensive work on computer simulation and on experimental study of solidification can be found in the literature for each of these processes; we consider here as an example only the case of the continuous, or direct-chill, casting.

The basic heat-flow equation for the continuous process for fixed coordinates and steady state is

$$\frac{\partial T}{\partial t} = 0 = \alpha \nabla^2 T - R \frac{\partial T}{\partial z} \quad (1\text{-}30)$$

where $z$ is the distance along the long axis of the ingot. The second term on the right accounts for heat transferred by movement downward of the ingot at velocity $R$, and $\partial T/\partial t$ is zero at fixed location and steady state. Solution of the heat-flow problem is complicated by the fact that the external boundary conditions (heat-transfer coefficient at the ingot surface) vary with location. Near the upper part of the mold, the solid is in contact with the mold, and the interface coefficient $h$ is high. Subsequently, it pulls away, and $h$ drops markedly. Below the mold, water strikes the metal directly; and the heat-transfer coefficient is again high but variable depending on spray characteristics, flow rate, and metal temperature. At still lower positions, $h$ is often intentionally reduced to minimize ingot cracking problems (by having no water in these locations).

Certain simplifications to the heat-flow problem in continuous casting are often possible. As example, at sufficiently high casting velocity $V$, heat flow in the $z$ direction becomes negligible with respect to that in the directions of the ingot thickness and width ($x$ and $y$ directions). Also, for a slab casting ($x \ll y$), heat flow in the $y$ direction is negligible. Thus, the problem is reduced to one of simple unidirectional heat flow. Mizikar[22] has treated slab casting of steel in this way; his results are plotted in Fig. 1-19a, assuming a casting speed of 60 in/min. Note that the liquid-solid zone extends to 50 ft below the mold. The vertical distance between the liquidus and solidus at the centerline is nearly 40 ft. This length is the order of those typically obtained in ferrous continuous casting. It is a very long mushy zone indeed and is a potential source of macrosegregation, centerline shrinkage, and liquid-metal breakouts.

Neglecting longitudinal heat flow is a reasonable approximation for ferrous continuous casting as usually practiced commercially.[23] In nonferrous casting, however, at the speeds typically employed, mushy zones are much shallower; and heat flow in this direction must be considered. Figure 1-19b shows an example of calculated liquidus and solidus lines for a commercial magnesium-alloy continuous casting solidified at a casting speed of 2 in/min.[24] These results were also obtained by numerical solution of Eq. (1-30) with approximate boundary conditions. Qualitatively similar liquid pool depths and mushy-zone shapes have been reported for aluminum-alloy continuous castings.[20,25]

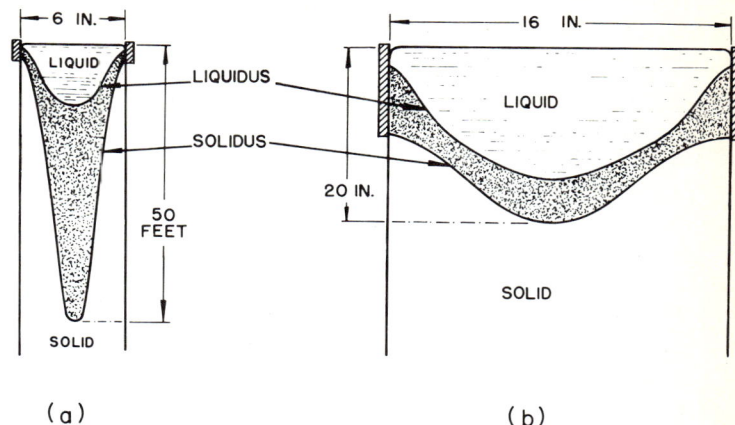

**FIGURE 1-19**
Calculated liquidus and solidus isotherms in continuous casting. (*a*) Steel slab casting speed 60 in/min (*after Mizikar*[22]); (*b*) magnesium alloy, cylindrical ingot, casting speed 2 in/min (*after Adenis et al.*[24]).

# REFERENCES

*1* LAUDICE, R. A.: Techniques of Crystal Growth, in H. S. Reiner (ed.), "Crystal Growth," Pergamon Press, New York, 1967.
*2* MOLLARD, R. F., and FLEMINGS, M. C.: *Trans. AIME*, **239**:1534 (1967).
*3* SIMPSON, B.: "Development of the Metal Castings Industry," American Foundrymen's Assoc., Chicago, Ill., 1948.
*4* TAYLOR, H. F., FLEMINGS, M. C., and WULFF, J.: "Foundry Engineering," John Wiley & Sons, Inc., New York, 1959.
*5* CHVORINOV, N.: *Giesserei*, **27**:177 (1940).
*6* DUWEZ, P.: *Trans. ASM*, **60**:607 (1968).
*7* STRACHAN, R.: Doctoral Dissertation, Department of Metallurgy and Materials Science, M.I.T., Cambridge, Mass., 1967.
*8* PERRY, R. (ed.): "Chemical Engineers Handbook," McGraw-Hill Book Company, New York, 1950.
*9* CARSLAW, H. S., and JAEGER, J. C.: "Conduction of Heat in Solids," 2d ed., Oxford University Press, London,
*10* SCHWARTZ, C.: *Z. Angew. Math. Mech.*, **13**:202 (1933).
*11* CHIPMAN, J., and FONDERSMITH, C. R.: *Trans. AIME*, **125**:370 (1937).
*12* SPRETNAK, J. W.: *Trans. ASM*, **39**:569 (1947).
*13* MARBURG, E.: *Trans. AIME*, **197**:157 and 1553 (1953).
*14* JONES, H.: *J. Inst. Metals*, **97**:38 (1969).
*15* ADAMS, C. M.: "Liquid Metals and Solidification," p. 187, American Society for Metals, Cleveland, Ohio, 1958.
*16* MUEHLBANER, J. C., and SUNDERLAND, J. E.: *Appl. Mech. Rev.*, **18**:951 (1965).

17 BANKOFF, S. G.: *Advan. Chem. C.*, **5**:75 (1964).
18 HILLS, A. W. D.: *Trans. AIME*, **245**:1471 (1969).
19 CAMPAGNA, A.: Sc.D. thesis, Department of Metallurgy and Materials Science, M.I.T., Cambridge, Mass., 1970.
20 RUDDLE, R. W.: "The Solidification of Sand Castings," 2d ed., no. 7, Institute of Metals, London, 1957.
21 HENZEL, JR., J. G., and KEVERIAN, J.: *J. Metals*, **17**:561 (1965).
22 MIZIKAR, E. A.: *Trans. AIME*, **239**:1747 (1967).
23 HILLS, A. W. D.: *J. Iron Steel Inst.*, **203**:18 (1968).
24 ADENIS, D. J. P., COATS, K. H., and RAGONE, D. V.: *J. Inst. Metals*, **91**:395 (1962–1963).
25 PEEL, D. A., and PENGELLY, A. E.: Mathematical Models in Metallurgical Process Development, *Iron and Steel Institute Rep. No.* 123, 1968.
26 NEREO, G. E., POLICH, R. F., and FLEMINGS, M. C.: *Trans. AFS*, **73**:1 (1965).

# PROBLEMS

1-1 A floating zone is passed up the length of an aluminum single crystal with 0.3 cm radius. Thermal conductivity of molten aluminum is approximately 0.20 cal/(cm)(°C)(s). Coefficient of heat transfer between aluminum and the furnace atmosphere $h$ is $1.2 \times 10^{-3}$ cal/(cm$^2$)(°C)(s). Other data are given in Appendix B.
   (a) What is the maximum temperature gradient achievable at the liquid-solid interface in the solid? In the liquid?
   (b) Would these temperature gradients be significantly different at a typical crystal-growing velocity of $5 \times 10^{-3}$ cm/s?
   (c) How might you alter the crystal or crystal-growing apparatus to increase these gradients?

1-2 Plot the longitudinal temperature profile during solidification of the single crystal of Example 1(b).

1-3 A large slab of aluminum, 25 cm thick, is poured in a sand mold with no superheat.
   (a) How long will it take for the slab to solidify? (Assume negligible resistance to heat flow at the mold-metal interface and within the solidified metal.)
   (b) Show schematically cooling curves for a thermocouple 6.25 cm from the surface and one at the center of the casting. (Remember, the solidified portion of the casting remains very close to $T_M$ until the casting is completely solid.)

1-4 An approximate method of calculation of solidification time of a sand casting poured with superheat is to add the heat content associated with the superheat to the heat of fusion. How long would it take the slab casting of Prob. 1-3 to solidify if it were poured with 100°C superheat?

1-5 Experimentally, it is found that when a casting such as that of Prob. 1-4 is cast, no significant solidification takes place until the superheat is completely dissipated. Explain.

1-6  A long cylinder of aluminum, 25 cm diameter, is poured with no superheat in a sand mold.
   (a) Calculate solidification time neglecting divergency of heat flow.
   (b) Calculate solidification time taking account of divergency of heat flow.
   (c) Derive an equation describing thickness of solid as a function of time during solidification of this cylinder. Neglect divergency of heat flow. Plot the result schematically.
   (d) You should find from (c) above that solidification rate is infinite at the beginning and end of solidification. Explain the physical reason for this.

1-7  Is the error involved in using the Chvorinov approximation for solidification times greater for a sphere or for a cylinder of the same radius? For large castings or small? For metals of high or low melting point?

1-8  Aluminum is splat-cooled on a copper substrate with a mold-metal-interface heat-transfer coefficient of 1 cal/(cm²)(°C)(s).
   (a) Assuming heat transfer is interface-controlled, how thin must the sample be to achieve a cooling rate (in the liquid at just above the melting point) of $10^6$ °C/s?
   (b) For the thickness calculated above, is it reasonable to assume interface-controlled heat transfer?

1-9  One of the advantages claimed for magnesium die casting compared with aluminum is a shorter solidification time. Show that this is true for processes where heat transfer is controlled at the mold-metal interface.

1-10  (a) A slab of iron, 25 cm thick, is poured with no superheat into an iron mold. Calculate the solidification time assuming no mold-metal resistance to heat flow.
   (b) How much longer is required for solidification of the slab in a sand mold? Neglect resistance to heat flow within the solidified metal and at the mold-metal interface.
   (c) What percentage error did you introduce in (b) by neglecting resistance to heat flow within the solidified metal?

1-11  What is the mold-metal interface temperature for the iron slab poured in the metal mold of Prob. 1-10. In the sand mold?

1-12  Show that the mold-metal-interface temperature for iron poured at its melting point in an aluminum mold is below the melting point of aluminum. Why is aluminum seldom used as a mold material for iron?

1-13  For subsequent use in abrasive applications, aluminum oxide is melted and cast in ingot form in heavy iron molds. What is the solidification time of a slab of aluminum oxide 25 cm thick? Does the mold-metal interface temperature stay below the melting point of iron? Approximate thermal properties of the aluminum oxide are $T_M = 2320$°C, $k_s = 0.013$, $\rho_s = 3.6$, $C_s = 0.33$.

1-14  You are to help design a vertical continuous-casting machine to make 8-in-thick low-carbon steel slabs. Casting speed to be aimed for is 100 in/min. What is the *minimum-height* casting tower you must plan to build?
   Note: Heat flow in the longitudinal direction can be neglected. You can obtain the answer to this problem analytically by using data for pure iron or, more simply, by using Fig. 1-13.

1-15 What thermal properties of a mold material affect its heat-absorbing ability, i.e., the rate at which an infinitely thick mold can extract heat? Rate quantitatively the mold materials sand, plaster, aluminum, iron, and copper.

1-16 It has been suggested that the ideal machine for making steel die castings would be one in which the metal is propelled at exceedingly rapid velocity into the mold cavity. It has even been suggested that if this were done, die materials of relatively low melting point (e.g., copper) might be employed. Examine this assumption critically, considering specifically the two major problems to be expected in die casting steel: lack of complete filling of detail and mold erosion, primarily from overheating.

1-17 What casting process or processes would you recommend for (a) gold rings, (b) costume jewelry to be plated, (c) soil pipe, (d) 100 aluminum ashtrays, (e) 500,000 aluminum ashtrays, (f) 100 ductile iron crankshafts, (g) 500,000 ductile iron crankshafts?

1-18 Draw schematic curves comparable to those of Fig. 1-16 for Al–4.5% Cu solidifying in a sand mold or die casting, with negligible thermal gradient in the metal.

# 2
# PLANE FRONT SOLIDIFICATION OF SINGLE-PHASE ALLOYS

## INTRODUCTION

Controlled plane front solidification of alloys is used in practice to grow single crystals, refine materials (e.g., *zone refining*), and obtain controlled uniform or non-uniform composition within the material grown. The most important commercial applications of this type of solidification are for growth of crystals for semiconductors. Another important application is in growth of oxides for laser systems and other optical applications; examples of materials are aluminates, tantalates, and niobates. Growth of oxide crystals for jewels is another, much older commercial application of single crystal growth. Melt-grown, plane front solidified crystals of metals and alloys find extensive use in research, but there is as yet no commercial application for these materials.

Solidification of most metals and inorganic nonmetals from their melts is closely approximated by the assumption of equilibrium at the interface during growth. That is, there may be large concentration gradients in the solid and liquid during solidification, but there is only a negligible barrier to transport of atoms across the interface. Consider a single crystal of alloy $C_0$ (Fig. 2-1), growing with a plane front. Equilibrium at the solid-liquid interface can be attained at temperatures below the

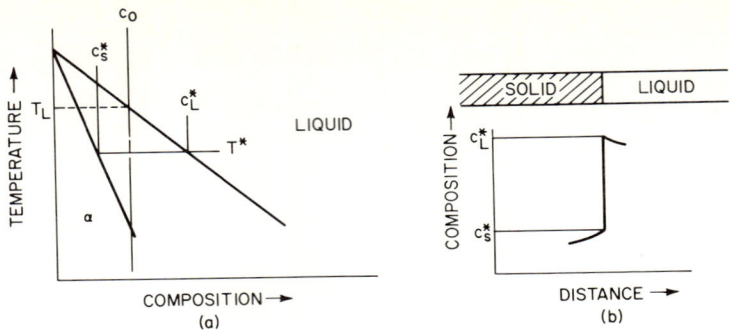

**FIGURE 2-1**
Solidification of an alloy with equilibrium at the liquid-solid interface. (a) Phase diagram; (b) composition profile across the interface.

liquidus temperature $T_L$. If solidification is occurring at temperature $T^*$, the condition of equilibrium at the interface requires that the liquid and solid compositions at the interface $C_L^*$ and $C_s^*$, respectively, be fixed by the phase diagram. Compositions away from the interface may be very different, but the condition of equilibrium at the interface requires that if either $T^*$, $C_L^*$, or $C_s^*$ is specified, the other two are fixed by the phase diagram.

In describing solidification under these conditions, it is convenient to define an *equilibrium partition ratio k*, where

$$k = \frac{C_s^*}{C_L^*} \qquad (2\text{-}1)$$

When the liquidus and solidus are straight lines emanating from the composition of the pure solvent as in Fig. 2-1, $k$ is a constant. In deriving various expressions below, $k$ is assumed constant to simplify the mathematics. When $k$ varies significantly, solutions to the various differential expressions given are readily obtained by numerical methods. Further, in most of the following discussion, $k$ is assumed to be less than unity; that is, the phase diagram is as in Fig. 2-1 with solidus and liquidus sloping downward. However, the expressions are valid for alloys in which $k$ is greater than unity.

In the following sections, solute redistribution in *normal solidification* is described first. Normal solidification is the term used to describe solidification when an entire charge is melted and solidified with plane front from one end. Solute redistribution in other types of crystal growth (*zone solidification* and *crystal pulling*) is then considered. All of the quantitative treatments given employ the assumptions of equilibrium at the liquid-solid interface and no significant undercooling before nucleation or from effect of curvature of the liquid-solid interface.

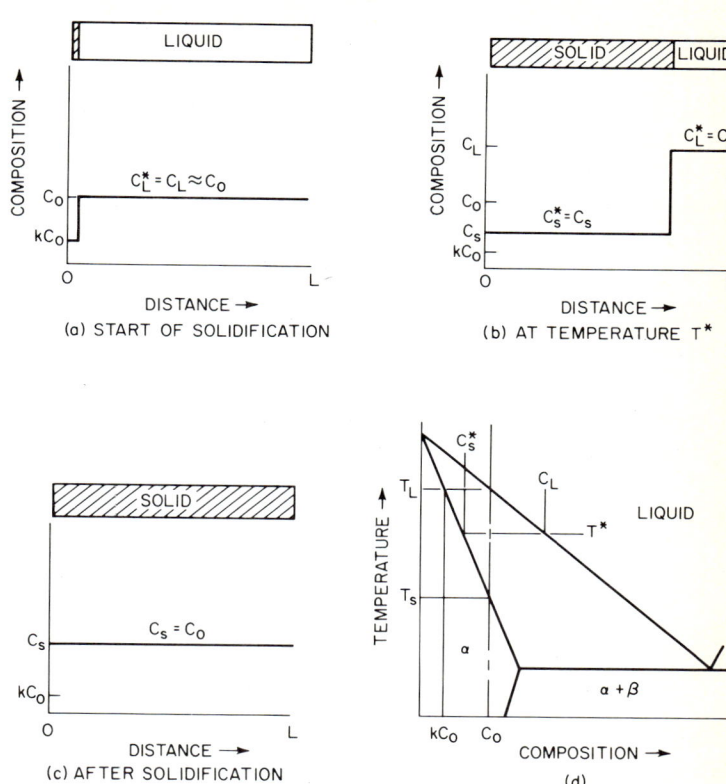

FIGURE 2-2
Solute redistribution in equilibrium solidification of an alloy of composition $C_0$. (a) At start of solidification; (b) at temperature $T^*$; (c) after solidification; (d) phase diagram.

## EQUILIBRIUM SOLIDIFICATION

Solidification in crystal growth is almost never slow enough to approach *equilibrium solidification*, although such solidification is possible and would result when $L^2 \ll D_S t$, where $L$ is the length of the growing crystal, $D_S$ is the diffusion coefficient of the solute in the solid, and $t$ is time. In addition to the assumptions given above, those of equilibrium solidification are simply complete diffusion in the liquid state and complete diffusion in the solid state.

Consider a crucible of liquid alloy of length $L$ and initial composition $C_0$ freezing from one end. The first solid begins to form at $T_L$ and is of composition $kC_0$, lower in solute than the initial liquid composition. The balance of the solute is rejected from the solid-liquid interface and diffuses into the liquid, Fig. 2-2a. During subsequent cooling and solidification, both the liquid and solid become enriched in solute; at temperature $T^*$, the solid of composition $C_s^*$ is forming in equilibrium at the liquid-solid interface with liquid of composition $C_L^*$. Because diffusion in the solid and liquid is complete, the entire solid becomes of uniform composition $C_s = C_s^*$ and the entire liquid of uniform composition $C_L = C_L^*$, Fig. 2-2b. At temperature $T^*$, a general materials balance (conserving solute atoms) is written:

$$C_s f_s + C_L f_L = C_0 \qquad (2\text{-}2)$$

where $f_s$ and $f_L$ are weight fractions solid and liquid, respectively. This is simply the *equilibrium lever rule*, which can be readily solved for fraction solidified at a given temperature since $f_s + f_L$ equals unity. Note that in spite of the equilibrium nature of solidification, substantial solute redistribution occurs during solidification; the material is homogeneous only before and after solidification.

## NO SOLID DIFFUSION

A case of much more practical interest than the foregoing is one where all the assumptions made above are retained except that it is assumed that no diffusion at all takes place in the solid. Consider again a crucible of liquid alloy of length $L$ and initial composition $C_0$ freezing from one end. As in equilibrium solidification, the first small amount of solid to form is of composition $kC_0$ at temperature $T_L$.

During subsequent cooling and solidification, the liquid becomes richer in solute and so the solid that forms is of higher solute content at later stages of solidification. However, since there is no diffusion in the solid state, the composition of the solid formed in the initial stages of freezing remains unchanged. At temperature $T^*$, solid of composition $C_s^*$ is freezing from liquid of composition $C_L^*$ and the solute distribution along the length of the growing crystal is as in Fig. 2-3. A quantitative expression is easily obtained by equating the solute rejected when a small amount of solid forms with the resulting solute increase in the liquid. This balance is

$$(C_L - C_s^*)\, df_s = (1 - f_s)\, dC_L \qquad (2\text{-}3)$$

Now, substituting the equilibrium partition ratio and integrating from $C_s^* = kC_0$ at $f_s = 0$ yields the composition of the solid at the liquid-solid interface $C_s^*$ as a function of fraction solid

$$C_s^* = kC_0(1 - f_s)^{(k-1)} \qquad (2\text{-}4a)$$

or, in terms of liquid composition and fraction liquid,

$$C_L = C_0 f_L^{(k-1)} \qquad (2\text{-}4b)$$

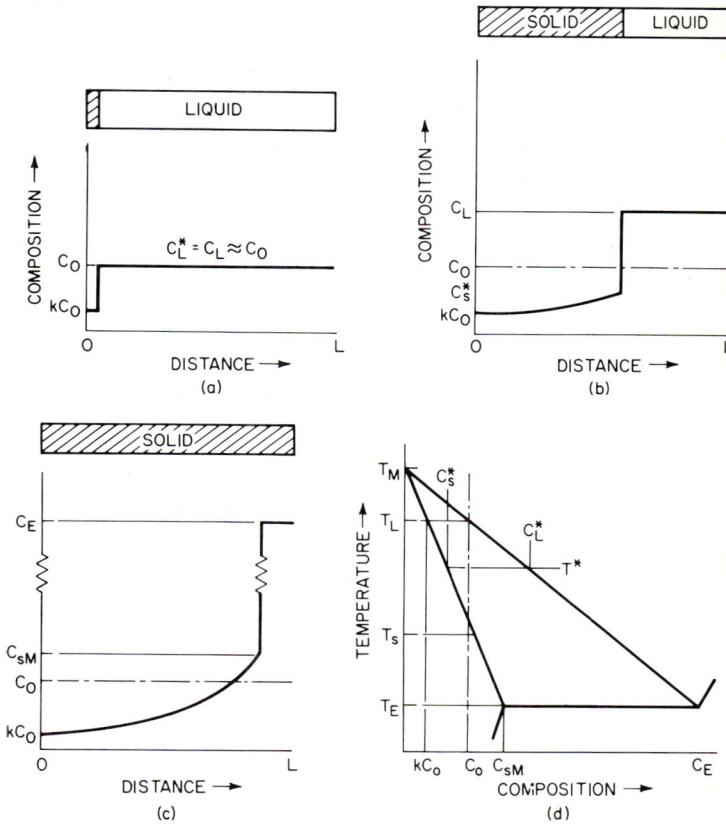

**FIGURE 2-3**
Solute redistribution in solidification with no solid diffusion and complete diffusion in the liquid. (*a*) At start of solidification; (*b*) at temperature $T^*$; (*c*) after solidification; (*d*) phase diagram.

Equations (2-3) and (2-4) have been derived repeatedly, for example, by Gulliver,[1] Scheil,[2] and Pfann,[3] and are termed the *nonequilibrium lever rule*, or the *Scheil equation*. These equations closely describe solute redistribution in crystal growth under a wide range of experimental conditions. Experimental examples will be given below after it is shown that this equation is a limiting case of a more general treatment that also considers the effect of limited liquid diffusion.

It is of interest to compare the solute distributions in Fig. 2-3, obtained from the normal segregation equation, with those of Fig. 2-2, obtained from the equilibrium lever rule. The starting condition is exactly the same, with the initial solid being of composition $kC_0$ in both cases. During solidification, at temperature $T^*$, the interface composition $C_s^*$ is the same for both cases, as required by the assumption of equilibrium at the interface. Average composition of the solid is less in the latter case, and there is more liquid remaining in this case. As long as the partition ratio $k$ is constant, some liquid will remain until an invariant temperature (e.g., *eutectic*) is reached and the remainder of the liquid then solidifies at eutectic composition. If $k$ is not constant, solute redistribution is still described by the differential Eq. (2-3), but in general solution must be by numerical methods. In this case, liquid is exhausted either when the eutectic is reached or when $k$ goes to unity.

## LIMITED LIQUID DIFFUSION, NO CONVECTION

Another practically important limiting case of normal solidification occurs when all the assumptions of the preceding case apply except that diffusion is limited in the liquid and there is no convection. Here, solidification begins exactly as in the previous examples, with the initial solid forming of composition $kC_0$. The solute rejected into the liquid is transported only by diffusion and so an enriched *solute boundary layer* forms and gradually increases in solute. If the crystal is sufficiently long, a *steady state* is approached, which is as sketched in Fig. 2-4a. At this steady state, the composition of the solid forming is exactly the overall alloy composition $C_0$. Equilibrium at the interface then requires that the composition in the liquid at the interface be $C_0/k$ and that solidification be occurring at the solidus temperature $T_s$. The solute distribution in the boundary layer in the steady-state region is given by the differential equation

$$D_L \frac{d^2 C_L}{dx'^2} + R \frac{dC_L}{dx'} = 0 \qquad (2\text{-}5)$$

where $x'$ is distance from interface, $D_L$ is diffusion coefficient of solute in the liquid, and $R$ is rate of movement of interface. Boundary conditions are that $C_L = C_0/k$ at $x' = 0$, and $C_L = C_0$ at $x' = \infty$. Also, the requirement of solute conservation at the interface gives directly the gradient of composition in the liquid at the interface:

$$\left( \frac{dC_L}{dx'} \right)_{x'=0} = -\frac{R}{D_L} C_L^*(1 - k) \qquad (2\text{-}6)$$

The solution to Eq. (2-5), as given by Tiller, Jackson, Rutter, and Chalmers[4] and as readily obtained with the aid of Eq. (2-6), is

$$C_L = C_0 \left( 1 + \frac{1-k}{k} e^{-(R/D_L)x'} \right) \qquad (2\text{-}7)$$

The quantity $D_L/R$ may be considered a *characteristic distance*, the distance at which the quantity $C_L - C_0$ falls to $1/e$ of the maximum $C_0/k - C_0$. Note that this charac-

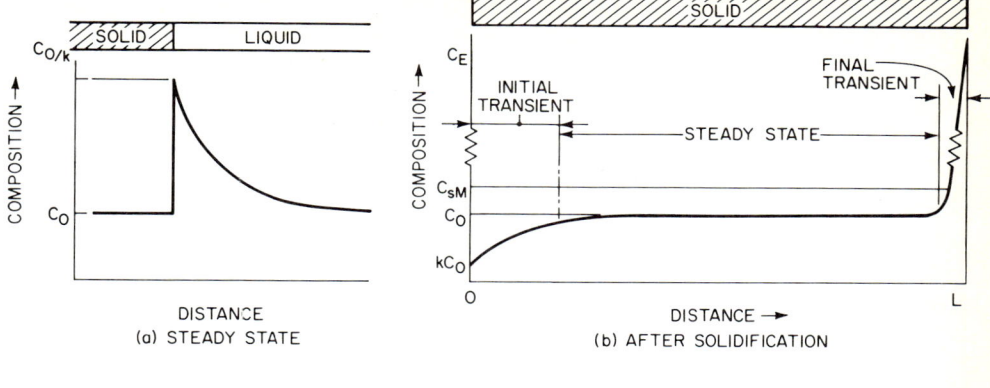

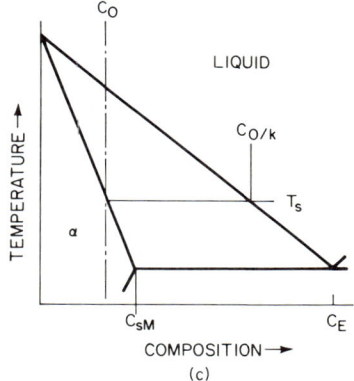

**FIGURE 2-4**
Solute redistribution in solidification with limited liquid diffusion and no convection. (*a*) Composition profile during steady-state solidification; (*b*) composition profile after solidification; (*c*) phase diagram.

teristic distance is dependent only on $D_L$ and $R$. Equation (2-7) applies in many real cases of single crystal growth even though some convection is present. The requirement for its applicability is that the characteristic distance $D_L/R$ must be small compared with the momentum boundary layer.

This type of solidification results in a crystal of nearly uniform composition except for the initial and final transients, as shown in Fig. 2-4*b*. The initial transient is formed while the solute boundary layer builds up to its maximum steady-state value. Calculation of solute redistribution during this period is done using the time-dependent form of Eq. (2-5):

$$\frac{\partial C_L}{\partial t} = D_L \frac{\partial^2 C_L}{\partial x'^2} + R \frac{\partial C_L}{\partial x'} \qquad (2\text{-}8)$$

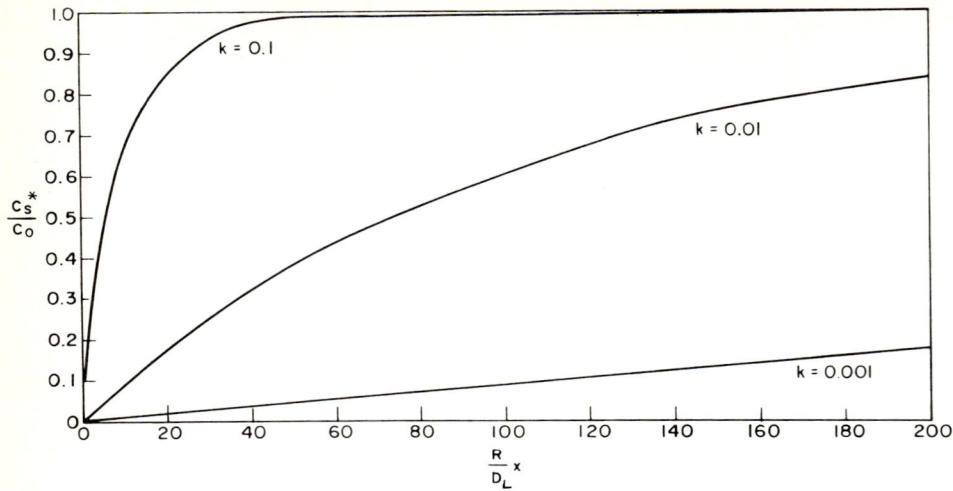

**FIGURE 2-5**
Solute distribution in initial transient. (*After Smith, Tiller, and Rutter.*[5])

where $t$ = time. The solute conservation Eq. (2-6) applies here as in steady-state solidification, and boundary conditions for the initial transient are $C_L = C_0$ at $t = 0$ for $x' > 0$, and $C_L = C_0$ at $x' = \infty$ for $t > 0$. A rigorous solution for this case has been obtained by Smith et al.,[5] which is rather cumbersome, however. For small values of $k$, this equation reduces to[6]

$$C_s^* = C_0[1 - (1 - k)e^{-(kRx/D_L)}] \qquad (2\text{-}9)$$

Results of the analysis of Smith et al. are plotted in Fig. 2-5, and it can readily be shown these are closely similar to those that would be obtained from Eq. (2-9). Solute content in the solid builds up gradually to its steady-state value $(C_s^*/C_0) = 1$ as solidification proceeds. The distance $x$ required to reach essentially the steady-state value depends on $R/D_L$ and $k$. From Eq. (2-9), it is seen that for small $k$, a characteristic distance for the length of this transient is $D_L/Rk$. At this distance, the composition of the solid forming has risen to $1 - (1/e)$ of its maximum (that is, 67 percent of its steady-state value).

The final transient is much smaller than the initial transient since it results simply from impingement of the solute boundary layer on the extremity of the crucible. Thus, its length is the order of the characteristic distance of the solute boundary layer, or $D_L/R$. An exact solution for this case is given by Smith et al.[5] Figure 2-4b shows a schematic representation of this final transient for the alloy used as example. Solute concentration in the final transient increases continuously from

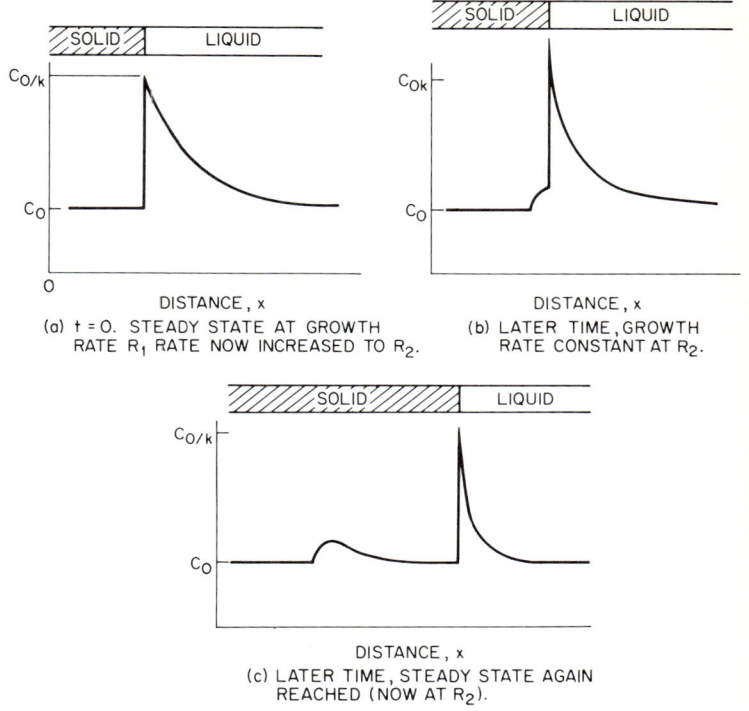

**FIGURE 2-6**
Solute-rich band resulting from increase in solidification velocity.

$C_0$ to $C_E$ at the ingot end. At solute concentrations greater than $C_{SM}$, the ingot is two-phase, as discussed in Chap. 4.

Another type of transient, which is of considerable engineering importance, is that resulting when interface velocity changes during solidification. This is shown schematically in Fig. 2-6 for a sudden increase in growth velocity from $R_1$ to $R_2$. The steady-state profiles of solute distribution in the liquid are different for $R_1$ and $R_2$ only in that the characteristic distance is less for $R_2$ than for $R_1$. Thus, the excess solute that was initially in the boundary layer must appear as a solute-rich *band* in the vicinity of the region where the velocity change occurred. Smith et al.[5] have again obtained an exact solution for this case using the differential equation (2-8) and with the steady-state solute distribution at $R_1$ as initial boundary condition. Results are given in Fig. 2-7 for several values of $R_2/R_1$ and $k$. As in the initial transient, the characteristic length of the perturbations that form is much greater than $D_L/R$ for small $k$ and increases with decreasing $k$. Also, as would be expected intuitively, the

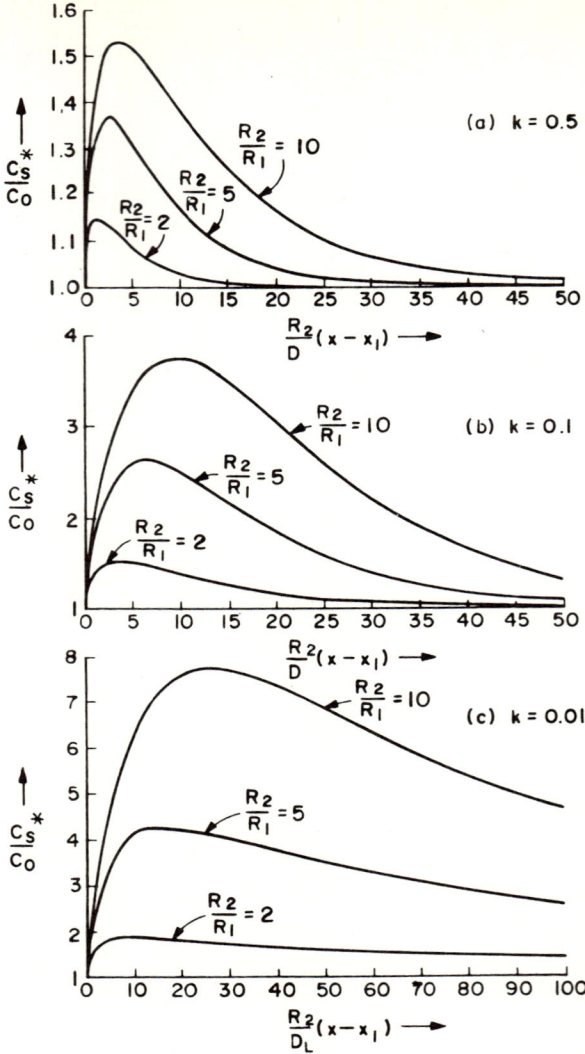

**FIGURE 2-7**
Transient solute distribution in solid resulting from an instantaneous change in growth velocity from $R_1$ to $R_2$ at location $x_1$. (*After Smith, Tiller, and Rutter.*[5])

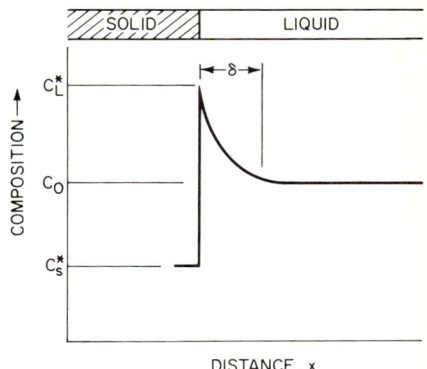

**FIGURE 2-8**
Solute profile in the liquid for solidification with convection.

maximum solute content in the banded region increases with increasing velocity change. Similar plots for a sudden decrease in velocity ($R_2/R_1 < 1$), of course, would show a solute-poor band.

## EFFECT OF CONVECTION

In most real cases of crystal growth, some convection is present; and solute redistribution in these cases has been treated by Burton, Prim, and Slichter,[7] and by Wagner[8] as well as others. The following analysis is after Burton, Prim, and Slichter. A diffusion boundary layer of thickness $\delta$ is assumed, outside of which liquid composition is maintained uniform by convection and inside of which mass transport is by diffusion only, Fig. 2-8. If solidification is taking place in a very large liquid bath, the bulk liquid composition is not altered by the solid forming and remains constant at $C_0$. After steady state is reached, the differential equation (2-5) applies with the boundary conditions that $C_L = C_L^*$ at the liquid-solid interface (at $x' = 0$), and that $C_L = C_0$ at $x' = \delta$. The solution is

$$\frac{C_L^* - C_s^*}{C_0 - C_s^*} = e^{R\delta/D_L} \qquad (2\text{-}10)$$

For convenience, we now define an *effective partition ratio $k'$*, which is the solid composition forming $C_s^*$ divided by the bulk liquid composition (in this case $C_0$). Then, substituting in Eq. (2-10),

$$k' = \frac{k}{k + (1 - k)e^{-(R\delta/D_L)}} \qquad (2\text{-}11)$$

This expression is of considerable engineering usefulness because it relates the composition of the solid forming in crystal growth to alloy composition and growth

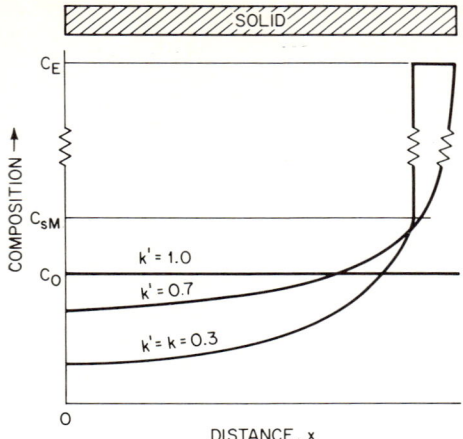

FIGURE 2-9
Final solute distributions for solidification with limited liquid diffusion and different amounts of convection.

conditions. It can be used to describe solute redistribution in crucibles of finite extent provided only that the thickness $\delta$ of the boundary layer is small compared with the length of the crucible. When this is true, a dynamic equilibrium is attained between the bulk melt and growing solid and equations identical to Eqs. (2-4) are readily derived, except that the equilibrium partition ratio $k$ is replaced by the effective partition ratio $k'$:

$$C_s^* = k'C_0(1 - f_s)^{(k'-1)} \qquad (2\text{-}12a)$$
$$C_L = C_0 f_L^{k'-1} \qquad (2\text{-}12b)$$

Here, $C_L$ is the bulk liquid composition and $k' = C_s^*/C_L$; Eqs. (2-12a) or (2-12b) constitute a modified *normal segregation equation*.

Figure 2-9 shows some calculated distributions of solute for the alloy of the preceding examples, taking $k'$ equal to $k$, unity, and an arbitrary value between the minimum ($k$) and the maximum (unity). As seen from Eq. (2-10), the minimum value occurs when $R\delta/D_L \ll 1$, that is, at slow growth rate, high liquid diffusivity, and maximum stirring, and so $\delta$ is a minimum. At this limit, solute distribution is described by the special case given earlier where infinite diffusivity in the liquid was assumed. The maximum value of $k$ (equal to unity) is obtained for $R\delta/D_L \gg 1$. Under these conditions, any convection present has negligible effect on solute distribution and this limit is therefore described by the special case (given earlier) of negligible liquid convection. Equations (2-12) apply only in the region of single-phase growth and exclude initial and final transients. When $k' = k$, bulk liquid composition reaches the eutectic $C_E$ when solid composition reaches its maximum $C_{SM}$. Thereafter, solid forming is of uniform composition $C_E$. For $k < k' < 1$, a region of two-phase solid of continuously varying composition forms at the end of solidification; this is discussed in Chap. 4.

The curves in Fig. 2-9 are drawn such that the initial composition to form is $k'C_0$; that is, the *initial transient* where the solute builds up from $kC_0$ to $k'C_0$ is omitted. For usual conditions of crystal growth, this transient is usually extremely small compared with the length $L$ of the crystal to be grown. It can be seen from Eqs. (2-10) and (2-11) that changes in growth velocity $R$ or solute boundary layer $\delta$ will have a marked influence on solute distribution of the growing crystal; $C_s^*$ increases with increasing $R$ and $\delta$. This problem has been treated by Hurle, Jakeman, and Pike.[9] When the perturbation occurs over a distance which is large compared with the boundary layer $\delta$, the problem is treated simply by using Eq. (2-11) if $R$ and $\delta$ are known as a function of position.

Experiments have been conducted by many workers to compare calculated effective partition ratios with those experimentally determined. Quantitative interpretation of some of these experiments is made difficult by the fact that in some cases solidification was probably not plane front (i.e., that *cells* or *dendrites* may have formed). In other cases, *banding* was probably present resulting from fluctuations of $R$ or $\delta$ during growth. As shown by Hurle and Jakeman,[10] banding will lead to spurious calculated values of effective partition ratio $k'$ or boundary-layer thickness $\delta$.

Figure 2-10 summarizes some results of Dean, Kerr, and Hellawell[11] on normal solidification of a Pb–1% Sb alloy. Solute-distribution curves such as the one in Fig. 2-10a were obtained, which conform closely to theoretical equation expectation, with the effective partition ratios as in Fig. 2-10b. At the lower growth rates and higher gradients, $k'$ varied according to the Burton-Prim-Slichter equation, assuming $\delta = 0.05$ cm. At higher growth rates or lower temperature gradients, measured values of $k'$ were much higher than expected from the Burton-Prim-Slichter relation. This was shown metallographically to be caused by the fact that solidification was no longer with a plane front, but that cells or dendrites had formed. Figure 2-10c, for example, shows that when the structure solidified with plane front, $\delta$ increased very slightly with decreasing temperature gradient. Presumably this was because the lower thermal gradient resulted in less vigorous thermal convection and therefore slightly larger $\delta$. However, below about 30°C/cm temperature gradient at the growth rate employed, cells or dendrites formed and as a result $k'$ increased rapidly. In both cases, $k'$ was determined from plots such as those of Fig. 2-10a. In material solidified with plane front, solid composition was uniform, of course, at a given position along the ingot length. In the cellular or dendritic material, it was not uniform and the composition measured represented a local average.

Table 2-1 lists some typical values of $\delta$ compiled by Zief and Wilcox[12] from various investigators. As in the above example, these were calculated using the Burton-Prim-Slichter relation and, in the case of normal solidification, using Eq. (2-12).

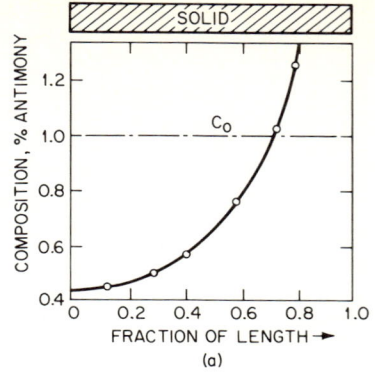
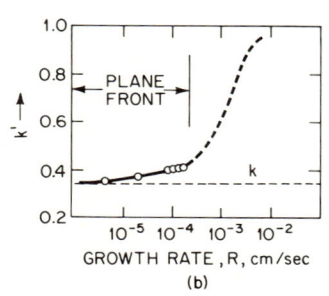
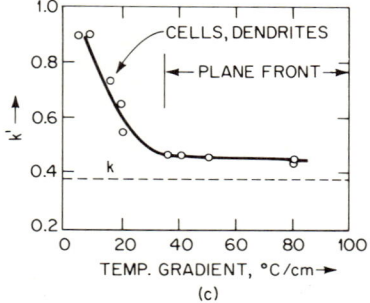

FIGURE 2-10
Solute redistribution in Pb–1% Sb alloy. (a) Example of experimentally determined composition along length of ingot solidified with a plane front; (b) $k'$ versus $R$ for ingots solidified with a temperature gradient of 80°C/cm; (c) $k'$ versus temperature gradient for ingots solidified at $R = 10^{-4}$ cm/s. (After Dean, Kerr, and Hellawell.[11])

## CZOCHRALSKI GROWTH (CRYSTAL PULLING)

Most semiconductor crystals are made, not by normal solidification, but by Czochralski growth, Fig. 1-1. In this case, the supply of liquid is very large compared with the size of the crystal grown and some thermal convection is always present. Thus, the bulk liquid remains of uniform constant composition at the initial melt composition $C_0$. Provided growth rate and convection are constant, the composition of the growing crystal is also uniform; its composition is given by the Burton-Prim-Slichter relation, Eq. (2-11). Figure 2-11, from the work of Bridgers,[24] shows typical experimental results for $k'$ in Czochralski growth as a function of growth rate $R$. The different slopes of the curves are due to the very different equilibrium partition ratios of the

## CZOCHRALSKI GROWTH (CRYSTAL PULLING)  45

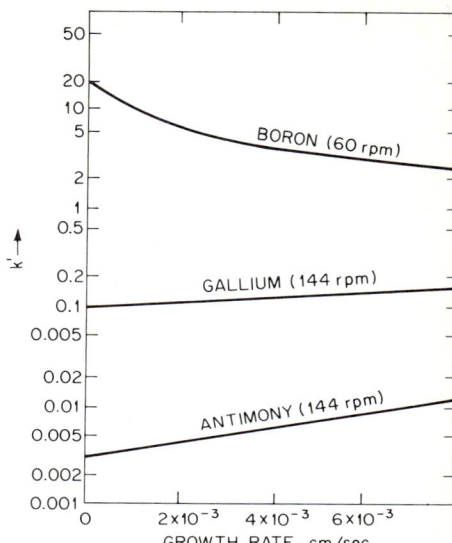

**FIGURE 2-11**
Effective partition ratio $k'$ versus growth rate for Czochralski growth of germanium with different solutes. (*From Bridgers.[24]*)

Table 2-1  TYPICAL BOUNDARY-LAYER THICKNESSES IN CRYSTAL GROWTH†

| Solvent | Solute | $\delta$, cm | Reference |
|---|---|---|---|
| | Normal solidification | | |
| Pb | 1% Sb | 0.05 | 11 |
| Mg | 0.0016% Fe | 0.007 | 13 |
| Anthracene | 0.1% tetracene | 0.2 | 14 |
| $H_2O$ | 2–10 g/100 ml urea | 0.02 | 15 |
| $H_2O$ | 1–5 g/100 ml urea‡ | 0.009 | 15 |
| | Czochralski growth | | |
| Ge | $<10^{-3}$% Ta | 0.04 | 16 |
| Ge | Sb | 0.01 | 17 |
| Ge | Ga | 0.01 | 17 |
| Si | Al | 0.03 | 18 |
| NaCl | 0.5% I | 0.0005 | 19 |
| | Zone melting | | |
| Pb | 10% Sn | 0.08 | 20 |
| Pb | 10% Sn‡ | 0.016 | 20 |
| Fe | 0.2% Ta | <0.015 | 22 |
| $GaCl_3$ | $1.4 \times 10^{-5}$ $FeCl_3$ | 0.008 | 23 |
| Napthalene | 10% benzoic acid | 0.07 | 21 |

† Boundary layers $\delta$ were calculated from experimental solute distribution using Eq. (2-11) (from Wilcox[21]).
‡ Crystals grown with forced convection.

alloy elements [$k$(Sb) = 0.003; $k$(Ga) = 0.01; $k$(B) = 17]. Table 2-1 shows some calculated values of $\delta$.

## ZONE MELTING

*Zone melting* (zone solidification), described in Chap. 1, is widely employed commercially because a greater degree of purification can be obtained by this technique than by normal solidification and because it can be conducted without a crucible. Figure 2-12 shows schematically the solute-distribution curve during and after solidification for the alloy used in the previous examples. Only a small portion of the bar is melted, and this molten *zone* is moved from one end of the bar to the other. Figure 2-12 shows the solute distribution during and after one pass, but the process can be repeated many times with additional purification obtained each time.

For calculation of the solute distribution in the first pass, assume initially no diffusion in the solid and complete diffusion in the liquid. Then, solute distribution until the final transient is reached is calculated from a simple materials balance of solute entering and leaving the zone:

$$(C_0 - C_s^*)\, dx = l\, dC_L^* \tag{2-13}$$

where $l$ = zone length. Substituting the equilibrium partition ratio and integrating from $C_s^* = kC_0$ at $x = 0$ yields the solute-distribution equation derived by Pfann:[3]

$$\frac{C_s^*}{C_0} = 1 - (1-k)e^{-(kx/l)} \tag{2-14}$$

If convection is present but diffusion not complete in the liquid (and if, as is usually the case, the boundary layer $\delta$ is much smaller than $l$), then Eq. (2-14) can be rewritten directly in terms of the effective partition ratio $k'$:

$$\frac{C_s^*}{C_0} = 1 - (1-k')e^{-(k'x/l)}. \tag{2-15}$$

The final transient is reached when the liquid zone impinges on the end of the ingot. Here, Eq. (2-15) no longer applies but solute distribution is given directly by the normal solidification equation (2-12) as long as only a single phase is forming. The bulk liquid composition in this case, of course, is the bulk composition of the zone at the time it reaches the ingot end. When the second phase begins to form near the end of solidification, solute redistribution is as given by Eq. (4-20).

There is no simple way to calculate solute distribution after multiple-pass zone melting, but a large number of computed curves are given by the inventor of this

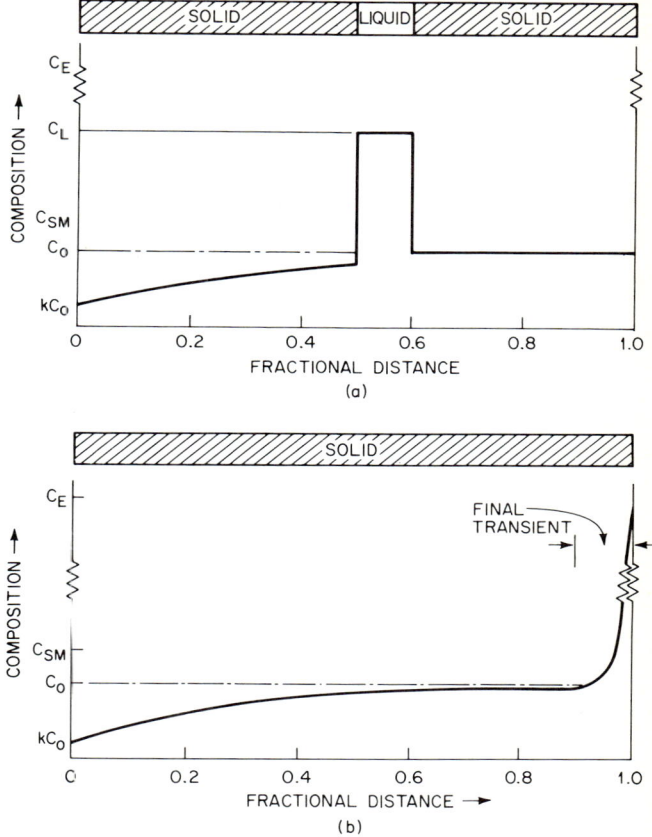

FIGURE 2-12
Solute redistribution in zone solidification (*a*) during solidification and (*b*) after solidification.

process, W. G. Pfann.[25] One set of these curves is shown in Fig. 2-13 for a single-phase alloy of very high effective partition ratio ($k' = 0.9524$) and a ratio of zone length to ingot length of 1:100. After 2,000 passes, only a very small fraction of the solute initially present remains in the first half of the bar. The rapid increase in solute at the last part of the bar to solidify is not to be confused with the final transient of Fig. 2-12. This transient comprises only one-hundredth of the bar length and is not shown.

When only one or a few passes are conducted in zone melting, and when the zone is small compared with the ingot length, the bulk of the ingot solidifies at steady

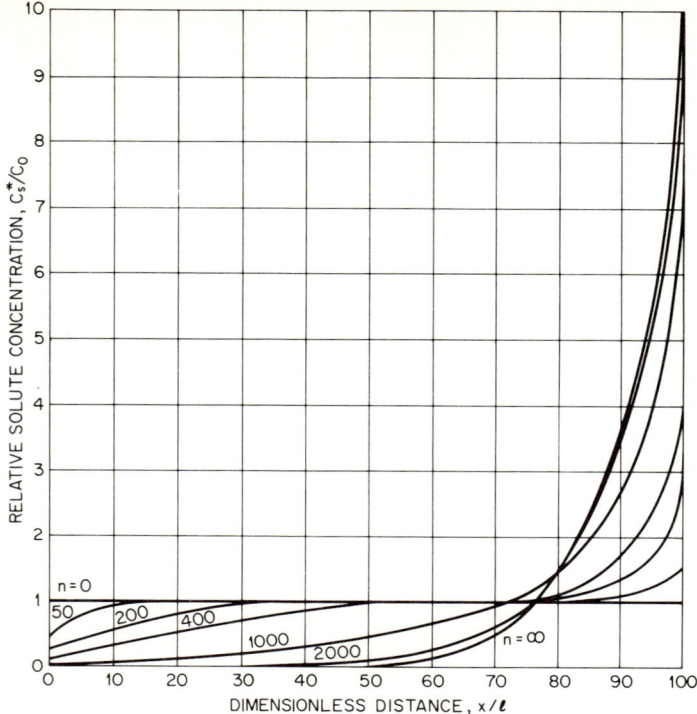

**FIGURE 2-13**
Relative solute concentration $C_s^*/C_0$ versus distance $x/l$ with number of passes $n$ as a parameter for $k = 0.9524$ and a ratio of zone length to ingot length of 1:100. (*After Reiss.*[26])

state, with the solid material in this zone of uniform composition $C_0$. When zone solidification is conducted in this way to achieve uniformity, it is termed *zone leveling*; when it is conducted to obtain maximum transport of solute, it is termed *zone purification*.

Zone melting is used most extensively commercially for metallic and semiconductor materials, for zone refining and zone leveling, and simply for single crystal growth. Much work, however, is also now being conducted on organic and inorganic materials primarily for purposes of purification. Pfann describes a large number of applications of the process, and other studies are summarized in a book edited by Zief and Wilcox.[12] Table 2-1 lists some values of $\delta$ calculated for typical zone-melting experiments, and Fig. 2-14 is an example of experimental results on zone melting in an organic material.

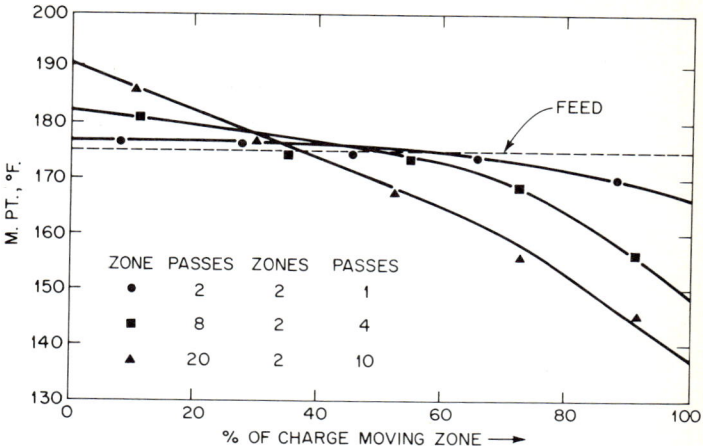

**FIGURE 2-14**
Melting-point-distribution profile for zone precipitation of microcrystalline wax using sec-butyl acetate as a solvent. (*After Eldib.*[27])

## CRYSTAL GROWTH WITH VOLATILE CONSTITUENTS

When impurity elements present in a crystal growth or refining operation are volatile, advantage can be taken of this fact to achieve enhanced purification. Conversely, special precautions must be taken to retain volatile-alloy elements (e.g., arsenic in GaAs) in solution during growth.

Assuming volatilization is limited by transfer across the liquid-vapor interface, the rate equation for the transfer is written

$$J = -\alpha \rho_L (C_L^e - C_L) \quad (2\text{-}16)$$

where  $J$ = flux across liquid-vapor interface, g/(cm²)(s)

$\alpha$ = evaporation coefficient, cm/s

$\rho_L$ = liquid density

$C_L^e$ = liquid composition that would be in equilibrium with the gas phase, wt fraction

$C_L$ = actual liquid composition, wt fraction

The rate constant $\alpha$ can be calculated from evaporation theory or measured experimentally. Table 2-2 gives some measured values for semiconductors.

Solute-distribution equations for the various types of solidification given earlier can now be readily rederived with the added effect of volatilization, represented by

Eq. (2-15). Solutions for a variety of cases are given in the literature.[25,28] One simple example is given here, that of vertical normal solidification (Bridgeman method) in which the top surface is exposed to zero partial pressure of solute. Solute volatilization is limited by transfer across the liquid-vapor interface, and other assumptions involved in deriving Eq. (2-3) apply (e.g., no solid diffusion). Then Eq. (2-3) becomes

$$\begin{array}{c} \text{Grams solute rejected} \\ \text{from solid} \end{array} = \begin{array}{c} \text{grams solute} \\ \text{added to liquid} \end{array} - \begin{array}{c} \text{grams solute} \\ \text{volatilized} \end{array}$$

$$(C_L - C_s^*)\, df_s \rho_L V = (1 - f_s)\, dC_L \rho_L V - JA\, dt \qquad (2\text{-}17)$$

where $V$ is crystal volume, $A$ is crystal cross section, and $t$ is time. Assuming constant liquid and solid densities,

$$\frac{df_s}{dt} = \frac{A}{V} R \qquad (2\text{-}18)$$

where $R$ is rate of interface movement. Substituting Eqs. (2-16) and (2-18) into Eq. (2-17) with $C_L^e = 0$ and integrating from $C_L = C_0$ at $f_s = 0$ yields

$$C_s^* = kC_0(1 - f_s)^{(k-1)+(\alpha/R)} \qquad (2\text{-}19)$$

and the resulting equation is seen to be the simple equation for normal solidification with no solid diffusion, modified only by the term $\alpha/R$ in the exponent. If diffusion is limited in the liquid but convection maintains the bulk liquid at uniform composition $C_L$, then the effective partition ratio $k'$ can be substituted for $k$.

When the volatile component is a desirable alloying element in the crystal to be grown, there are several techniques available to retain it in solution. One of these is simple encapsulation of the melt in a container of small volume so that the vapor pressure builds up to its equilibrium and no further volatilization occurs. Another

Table 2-2  EVAPORATION COEFFICIENTS FOR SOLUTE IN GERMANIUM†

| Solute | $\alpha$, cm/s |
|---|---|
| P | $5.4 \times 10^{-3}$ |
| As | $2.3 \times 10^{-4}$ |
| Sb | $1.35 \times 10^{-3}$ |
| In | $2.3 \times 10^{-4}$ |
| Ag | $1.2 \times 10^{-4}$ |
| Au | $4.8 \times 10^{-5}$ |

† From Brice.[28]

exceptionally useful technique is simply to cover the melt with a slag impervious to the volatile species.[29] Then, provided the vapor pressure of this element is not greater than ambient pressure, no loss occurs.

Of course, when the vapor pressure of the solute species is greater than the equilibrium vapor pressure of the melt, solute can be added to the crystal via this route. This has been done successfully both to produce concentration perturbations and to counteract normal segregation so as to produce a crystal of uniform solute content.[25]

## DIFFUSION IN THE SOLID

Throughout this chapter and in most analyses of solute redistribution in crystal growth, diffusion of solute in the solid is neglected. That this is an excellent assumption for usual types of crystal growth is readily shown by the following approximate analysis. This analysis places an upper limit on the diffusion into the solid that can occur by assuming that the diffusion does not significantly change the composition gradient in the solid at the liquid-solid interface; that is,

$$\left(\frac{\partial C_s}{\partial x}\right)_{x=x^*} \simeq \frac{dC_s^*}{dx^*} \qquad (2\text{-}20)$$

where $x^*$ is the position of the liquid-solid interface. Now, the solute redistribution equation for normal solidification can be calculated readily following steps similar to those used in deriving the preceding equation.[30] The result is

$$C_s^* = kC_0 \left[1 - \frac{f_s}{1 + (D_s k/RL)}\right]^{(k-1)} \qquad (2\text{-}21)$$

where $D_s$ is the solid-diffusion coefficient at the melting point and $L$ is crystal length. The dimensionless parameter $D_s k/RL$ is much smaller than unity in usual cases of crystal growth.

## THE FACET EFFECT

Until this point, it has been assumed that the equilibrium partition ratio $k$ applies at the liquid-solid interface regardless of crystal orientation or growth rate. So far as it has yet been proved experimentally, this assumption is valid for the growth rates encountered in usual crystal growing processes—except when facets form. Most metals and metal alloys do not form facets, but they are a common occurrence in many other materials, particularly those materials which have a high entropy of

**FIGURE 2-15**
The facet effect. (*a*) Flat facet perpendicular to the growth direction; (*b*) autoradiograph of a longitudinal section of an InSb crystal doped with $^{125}$Te. A (111) facet was present perpendicular to the growth direction, and $^{125}$Te concentration in the faceted region is much higher than in the periphery. (*From Bardsley, Mullin, and Hurle.*[32])

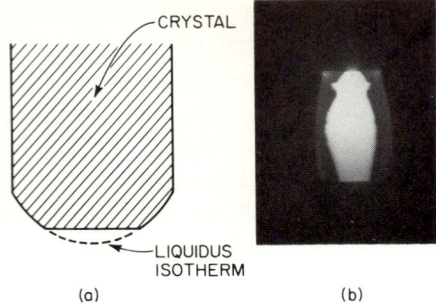

fusion. In faceted growth, crystallization occurs by the formation and lateral spreading of planes. The formation of new planes is by two-dimensional nucleation or by a screw-dislocation mechanism. The faceted interface is undercooled below its equilibrium melting temperature, and this undercooling is variable across the surface. As an example, undercooling of the facet shown schematically in Fig. 2-15 is greatest at the facet center. Kinetics of this type of growth are discussed in detail in Chap. 9.

The departure from equilibrium associated with faceting can result in either a decrease or an increase in measured partition ratio; that is, the composition of the

Table 2-3  FACET RATIO $k^*/k$ FOR VARIOUS SOLUTES IN TWO SEMICONDUCTORS[†]

| Germanium solute | $k^*/k$ |
|---|---|
| P | 2.5 |
| As | 1.8 |
| Sb | 1.45 |
| Ga | 0.85 |
| In | 1.4 |
| Tl | 1.2 |
| **Indium-antimonide solute** | |
| Cu | 1.3 |
| Zn | 1.3 |
| Cd | 3.3 |
| Si | 1.7 |
| Ge | 1.6 |
| Sn | 3.9 |
| S | 3.2 |
| Se | 6 |
| Te | 9 |

† From Bardsley, Mullin, and Hurle.[32]

solid forming may be either less than or greater than $kC_L^*$. The new kinetically influenced partition ratio given by the actual ratio of $C_s^*/C_L^*$ is herein termed $k^*$; it is equal to $k$ when the kinetic effect is small. The effect of faceting on segregation was first shown by Hulme and Mullin[31] in a study on tellurium-doped indium-antimonide. The presence of (111) facets was correlated with anomalously high tellurium incorporation in the faceted regions of growth. The effect is seen in the autoradiograph of the indium-antimonide crystal of Fig. 2-15. The central region grew with a faceted interface, and tellurium concentration in this region was approximately nine times that in the outer (nonfaceted) regions. The liquid composition at the interface $C_L^*$ is very nearly the same on and off the facet. Thus, the ratio of 9 also represents the ratio of the effective distribution coefficient on the (111) facet to that off the (111) facet, or approximately $k^*/k$, termed here the *facet ratio*. Table 2-3 summarizes some measurements of the facet ratio in various semiconductors. Qualitatively similar results have been described by Cockayne[33] for melt-grown oxide crystals.

## GROWTH OF SINGLE CRYSTALS OF HIGH PERFECTION

In recent years much effort has been expended in developing techniques for producing single crystals of desired orientations, that is, of high purity, high homogeneity, and high crystalline perfection. Orientation control is generally obtained by starting with a properly oriented *seed* crystal. Purity is obtained by the usual chemical methods, i.e., by zone refining (without crucible and in vacuum when feasible) and generally avoiding contamination during melting and growth.

Homogeneity on a macroscopic scale is obtained in Czochralski growth, zone leveling, and normal solidification when the effective partition ratio is near unity. In each of these processes, growth rate and convection must be constant during growth to achieve homogeneity. Homogeneity on a microscopic scale is much more difficult to achieve. Minor fluctuations in growth rate bring about solute *banding*, as described earlier. Avoiding such fluctuations is not simple, and despite years of intensive effort in the semiconductor field, the problem still plagues those responsible for the control of the quality of transistors and those who produce high-quality crystals of other materials.

Some causes of growth perturbations are not difficult to eliminate; these include such external causes as variations in power input, crystal-pulling speed, and vibrations. A much more difficult source to eliminate is that resulting from thermal convection. Thermal gradients are always present in crystal growth, and these give rise to convection patterns which, in typical crystal-growing furnaces, are turbulent. The turbulent flow then results in local temperature fluctuations, which in metal and semiconductor growth have been measured to be in the range of a few tenths of a degree to several tens of degrees, depending on thermal gradient, material, and crucible geometry.[34-37] The temperature fluctuations, when they reach the liquid-solid interface, perturb the growth rate and result in bands. Figure 2-16 is an example of the bands that result from thermal convection in indium-antimonide and shows also one way to eliminate the bands, i.e., by imposition of a steady magnetic field. Flow of conductive liquids is strongly damped by a magnetic field, and quite modest magnetic fields are sufficient for metals and semiconductors to eliminate the turbulence.[37,38] Some quantitative aspects of fluid flow in solidification processing are considered in Chap. 7.

Thermal stresses resulting during and after crystallization affect the structure and performance of melt-grown crystals. In semiconductors, they are an important source of dislocations; crystalline perfection in these materials is significantly improved by employing *after-heaters* to reduce thermal gradients in the solid. Similarly, excessive thermal stresses introduce slip or cracking in oxide single crystals, either of which interferes with optical perfection. Here, also, after-heaters are often used to reduce thermal gradient in the solid.

Dash[39] was the first to grow crystals (of silicon) completely free of dislocations. He did this by starting with as perfect a seed crystal as possible and then growing through a long narrow neck into a larger volume of material which comprised the final crystal. Any dislocations initially present in the seed or formed during the first part of the process grew out before they reached the main body of the melt. Dash and others found that once a perfect single crystal was obtained, modest liberties could be taken with growth conditions without introducing new dislocations. Materials other than silicon have now been made dislocation-free using similar techniques. Success has not been obtained with metals in this regard, however, presumably because metals are so readily deformed plastically. Variations in solute content introduced by banding are also a source of introduction of dislocations. Goss et al.[40] showed that fluctuations in silicon content in Ge–Si crystals produced sharp changes in the lattice constant that were taken up by grids of edge dislocations lying parallel to the liquid-solid interface. Variations in solute content from the *facet effect* result in similar introduction of defects. Incorporation of inclusions and precipitation of impurity elements during cooling after solidification are still other sources of dislocations. During growth, dislocations sometimes amalgamate by processes similar to polygonization to form subboundaries. Jackson has summarized in greater detail our current understanding of how dislocations and other crystalline defects form during crystal growth.[41]

The growth of quality single crystals requires that the liquid-solid interface be planar, at least locally. That is, the formation of cells or dendrites must be avoided, and this is done by growing the crystal with sufficiently high thermal gradient in the liquid $G_L$ and with sufficiently low growth rate $R$. Under these conditions, *breakdown* of the plane front interface is avoided (this important topic is considered in Chap. 3).

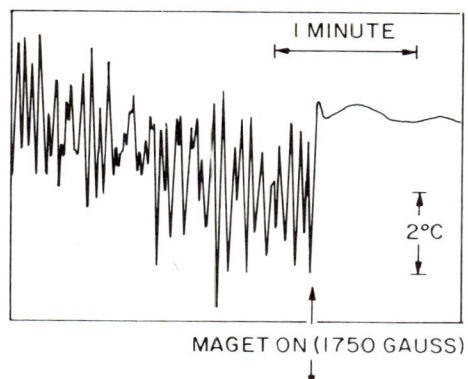
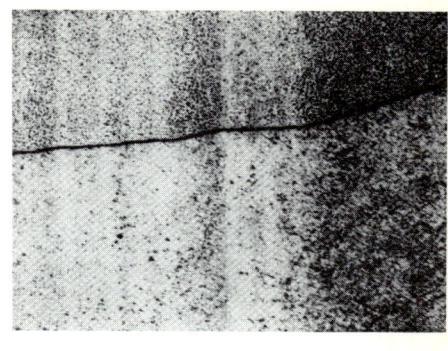

FIGURE 2-16
Thermal convection and banding in Te-doped InSb. *Left:* Temperature fluctuations in front of the liquid result from thermal convection and are eliminated by imposition of a steady magnetic field; *right:* Te-rich bands which result from the temperature fluctuations. (*From Utech and Flemings.*[37])

# REFERENCES

*1* GULLIVER, G. H.: "Metallic Alloys" (Appendix), Charles Griffin & Co., Ltd., London, 1922.
*2* SCHEIL, E.: *Z. Metallk.*, **34**:70 (1942).
*3* PFANN, W. G.: *Trans. AIME*, **194**:747 (1952).
*4* TILLER, W. A., JACKSON, K. A., RUTTER, J. W., and CHALMERS, B.: *Acta Met.*, **1**:428 (1953).
*5* SMITH, V. G., TILLER, W. A., and RUTTER, J. W.: *Can. J. Phys.*, **33**:723 (1955).
*6* POHL, R. G.: *J. Appl. Phys.*, **25**:668, 1170 (1954).
*7* BURTON, J. A., PRIM, R. C., and SLICHTER, W. P.: *J. Chem. Phys.*, **21**:1987 (1953).
*8* WAGNER, C.: *J. Metals*, **6**:154 (1954).
*9* HURLE, D. T. J., JAKEMAN, E., and PIKE, E. R.: *J. Crystal Growth*, **3**:633 (1968).
*10* HURLE, D. T. J., and JAKEMAN, E.: *J. Crystal Growth*, **5**:227 (1969).
*11* DEAN, F. V., KERR, J. R., and HELLAWELL, A.: *J. Inst. Metals*, **90**:234 (1962).
*12* ZIEF, M., and WILCOX, W. R.: "Fractional Solidification," Marcel Dekker, Inc., New York, 1967.
*13* YUE, A. S.: *J. Inst. Metals*, **91**:166 (1962).
*14* SLOAN, G. J.: unpublished, E. I. DuPont de Nemours & Co., Wilmington, Del., 1964.
*15* BEUTEL, J.: "Water Recovery by Reiterative Freezing," Minneapolis-Honeywell Regulator Co., Minneapolis, Minn., 1962.
*16* TAGIROV, V. I., and KULIEV, A. A.: *Soviet Phys.-Solid State* (*English Trans.*), **3**:1944 (1962); *Fiz. Tverd. Tela*, **3**:2669 (1963).
*17* BURTON, J. A., KOLB, E. D., SLICHTER, W. P., and STRUTHERS, J. D.: *J. Chem. Phys.*, **21**:1991 (1953).

*18* KODERA, H.: *J. Appl. Phys. (Japan)*, **2**:212 (1963).
*19* ANDREEV, G. A., and ALEKSANDROV, B. N.: *Soviet Phys.-Solid State (English Transl.)*, **7**:135 (1965); *Fiz. Tverd. Tela*, **7**:177 (1965).
*20* JOHNSTON, W. C., and TILLER, W. A.: *Trans. AIME*: **221**:331 (1961).
*21* WILCOX, W. R.: "Fractional Crystallization from Melts," doctoral dissertation, University of California, Berkeley, 1960; also *Lawrence Rad. Lab. Rept.* UCRL-9213, Berkeley, 1960.
*22* OLIVER, B. F.: *Trans. AIME*, **230**:1352 (1964).
*23* KERN, W.: *J. Electrochem. Soc.*, **110**:60 (1963).
*24* BRIDGERS, H. E.: *J. Appl. Phys.*, **27**:746 (1956).
*25* PFANN, W. G.: "Zone Melting," 2d ed., John Wiley & Sons, Inc., New York, 1966.
*26* REISS, H.: *Trans. AIME*, **200**:1053 (1954).
*27* ELDIB, I. A.: *Ind. Eng. Chem., Process Des. Develop.*, **1**:2 (1962).
*28* BRICE, J. C.: The Growth of Crystals from the Melt, in "Selected Topics in Solid State Physics," vol. 5, John Wiley & Sons, Inc., New York, 1965.
*29* BARDSLEY, W., private communication, Malvern, England, 1970.
*30* BRODY, H. D., and FLEMINGS, M. C.: *Trans. Met. Soc. AIME*, **236**:615 (1966).
*31* HULME, K. F., and MULLIN, J. B.: *Phil. Mag.*, **4**:1286 (1959).
*32* BARDSLEY, W., MULLIN, J. B., and HURLE, D. T. J.: "Solidification of Metals," p. 93, *Iron and Steel Inst. Publ. No.* 110, London, 1968.
*33* COCKAYNE, B.: *J. Crystal Growth*, **3**:60 (1968).
*34* COLE, G. S., and WINEGARD, W. C.: *Can. Met. Quart.*, **1**:29 (1962).
*35* KOMAROV, G. V., and REGEL, A. R.: *Soviet Phys.-Solid State (English Transl.)*, **5**:563 (1963); *Fiz. Tverd. Tela*, **5**:773 (1963).
*36* MULLER, A., and WILHELM, M.: *Z. Naturforsch*, **19a**:254 (1964).
*37* UTECH, H. P., and FLEMINGS, M. C.: *J. Appl. Phys.*, **37**:2021 (1966).
*38* UTECH, H. P., and FLEMINGS, M. C.: in H. S. Peiser (ed.), "Crystal Growth," p. 651, Pergamon Press, New York, 1967.
*39* DASH. W. C.: *J. Appl. Phys.*, **29**:736 (1958).
*40* GOSS, A. J., BENSON, K. A., and PFANN, W. G.: *Acta Met.*, **4**:332 (1956).
*41* JACKSON, K. A.: in "Solidification," p. 121, American Society for Metals, Metals Park, Ohio, 1971.

## PROBLEMS

2-1 A Ge–Ga ingot containing 10 ppm Ga is solidified at $R = 8 \times 10^{-3}$ cm/s with negligible convection. Show schematically, the composition along the length of the fully solidified ingot, giving the initial composition and lengths of the initial and final transients. Assume $D_L = 5 \times 10^{-5}$ cm²/s, $k = 0.1$.

2-2 The rate of solidification of the above ingot is increased abruptly by a factor of 5 and held constant at this new velocity. Sketch the composition variation along the resulting band, giving the maximum composition and approximate thickness of the band.

## PROBLEMS

**2-3** A Ge–Ga ingot 10 cm long containing 10 ppm Ga is zone-leveled using one pass and a zone length of 0.5 cm. What fraction of the ingot is within $\pm 5$ percent of the average composition? Assume $k' = k$.

**2-4** A Ge–Ga crystal is grown by normal freezing with forced convection so that $\delta = 0.005$ cm. Initial composition is 10 ppm Ga. Assume $D_L = 5 \times 10^{-5}$ cm$^2$/s, $k = 0.1$.
  (a) For a solidification rate of $8 \times 10^{-3}$ cm/s, what will be the composition of the solid forming when the crystal is 50 percent solidified?
  (b) How much lower would the solidification rate need to be to make it reasonable to assume complete liquid diffusion?
  (c) How much higher would the solidification rate need to be to obtain a crystal of essentially uniform composition?

**2-5** An Al–1% Cu alloy is normally solidified with $k' = k$. The phase diagram for this alloy is schematically as in Fig. 2-3 with $C_E = 33\%$ Cu, $C_{sM} = 5.65\%$ Cu, $T_M = 660°C$, and $T_E = 548°C$.
  (a) How much eutectic will be present in the finally solidified bar assuming no solid diffusion?
  (b) At a solidification velocity of $10^{-3}$ cm/s, do you expect significant diffusion in the solid? [Note: $D_s = 0.29\ e^{-31,120/RT}$ cm$^2$/s; $R = 1.99$ cal/(mol)(°K).]

**2-6** List as many ways as you can think of whereby you could obtain controlled-composition fluctuations in an ingot solidified "normally."

**2-7** Describe a crystal-growing apparatus you might build to grow an Al–2% Cu alloy crystal, $\frac{1}{4}$ in. diameter, with minimum convection.

**2-8** Show schematically how partial volatilization of the solute element would change the solute-distribution curve of Fig. 2-3c, assuming (a) a large furnace chamber and (b) a small closed furnace chamber.

**2-9** Describe the steps you would follow to grow a nickel single crystal with as low a dislocation count as possible.

**2-10** A small Ge–Ga crystal is grown with a plane front by the Czochralski technique with forced convection so that $\delta = 0.005$ cm. Melt composition is 10 ppm G. Assume $D_L = 5 \times 10^{-5}$ cm$^2$/s, $k = 0.1$. Plot the composition along the length of the crystal for (a) a very slow rate, (b) $8 \times 10^{-3}$ cm/s, and (c) a very fast rate.

**2-11** What is the thickness of the diffusion boundary layer of a crystal growing at a velocity of 2.5 cm/h? If the crystal is 10 cm long, show that the initial and final transients are of negligible thickness.

**2-12** Plot schematically a solute-distribution curve for normal solidification comparable to that of Fig. 2-3a but for an alloy in which $k > 1$.

**2-13** Plot schematically a solute-distribution curve for a Si–50% Ge alloy (see Fig. 8-5). Indicate how you could obtain this curve quantitatively.

# 3

# CELLULAR SOLIDIFICATION

## CONSTITUTIONAL SUPERCOOLING AND CELL FORMATION

In the preceding chapter numerous examples have been given of how a solute-rich boundary layer builds up in front of a solidifying planar interface. Several early workers, among them Papapetrou,[1] understood qualitatively how such a solute buildup could lead to instability of the plane front. Not until the important studies of Chalmers and coworkers,[2,3] however, were the ideas quantified and applied to metal crystal growth from the melt.

Figure 3-1 shows qualitatively how the driving force for instability of the plane front develops. A solute-rich layer is present in front of a growing interface, in which liquid composition is a maximum $C_L^*$ at the interface and decreases with increasing distance from the interface. Now, with the aid of the phase diagram, it is a simple task to plot the equilibrium liquidus temperature of the liquid as a function of distance from the interface; this is done in Fig. 3-1c and d. The equilibrium liquidus temperature increases with distance from the interface because the lower the solute content, the higher the liquidus temperature. Next, the actual temperature in the growing crystal is superimposed on the same graph. Since equilibrium is assumed at the solid-

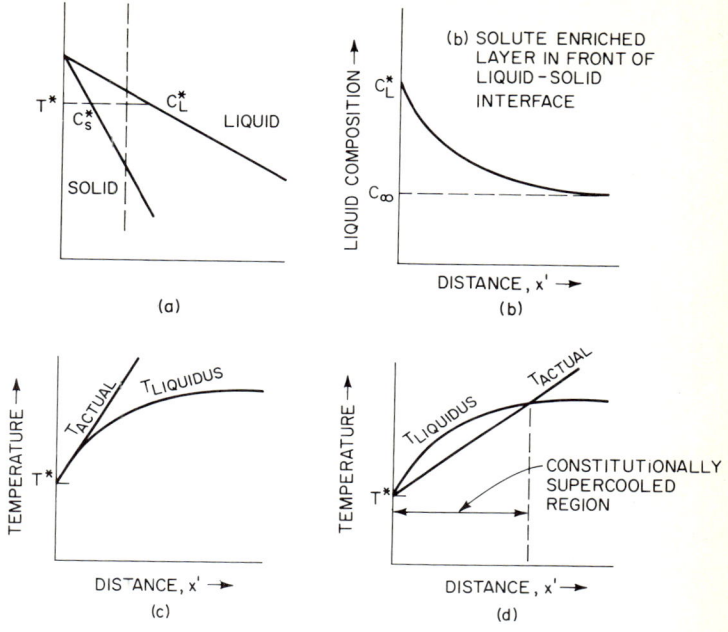

**FIGURE 3-1**
Constitutional supercooling in alloy solidification. (*a*) Phase diagram; (*b*) solute-enriched layer in front of liquid-solid interface; (*c*) stable interface; (*d*) unstable interface.

liquid interface, this curve must pass through $T^*$ at the interface $x' = 0$, but otherwise its shape is dictated by heat flow.

The curves in Fig. 3-1c show a condition where the interface is exactly at the equilibrium liquidus temperature and where every point in front of the interface is at a temperature above the liquidus. This represents the condition necessary for stable plane front solidification. If an instability causes a protuberance to form on the flat interface, it will find itself in a superheated environment and will melt back. Figure 3-1d, on the other hand, represents an unstable case; here, liquid immediately in front of the interface is at an actual temperature that is below its equilibrium liquidus temperature. It is, therefore, supercooled. Chalmers and coworkers termed this *constitutional supercooling*; the word *constitutional* indicates that the supercooling arises from a change in composition, not temperature. According to the constitutional supercooling theory, this supercooling results in instability of the plane front since any protuberance forming on the interface would find itself in supercooled liquid and therefore would not disappear. A quantitative treatment is given below. The agreement of this simple theory with experiment is excellent even though it neglects a number of factors that have been included in the later theoretical work on interface stability by Mullins and Sekerka[4-7] and others.

To develop quantitatively the constitutional supercooling criterion, we need consider heat and mass flow only at the interface. The gradient of solute in the liquid at the interface is as given by Eq. (2-6):

$$\left(\frac{dC_L}{dx'}\right)_{x'=0} = -\frac{R}{D_L} C_L^*(1-k) \tag{3-1}$$

Assuming equilibrium at the flat interface, the slope of the curve of the equilibrium liquidus temperature $T_L$ versus distance from the interface $x'$ is related to that of liquid composition $C_L$ by the slope of the liquidus line $m_L$:

$$\left(\frac{dT_L}{dx'}\right)_{x'=0} = m_L \left(\frac{dC_L}{dx'}\right)_{x'=0} \tag{3-2}$$

Constitutional supercooling is absent when the actual temperature gradient in the liquid at the interface $G_L$ is equal to or greater than $(dT_L/dx')_{x'=0}$. Combining this statement with Eq. (3-1) and (3-2) and letting $C_s^* = kC_L^*$ gives the general constitutional supercooling criterion;[8,9] that is, a plane front is stable when

$$\frac{G_L}{R} \geq -\frac{m_L C_s^*(1-k)}{kD_L} \tag{3-3}$$

Equation (3-3) is applicable regardless of the presence or absence of convection since a laminar layer exists next to the solidifying interface regardless of degree of convection. $C_s^*$ is given by the Burton-Prim-Slichter relation [Eq. (2-10)]. At steady state, with no convection, $C_s^* = C_0$ and Eq. (3-3) becomes the constitutional supercooling criterion originally derived by Chalmers et al.:

$$\frac{G_L}{R} \geq -\frac{m_L C_0(1-k)}{kD_L} \tag{3-4}$$

When convection is sufficiently vigorous that, from a solute-redistribution standpoint, diffusion in the liquid is complete, there still exists a laminar layer and its small but finite concentration gradient. In this case,

$$\frac{G_L}{R} \geq -\frac{m_L C_\infty(1-k)}{D_L} \tag{3-5}$$

where $C_\infty$, the bulk liquid composition, is equal to $C_0$ for a small amount of solidification from a large melt.

A very large amount of qualitative and quantitative confirmation of the constitutional supercooling theory has been obtained since its formulation in 1953. The most direct qualitative confirmation is on transparent organic materials where flat

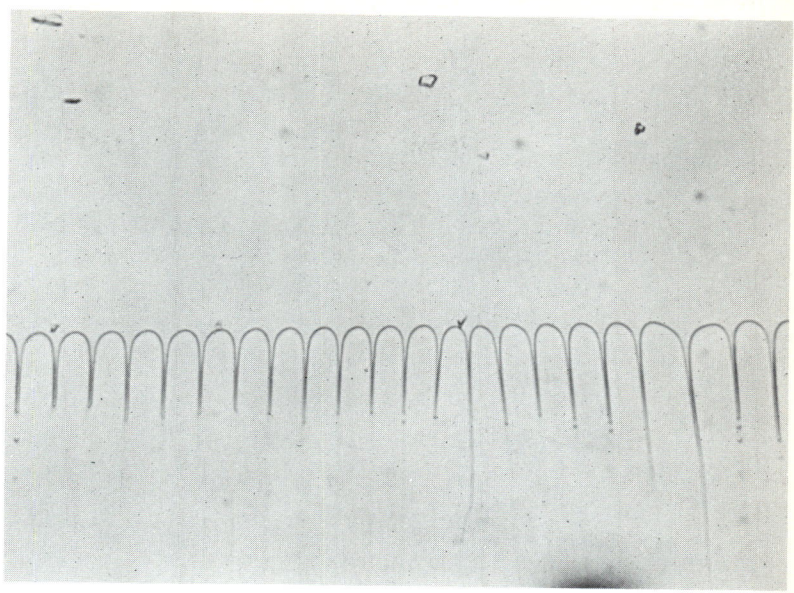

FIGURE 3-2
Cellular interface in a transparent organic (carbon tetrabromide). (Magnification × 60.) (*From Jackson and Hunt.*[10])

interfaces are seen to become rippled and then *cellular* as the ratio $G_L/R$ is decreased. An example, from work of Jackson and Hunt[10] is shown in Fig. 3-2. Many studies on metal systems have provided quantitative confirmation of the theory.[11-16] One of the earliest of these was by Walton et al.,[11] who solidified Sn–Pb alloys in a horizontal boat under various thermal conditions and examined the liquid-solid interface after decanting; their results are plotted in Fig. 3-3 as liquid composition versus $G_L/R$. Points at the upper left denote the cellular interface; those at the lower-right the plane front. A straight line is seen to separate the plane front reasonably well from the nonplane front experiments. Assuming negligible liquid convection, Eq. (3-5) applies and this line is then expected to have the slope $m_L(1-k)/kD_L$. Substitution of handbook values for $m_L$ and $k$ yields a calculated value for the diffusion coefficient $D_L$ of $2 \times 10^{-5}$ cm²/s. A more recent similar study of Cole and Winegard[12] yields a value of $D_L$ of $4.5 \times 10^{-5}$ cm²/s. Both results are close to that expected from published diffusion data.

Recent work on the same alloy system but in an intentionally stirred melt showed similar results.[14] Here, the applicable stability equation is Eq. (3-3). Results are plotted in Fig. 3-4 according to that expression except with the solid temperature gradient $G_s$ in place of $G_L$. The experiments were conducted at constant $G_s$. At low

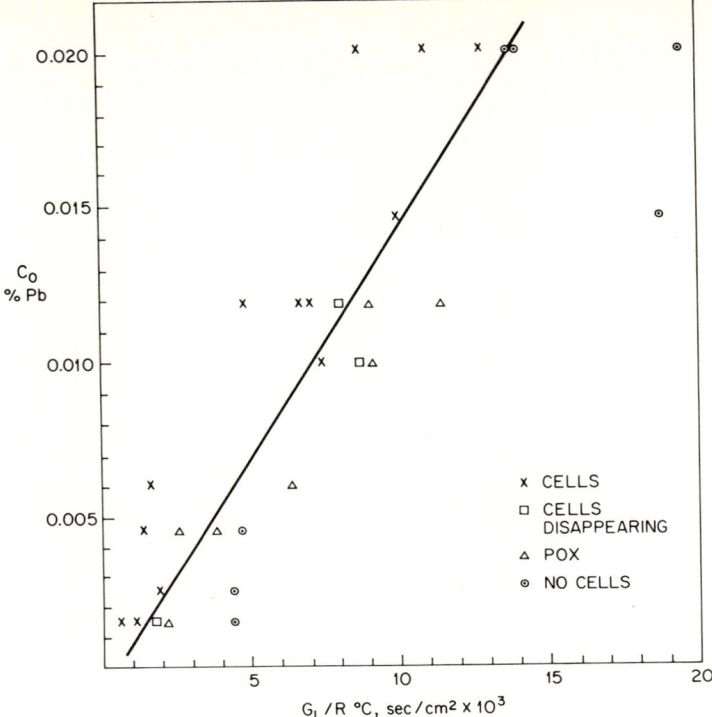

**FIGURE 3-3**
Conditions for interface stability in Sn–Pb alloys. (*From Walton et al.*[11])

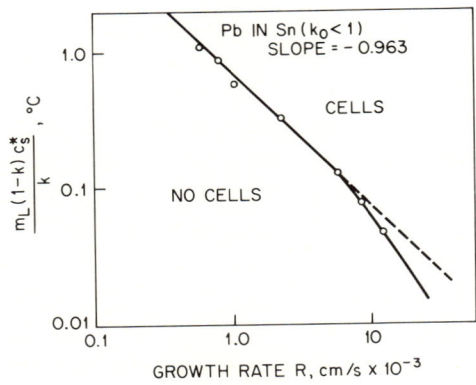

**FIGURE 3-4**
Stability in convecting Pb–Sn melts. Temperature gradient in solid $G_s$ was 18°C/cm. Czochralski growth. (*From Hunt et al.*[14])

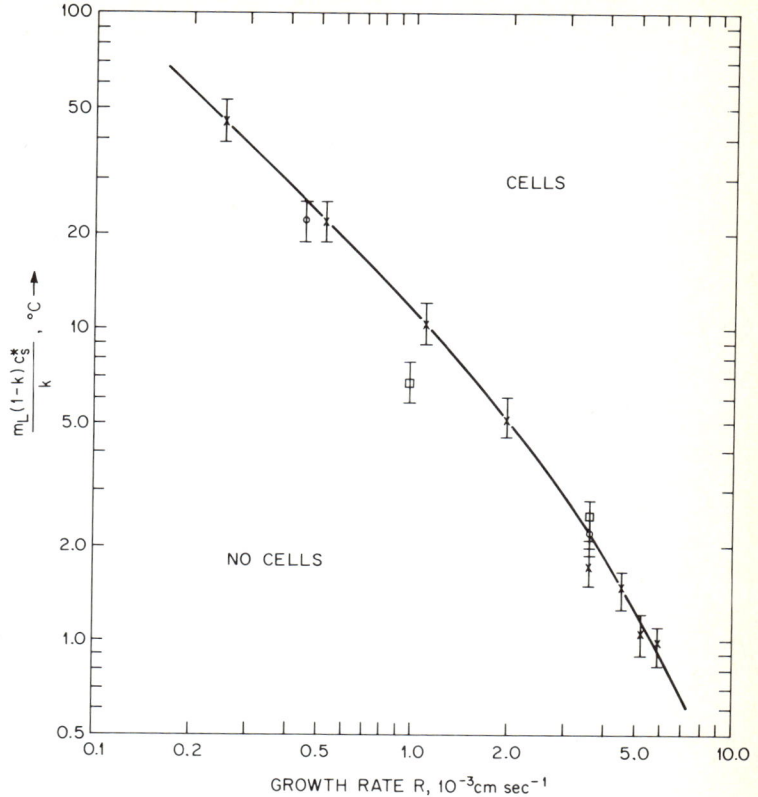

FIGURE 3-5
Stability in convecting Ge–Ga melts. Solid temperature gradient was 95°C/cm. Circles and squares represent two different crystal-rotation rates. Czochralski growth. (*From Bardsley et al.*[15])

growth rate, the curve is linear with very nearly a slope of unity, as expected from Eq. (3-3). The diffusion coefficient calculated from this portion of the curve is $1.12 \times 10^{-5}$ cm²/s, in approximate agreement with the values calculated from the experiments cited above. Deviation of the curve from linearity at higher growth rates is because $G_L$ is a function of growth rate at higher growth rates for constant $G_s$ [see Eq. (1-5)]. Closely similar results to those of Fig. 3-4 are shown in Fig. 3-5; but these are for gallium-doped germanium from the work of Bardsley, Callan, Chedzey, and Hurle.[15] Again, back calculation of the liquid diffusion coefficient gives a calculated value close to that obtained from separate measurements.

The various experiments outlined above and many similar ones indicate that the constitutional supercooling theory closely predicts conditions required to initiate breakdown of a plane front in metals and semiconductors solidifying without facet formation. Alternatively, the constitutional supercooling required to break down a plane front in such materials is so small that it is within experimental error. Of course, it must be recognized that experimental error in these studies is relatively large partly because the decision as to whether or not a front is exactly plane involves some degree of subjectivity.

## INTERFACE STABILITY THEORY

The constitutional supercooling theory deals with the question of which state, solid or liquid, is thermodynamically stable in front of an initially plane front interface. If liquid, the interface is assumed to remain plane; if solid, the interface is assumed to break down. However, thermodynamics represent only part of the story; the other part is concerned with what state is dynamically achievable, and it is to this that Mullins and Sekerka[4-7] and others have recently turned.

In plane front crystal growth, the problem considered is one where the interface is initially plane and a small perturbation is then imagined to form. Whether this perturbation will grow (as sketched in Fig. 3-6) or shrink depends on the interaction of the (now-perturbed) solute and thermal fields, on liquid-solid surface energy, and on interface kinetics. In the initial treatment of Mullins and Sekerka[4] and Sekerka,[5] the following assumptions were made: equilibrium at the liquid-solid interface, isotropic surface energy, and no convection. The stability equation then obtained is

$$\frac{G_L}{R} + \frac{\rho_L H}{2K_L} \geq -\frac{m_L C_0(1-k)}{kD_L} \frac{K_S + K_L}{2K_L} \mathscr{S} \qquad (3\text{-}6)$$

where $K_S$ and $K_L$ are the thermal conductivities of the solid and liquid, respectively, $\rho_L$ is liquid density, $H$ is heat of fusion, $\mathscr{S}$ is a dimensionless stability function, and other terms are as in the constitutional supercooling criterion [Eq. (3-4)]. The stability function $\mathscr{S}$ is related graphically to a dimensionless number $A$ in Fig. 3-7, where

$$A = -\frac{k^2 \gamma R T_M}{(1-k)\rho_L H D_L m_L C_0} \qquad (3\text{-}7)$$

$\gamma$ is liquid-solid surface energy, and $T_M$ is the equilibrium melting temperature of the pure solvent.

Equation (3-6) differs from the constitutional supercooling criterion for stability of an unstirred liquid only by the quantities $L/2K_L$, $(K_S + K_L)/2K_L$, and $\mathscr{S}$. When these quantities approach 0, 1, and 1, respectively, Eq. (3-6) becomes identical with the constitutional supercooling criterion. The reader can readily verify that for the experiments cited in this chapter, the interface stability criterion and constitutional

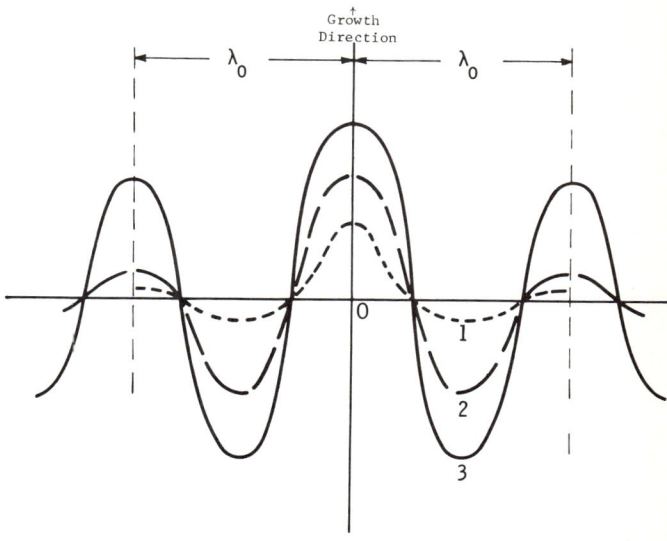

**FIGURE 3-6**
Perturbation forming at liquid-solid interface. Perturbation is stable. It grows in the overall growth direction and expands laterally with wavelength $\lambda_0$. (*From Sekerka.*[7])

supercooling criterion predict conditions for interface breakdown that are essentially identical. For example, in the experiments on tin of Walton et al.,[11] $L/2K_L$ was about 600°C/cm² whereas $G_L/R$ was in the range of 2000 to 20,000°C. The term $(K_S + K_L)/2K_L$ is within about 50 percent of unity, and $\mathscr{S}$ is within about 20 percent of unity.

Experiments have not yet been reported which could distinguish between these two criteria for plane front solidification. However, Eq. (3-6) does predict certain differences which should be experimentally observable. At sufficiently low values of $C_0$ (and hence low $G_L/R$), the term $L/2K_L$ becomes important and eventually overriding. Materials such as ice or other crystalline nonmetallics with relatively high $L/2K_L$ would be specially likely to show this effect. Also, in nonmetallics, the term $(K_S + K_L)/2K_L$ may differ by a factor of 2 or more from unity. Finally, at sufficiently high growth rates, the parameter $\mathscr{S}$ is seen to differ appreciably from unity, approaching zero as a lower limit, as the dimensionless number $A$ approaches unity. At this limit, a region of *absolute stability* is predicted. Unfortunately, for typical values of the various quantities comprising $A$ for metals, the velocity required for $\mathscr{S}$ to differ appreciably from unity is of the order of a centimeter per second or more, and it seems doubtful that the assumption of interface equilibrium, even for metals, is valid at these rates.

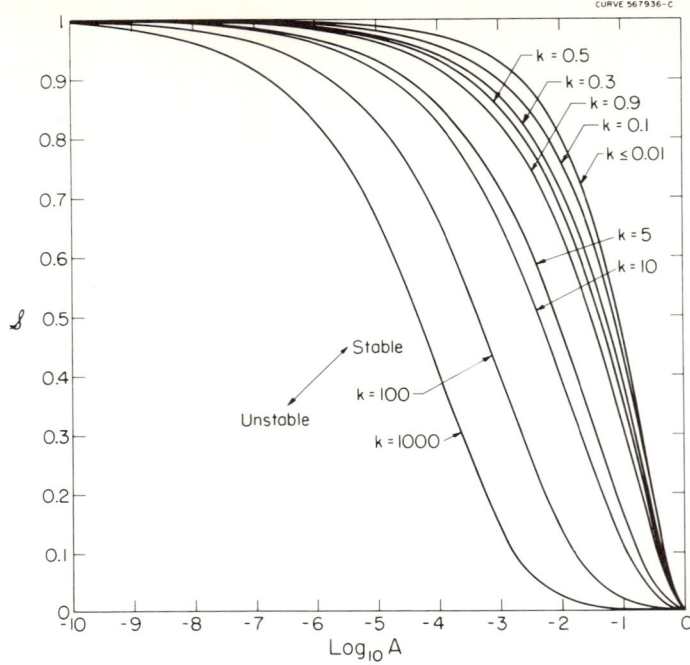

**FIGURE 3-7**
Dimensionless stability parameter $\mathscr{S}$ versus $\log_{10} A$. (*From Sekerka.*[5])

A different way of comparing interface stability theory with experiment is by use of Sekerka's time-dependent theory to predict the wavelength of the fastest growing instabilities which form on interface breakdown. Calculated curves, such as those in Fig. 3-6, are compared with the interface shapes observed experimentally. Hardy and Coriell[17] obtained good agreement between the theory and experiment in this way, using pure water. The water was supercooled below its equilibrium freezing temperature, and so the driving force for the instability was *thermal supercooling* rather than *constitutional supercooling*.

## CELL STRUCTURE

Experiments on transparent organic liquids show that as a planar interface becomes unstable, it first becomes gently undulatory, with protrusions later developing into the fully formed *cells* shown in Fig. 3-2. These observations are in qualitative agreement with predictions of the time-dependent interface stability theory. If grain boundaries are present, the grooves associated with these boundaries act as *built-in* distortions of the plane front and interface breakdown begins here, spreading outward to other portions of the crystal. Figure 3-8 is a schematic example of this initial breakdown from Schaefer and Glicksman,[18] based on their observations on solidification of a transparent organic alloy (impure succinonitrile).

One good way to observe cellular structures in metallic alloys is to interrupt the solidification process suddenly by a rapid quench. This method works best for alloys containing a relatively large amount of solute (that is, 1 percent or more). Figure 3-9 shows two structures in an Al–Cu alloy obtained in this way by Sharp and Hellawell.[19,20] These structures are closely similar to those that have been observed directly on transparent organic alloys. Transverse sections through fully developed cells, as shown at the bottom of Fig. 3-9, generally show them to be regularly spaced with six sides. An example is shown in Fig. 3-10; of course, this is the pattern expected to result from cylinders growing in a close-packed array.

As growth conditions depart from those required for plane front, a number of transition structures are observed before well-developed hexagonal cells, such as those of Fig. 3-10, are obtained. The structures obtained depend also on crystallographic orientation; examples are shown in Fig. 3-11. These are transverse sections of two Pb–Sb crystals grown with the ratio $G_L/R$ just slightly less than that required for stability. In the crystal growing in a $\langle 110 \rangle$ direction, elongated cells develop, with the segregate concentrated in the planar intercellular regions. In the crystal growing in a $\langle 100 \rangle$ direction, the segregate is concentrated in roughly cylindrical regions termed *nodes*. At low values of $G_L/R$, the regular structure of Fig. 3-10 develops regardless of orientation.

The liquid-solid interface for the two orientations discussed above is shaped as shown schematically in Fig. 3-12. Furrows develop on the interface in the case of the elongated cells, and regularly spaced depressions develop in the case of the nodes. A probable, although superficial, explanation of this effect of crystallography on interface morphology is as follows. In the case of the $\langle 100 \rangle$-growth direction in cubic metals, a perturbation on the interface will have four symmetrically placed regions of $\{111\}$ orientation; this is the closest packed plane and is expected to be slowest growing. Thus, these four slow growing regions "guide" the shape of the perturbation to be of roughly equal dimensions transverse to the heat-flow direction. The relatively small amount of segregate in dilute alloys with high $G_L/R$ then concentrates in the lowest positions of the interface as nodes. In the case of the $\langle 110 \rangle$-oriented crystal, on the other hand, a perturbation will have on it two, not four, such slow-growing planes, guiding the interface morphology to a furrowed structure and leading to *elongated cells*. (The effect of interface kinetics on growth morphology is discussed in Chap. 9.)

Structures between the extremes discussed above have been described.[21–24] As

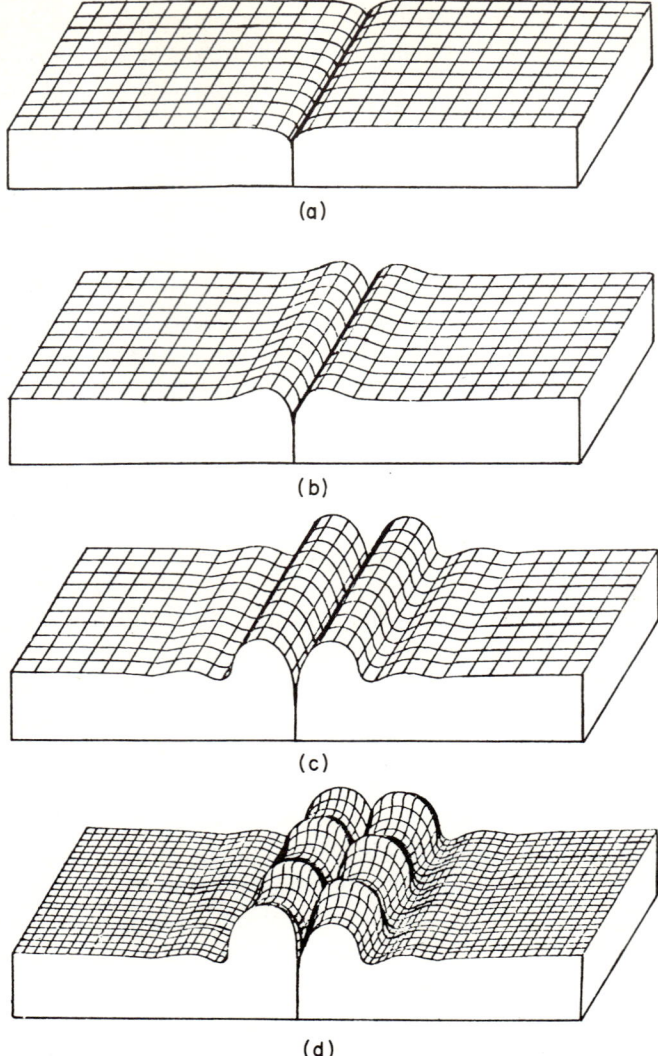

**FIGURE 3-8**
Modes of breakdown of planar interface near a grain boundary. (*a*) Grain boundary produces groove in equilibrated interface; (*b*) when growth starts, regions adjacent to grain boundaries grow most rapidly and form parallel ridges; (*c*) as ridges become large, secondary ridges appear beside them; (*d*) primary ridges break down into periodic rows of hills. (*From Schaeffer and Glicksman.*[18])

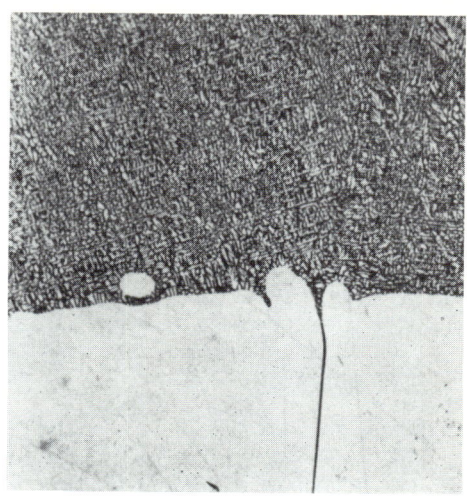

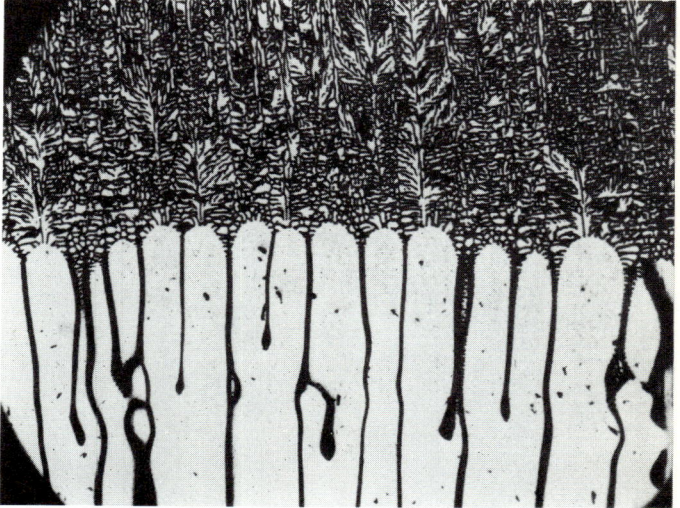

**FIGURE 3-9**
Interface breakdown in Al–Cu alloy. *Top:* Initial perturbations on plane front at grain boundary grooves; *bottom:* fully developed cells. (*From Sharp and Hellawell.*[19,20])

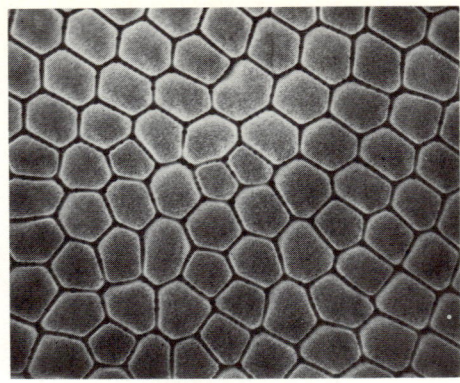

FIGURE 3-10
Transverse section of cellularly solidified Pb–Sb alloy. (Magnification ×48). (*From Morris and Winegard.*[21])

an example, the node structure of Fig. 3-11a assumes an irregular elongated-cell structure as $G_L/R$ is increased before finally becoming a "regular" structure. The elongated cells gradually segment as $G_L/R$ is increased, also becoming finally regular cells.[21] Since regular cells are obtained at high $G_L/R$ regardless of orientation, they are primarily a result of solute diffusion. However, crystallography exerts an influence in that, especially at high growth rates, the cellular direction deviates from the heat-flow direction toward the *dendrite* direction (that is, $\langle 100 \rangle$ for cubic metals[21,25]).

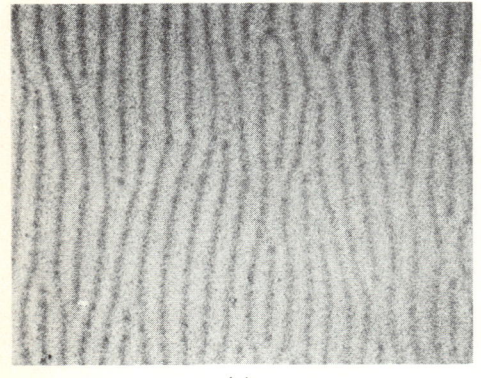
(a)

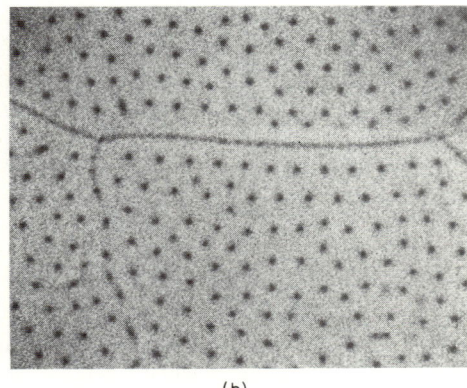
(b)

FIGURE 3-11
Variation in the initial breakdown morphology with crystal orientation in a Pb–Sb alloy; transverse sections (magnification ×24). (a) Elongated cells growing near a $\langle 110 \rangle$ direction; (b) nodes growing in a $\langle 100 \rangle$ direction. The continuous dark lines represent Sb segregation to striation boundaries that intersected the interface during growth. (*From Morris and Winegard.*[21])

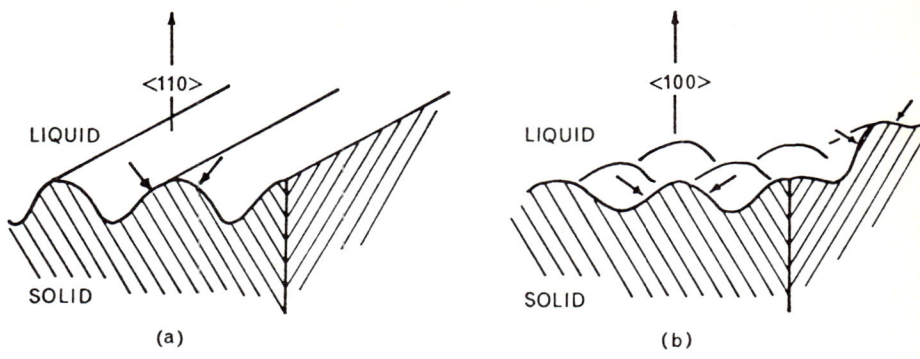

FIGURE 3-12
Schematic representation of developing perturbations for two crystalline orientations. Locations of regions of {111} orientation on the perturbations are shown by short arrows. (a) ⟨110⟩ growth direction (elongated cells develop); (b) ⟨100⟩ growth direction (nodes develop). (*From Morris and Winegard.*[21])

All the structures shown above have been of metallographic sections of solidified specimens. An alternative way of studying cellular structures that has been used by many investigators is to decant the bulk liquid during solidification. The liquid-solid interface is then examined directly. A disadvantage of this technique is that a thin film of liquid always remains on the surface.[26,27] Liquid in deep intercellular grooves is also not removed. Nonetheless, this technique shows directly the larger-scale structural features and has been used successfully by a number of investigators. An example of a structure obtained in this way is shown in Fig. 3-13.

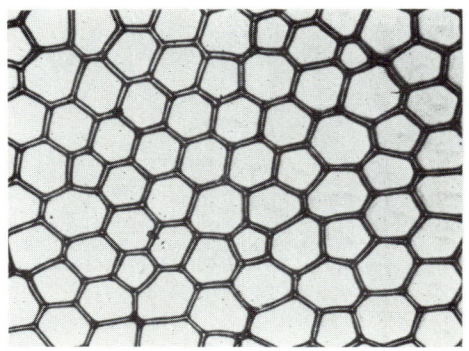
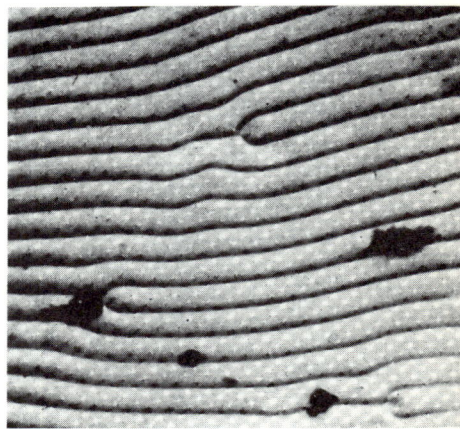

FIGURE 3-13
Decanted interface of cellularly solidified Pb–Sn alloys. (Magnification ×150.) (*From Chadwick.*[23])

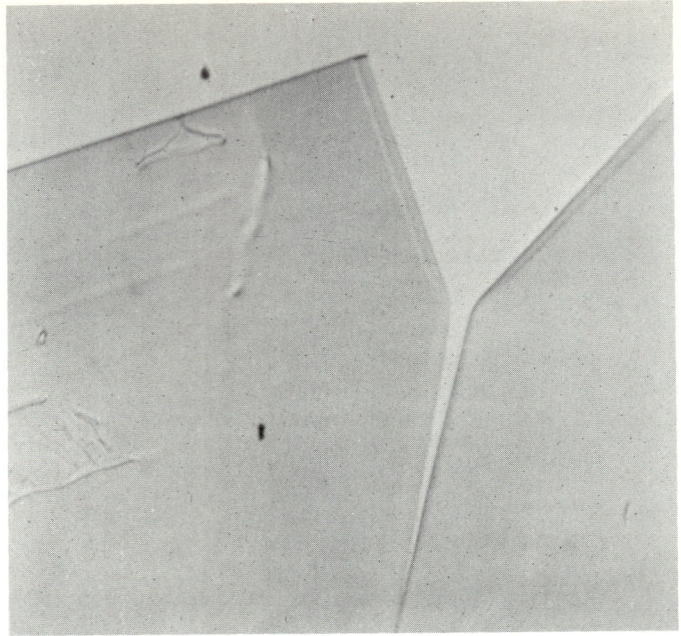

**FIGURE 3-14**
A transparent organic (salol) growing with faceted cells. (Magnification ×60.) (*From Jackson and Hunt.*[10])

SUCCESSIVE POSITIONS OF LIQUID-SOLID INTERFACE IN GE ALLOY, SHOWING DEVELOPMENT OF FACETED CELLS. (FROM BARDSLEY ET AL[28].)

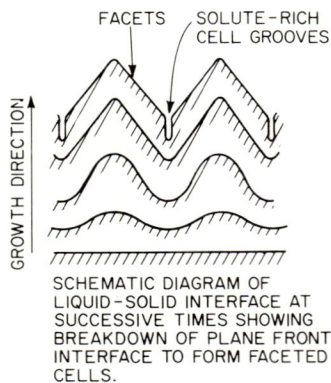

SCHEMATIC DIAGRAM OF LIQUID-SOLID INTERFACE AT SUCCESSIVE TIMES SHOWING BREAKDOWN OF PLANE FRONT INTERFACE TO FORM FACETED CELLS.

**FIGURE 3-15**
Development of cells in a faceting alloy. (*From Bardsley, Boulton, and Hurle.*[28])

 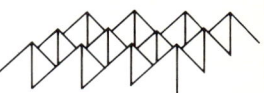

ELONGATED CELLS WHICH FORM WHEN GROWTH DIRECTION IS < 110 >. FACETS ON CELL SIDES ARE {111} PLANES INCLINED AT 54° 74' TO THE GROWTH AXIS. FACETS ON CELL ENDS ARE {111} PLANES PARALLEL THE GROWTH AXIS.

REGULAR CELLS WHICH FORM WHEN THE GROWTH DIRECTION IS < 100 >. FACETS ON CELL FACES ARE {111} PLANES 35°46' TO THE GROWTH DIRECTIONS.

**FIGURE 3-16**
Schematic illustration of cell morphology in germanium alloys.

The development of cells in materials whose liquid-solid interface develops facets is closely similar to that in the nonfaceted materials studied above. Figure 3-14 is an example of faceted cells in a transparent organic material, and Fig. 3-15 shows development of similar cells in a semiconductor from work by Bardsley, Boulton, and Hurle.[28] The latter authors and coworkers have studied cell formation in a variety of faceting alloys, including both semiconductors and melt-grown oxides.[15, 28-31]

Germanium single crystals, when grown in any orientation except the $\langle 111 \rangle$, break down to form cells as predicted by the constitutional supercooling criterion, i.e., at a very small amount of constitutional supercooling. The interface is first smoothly rippled (Fig. 3-15) and subsequently develops facets when portions of the interface reach the orientation of a $\{111\}$ plane. If sufficient solute is present, the deep-grain boundary grooves form at solute-enriched locations. Cell morphology is dependent on crystal orientation just as it is in metals. Elongated cells result when growth is in a $\langle 110 \rangle$ direction, and regular cells when growth is in a $\langle 100 \rangle$ direction, Fig. 3-16. Note the close similarity between Figs. 3-12 and 3-16; the formation of facets in the latter case is the only essential difference.

## FORMATION OF DENDRITES

When regular cells, such as those of Figs. 3-2 and 3-9, form and grow at relatively low rates, they grow perpendicular to the liquid-solid interface regardless of crystal orientation. When, however, growth rate is increased, crystallographic effects begin to exert an influence and the cell-growth direction deviates toward the preferred crystallographic growth direction (for example, $\langle 100 \rangle$ for cubic metals). Simultaneously, the cross section of the cell generally begins to deviate from its previously circular geometry owing to effects of crystallography. This structure has been described variously as a flanged structure, or *maltese cross*, in cubic materials; it is shown in Fig. 3-17. As growth rate increases still further, the cross structure first

# 74 CELLULAR SOLIDIFICATION

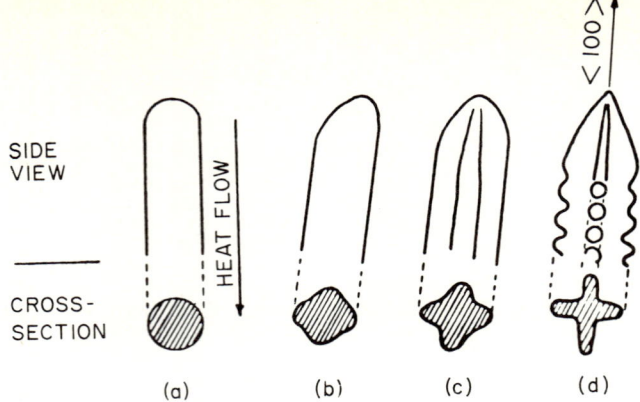

FIGURE 3-17
Sketch of the changing shape of the growth structure as the growth velocity is increased. (a) Regular cell growing at low velocity; (b) regular cell growing in $\langle 100 \rangle$ dendrite direction; (c) flanged cell; (d) dendrite exhibiting the start of periodic lateral branching. (*From Morris and Winegard.*[34])

becomes more apparent and then serrations begin to appear in the flanges of the cross; that is, secondary dendrite arms become discernible.[32-35]

The point at which the cellular structure in Fig. 3-17a becomes *dendritic* is a matter of terminology, and the terminology employed in the literature is not consistent. Some writers prefer to think of the structure as dendritic when it is growing in or near its crystallographic orientation, as in Fig. 3-17b. Others prefer to describe it as dendritic only when secondary branches can be discerned, as in Fig. 3-17d. The latter terminology is employed in this text.

Figure 3-18, which depicts a transparent organic liquid, shows one example of solidification where secondary arms are just discernible. Closely comparable structures have been described for a large number of metal alloys, including Pb–Sb alloys (Morris and Winegard,[34]) and aluminum alloys (Biloni et al.[36]). Figure 3-19 shows a range of structures in Fe–10% Ni alloy from cellular to dendritic. These structures were obtained in a unidirectionally solidified ingot, and so rate of heat removed decreased with the square of the distance from the chill. Near the chill, the structure is cellular although the shape of the cross section of the cells is seen to be influenced by crystallographic effects. At greater distances from the chill, the cross section becomes flanged and secondary dendrite arms become visible.

Another point on the terminology of dendritic structures will become more important in later chapters when we describe more complex dendritic structures: the difference between a *dendrite* and a *dendrite arm*. The central portion of the structure

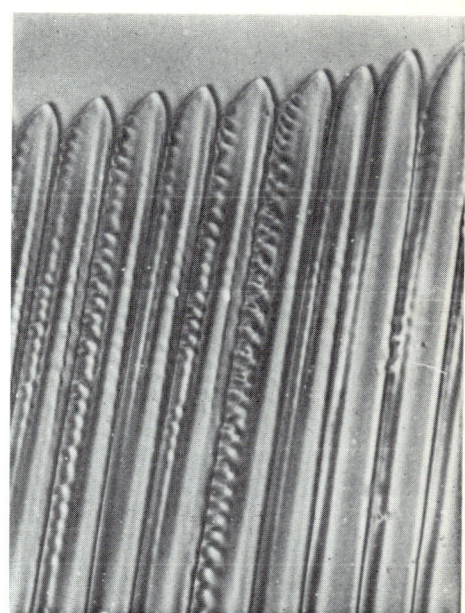

FIGURE 3-18
Dendrites with secondary arms just discernible. Transparent organic alloy (CBr$_4$). (*From Morris and Winegard.*[34])

in Fig. 3-17*d* which is growing in approximately the heat-flow direction is termed a *primary dendrite arm*; the rodlike protrusions perpendicular to the primary dendrite arm are *secondary dendrite arms*. In Fig. 3-18 and the upper part of Fig. 3-19, many such primary and secondary dendrite arms are seen. All the primary dendrite arms in these two figures have grown from the same nucleus and have nearly the same crystallographic orientation. They are thus a part of the same *grain*, or, in the terminology of this text, the same *dendrite*.

## CELLULAR-DENDRITIC TRANSITION

Quantitative description of effects of solidification variables on the cell-dendrite transition is one that has eluded investigators in spite of the many studies conducted over the last 15 years.[12,16,32,37] Qualitatively, secondary arms form because the approximately paraboloidal interface of the cell tip becomes unstable. The driving force for the instability is constitutional supercooling in the liquid just back from the cell (dendrite) tips. One would therefore expect that this instability should depend approximately on the same variables that enter into the constitutional supercooling theory for breakdown of a plane interface. Experimentalists who have studied the

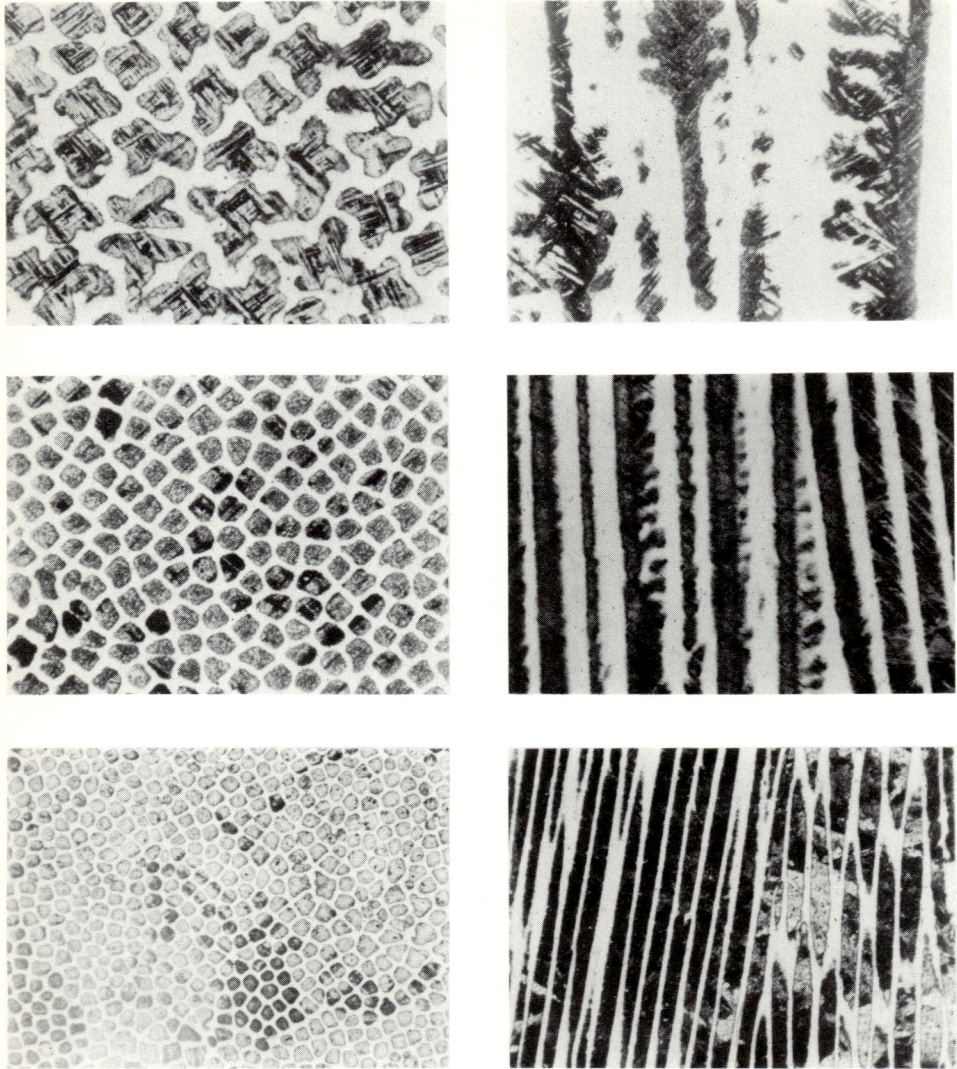

FIGURE 3-19
Dendrites in Fe–10% Ni alloy from directionally solidified ingot (magnification ×34). Samples from bottom to top at increasing distances from a chill. Gradient and growth rate decrease from bottom to top. *Left:* transverse sections; *right:* longitudinal sections.[46]

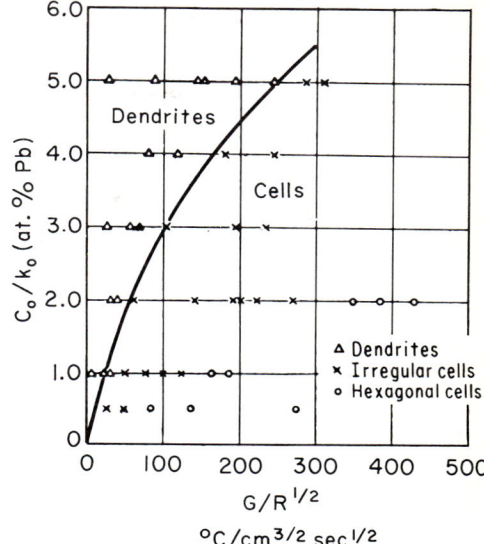

FIGURE 3-20
Growing conditions for the cell-to-dendrite transition in Sn–Pb alloys. (*From Plaskett and Winegard.*[32])

problem have plotted their data in various ways, generally as a function of $G/R^{1/2}$, where $G$ is the thermal gradient in the vicinity of the cell tips; Fig. 3-20 is an example from the work of Plaskett and Winegard.[32] As Davies[38] has pointed out, however, the data give little confidence that the transition is, in fact, a function of $G/R^{1/2}$. This entire topic remains a good one for theoretical and experimental study; it is discussed again in Chap. 5.

## SOLUTE REDISTRIBUTION IN CELLULAR SOLIDIFICATION

The problem of solute redistribution in cellular solidification is overwhelmingly complex if one approaches it with the same rigor applied to plane front solidification in Chap. 2. The multidimensional diffusion alone makes the problem difficult, and, in addition, there are the added complications of the effects of radius of curvature and solid-state diffusion. Detailed qualitative understanding of the problem is also made difficult because experimental data available are limited and, in some cases, have been obtained without adequate control of experimental variables (e.g., convection). Nonetheless, the data now available can be understood on the basis of the simple model discussed below.

The problem is shown schematically in Fig. 3-21a. Cells protrude into the liquid, and intercellular liquid becomes increasingly enriched with distance back from

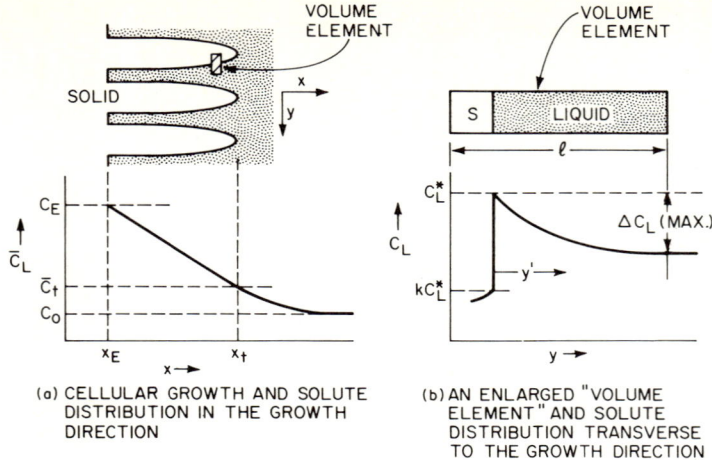

**FIGURE 3-21**
Solute redistribution in cellular growth.

(a) CELLULAR GROWTH AND SOLUTE DISTRIBUTION IN THE GROWTH DIRECTION

(b) AN ENLARGED "VOLUME ELEMENT" AND SOLUTE DISTRIBUTION TRANSVERSE TO THE GROWTH DIRECTION

the cell tips. Assuming a simple phase diagram containing a eutectic point, the maximum composition that can be attained by the intercellular liquid is $C_E$, as shown. The liquid does, in fact, often reach this maximum even when initial composition is a very small fraction of $C_E$.[33,36,39,40] The average liquid composition at any distance $x$ along the growth axis $\bar{C}_L$ then decreases to the average value $\bar{C}_t$ at the dendrite tips and, finally, to the bulk liquid composition at a distance far from the cell tips. Assuming no convection, this is $C_0$, initial liquid composition.

Now it is convenient to look at solute distribution in the $y$ direction (transverse to the growth direction) by taking a *volume element* which has a small thickness in the growth direction and which extends a distance $l$ from the center of a cell to the midpoint between two cell tips. The solute rejected as the cells thicken diffuses toward the center of the cells, and so the liquid in the $y$ direction is of nonuniform composition. The maximum composition difference at a given value of $x$ is shown as $\Delta C_L(\max)$ in Fig. 3-21b. Assuming equilibrium at the interface, the interface composition is $C_L^*$, the solid composition is $kC_L^*$, and the average composition $\bar{C}_L$ lies between $C_L^*$ and $C_L^* - \Delta C_L(\max)$.

If four assumptions are made in addition to those given above, the problem of solute redistribution in cellular solidification becomes simple.[41] These assumptions are: (1) isotherms are flat, perpendicular to the growth direction; (2) the cell spacing adjusts itself so that constitutional supercooling in intercellular regions is very small; (3) cell size is sufficiently coarse that the effect of radius of curvature on melting point can be neglected; and (4) solid-state diffusion is negligible.

The first two of the foregoing assumptions are equivalent to stating that at any distance $y$ behind the cell tips ($x < x_t$), the liquid is very nearly of uniform composition $C_L = C_L^*$, and that this liquid composition is given by the liquidus of the phase diagram. That is, assuming constant liquidus slope,

$$C_L - C_0 = \frac{1}{m_L}(T - T_L) \qquad \text{for } x \leq x_t \qquad (3\text{-}8)$$

where $T$ is temperature at $x$. Differentiating Eq. (3-8) with respect to $x$ and letting $G = \partial T/\partial x$, the thermal gradient in the temperature region where liquid and solid coexist ($x \leq x_t$), the concentration gradient $\partial C_L/\partial x$ is given by

$$\frac{\partial C_L}{\partial x} = \frac{G}{m_L} \qquad \text{for } x \leq x_t \qquad (3\text{-}9)$$

Solute diffuses down this concentration gradient and out ahead of the cell tips, and so the isoconcentrate at the cell tips is $C_t$, where $C_t > C_0$. At steady state, the requirement of no solute accumulation now gives

$$R(C_t - C_0) = -D_L \left(\frac{\partial C_L}{\partial x}\right)_{x=x_t} \qquad (3\text{-}10)$$

where $R$ is growth velocity. Combining Eqs. (3-9) and (3-10),

$$C_t = (1 - a)C_0 \qquad (3\text{-}11)$$

where

$$a = -\frac{D_L G}{m_L R C_0} \qquad (3\text{-}12)$$

The temperature of the dendrite tips $T_t$ is reduced to the liquidus temperature of the composition $C_t$, or, from Eqs. (3-8) and (3-10),

$$T_t = T_L - am_L C_0 \qquad (3\text{-}13)$$

A convenient way to compare the foregoing theory with experiment is to define an effective partition ratio for the tips $k_t$ as

$$k_t = \frac{kC_t}{C_0} \qquad (3\text{-}14)$$

For the case of no convection, $k_t$ varies in a simple way with the parameter $a$:

$$k_t = (1 - a)k \qquad (3\text{-}15)$$

The effective partition ratio $k_t$ can be calculated from experimentally measured cell-tip temperatures, as done by Kramer, Bolling, and Tiller.[40] Results of their study are in rough agreement with Eq. (3-15). Calculation of $k_t$ can also be done by quenching cell structures during solidification and then measuring compositions near the cell tips using an electron microprobe. The resulting composition $C_s^*$ is just equal to $kC_t$.

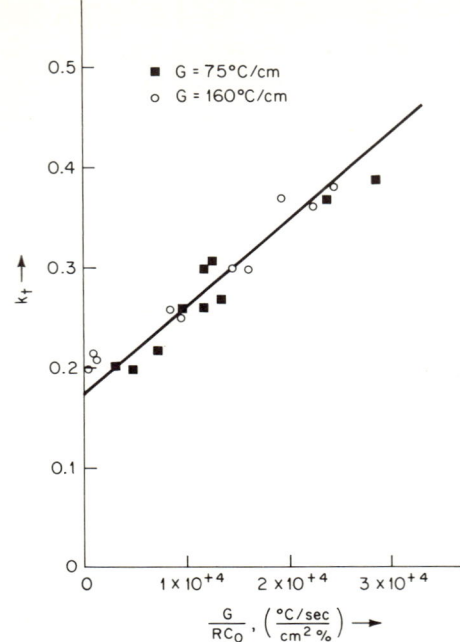

**FIGURE 3-22**
Partition ratio at cell tip $k_t$ versus $G/RC_0$. (*Data from Sharp and Hellawell.*[20])

Measurements were done in this way by Sharp and Hellawell[19] on several different alloys, and their results on aluminum–copper alloys are plotted in Fig. 3-22. In agreement with Eq. (3-15), $k_t$ increases linearly with $G/RC_0$. The slope of the curve yields a calculated value for $D_L$ of $4.5 \times 10^{-5}$ cm$^2$/s, which is in close agreement with that expected from the published literature.

It should be possible to compare experiment with theory on cellular solidification in other ways than by measurement of tip temperature or of effective partition ratio at the tip. Ultimately one would like to be able to predict qualitatively the overall local solute redistribution (intercellular microsegregation). In the case of the simple theory outlined above, this is done by writing an overall mass balance for solute in the volume element of Fig. 3-21. Assuming that solute flow exists only by diffusion and that solid and liquid densities are constant and equal, this is written

$$\text{Solute accumulation} = \text{solute in} - \text{solute out by diffusion} \qquad (3\text{-}16)$$

$$\frac{\partial \bar{C}}{\partial t} = \frac{\partial}{\partial x}\left(D_L f_L \frac{\partial C_L}{\partial x}\right)$$

where $\bar{C}$ is the average composition of the volume element at time $t$ and $f_L$ is the fraction of liquid. The net overall change of solute composition is the sum of the changes in liquid and solid:

$$\frac{\partial \bar{C}}{\partial t} = C_L(1-k)\frac{\partial f_L}{\partial t} + f_L \frac{\partial C_L}{\partial t} \qquad (3\text{-}17)$$

For steady-state solidification,

$$\frac{\partial f_L}{\partial x} = -\frac{1}{R}\frac{\partial f_L}{\partial t} \qquad (3\text{-}18)$$

$$\frac{\partial C_L}{\partial x} = \frac{\partial C_L}{\partial t}\frac{\partial t}{\partial x} = \frac{G}{m_L} \qquad (3\text{-}19)$$

Now, combining Eqs. (3-16) to (3-19) and integrating from $C_L = (1-a)C_0$ at $f_L = 1$ yields the *local solute-redistribution equation*[41]

$$C_S^* = kC_0 \left[ \frac{a}{k-1} + \left(1 - \frac{ak}{k-1}\right)(1-f_S)^{(k-1)} \right] \qquad (3\text{-}20)$$

where $C_S^* = kC_L$ is the solid composition at the liquid-solid interface when $f_S$ fraction of solid has formed in the element. Since there is no diffusion in the solid, $C_S^*$ is also the isoconcentrate enclosing $f_S$ weight fraction of solid in the volume element.

Equation (3-20) in no way depends on the geometry of the growing cell. For elongated *two-dimensional* cells, a suitable volume element would be as sketched in Fig. 3-21 and also at the upper left of Fig. 3-23. The fraction of solid within a given isoconcentrate in this volume element is proportional to the fractional distance $y/l$ along the element. For an ideal regular hexagonal array, an appropriate volume element would be as sketched at the upper right of Fig. 3-23, where the fraction of solid enclosed within a given isoconcentrate varies as $(r/l)^2$. For more irregular structures, the appropriate volume element is simply one which is large enough to be of the local average composition but small enough to be treated as a differential element (i.e., its cross section transverse to the heat-flow direction should be several cell spacings).

Figure 3-23 also shows microsegregation for an Al–4.5% copper alloy calculated according to Eq. (3-20). At the limit of $a = 0$, the result is exactly the same as obtained for single crystal growth, assuming no solid diffusion and complete liquid diffusion (Fig. 2-3c). The important difference here is that the segregation occurs over distances the order of the cell spacing, not over the crystal as a whole. When $a = 0$, about 9 wt% eutectic ($f_E = 0.09$) is expected in intercellular regions. Even at high values of $a$, substantial eutectic is predicted. In fact, as long as cells form at all [that is, until $a = -(1-k)/k$], some eutectic is predicted in the intercellular regions, no matter how low the original composition is. This prediction is in agreement with results of many experiments. As an example, Kramer, Bolling, and Tiller[40] found eutectic

**82** CELLULAR SOLIDIFICATION

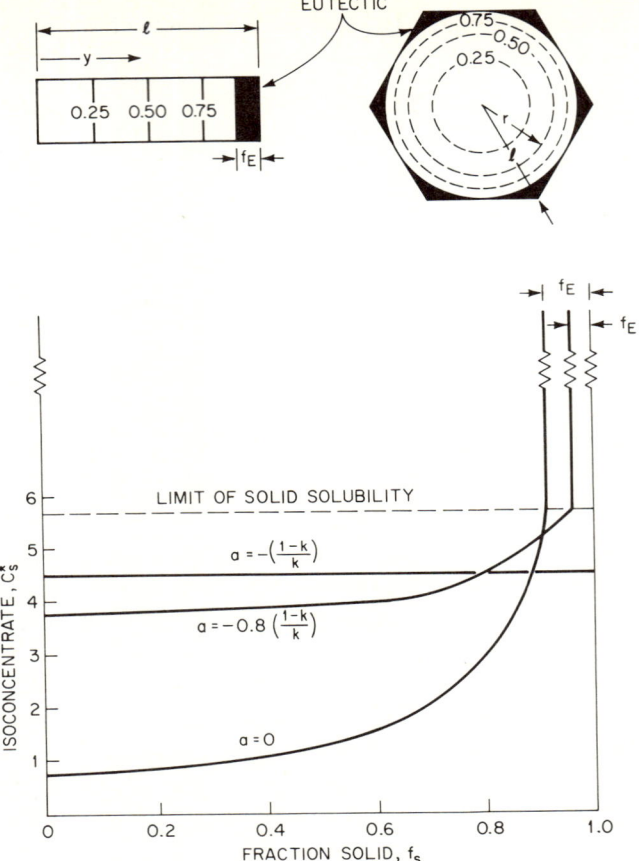

**FIGURE 3-23**
Microsegregation in cellular solidification. At the top are shown two idealized *volume elements*, the one on the left for elongated (two-dimensional) cells and the one on the right for regular cells. Dotted lines show isoconcentrates at the values of $f_s$ noted. The graph plots microsegregation for an Al–4.5% Cu alloy according to Eq. (3-20).

present in Sn–Pb alloys at initial compositions as low as 0.2% Pb, although the limit of solid solubility of Pb in Sn is 2.5% and the eutectic composition is 38% Pb. Biloni and coworkers[39] have found second phases present in several equally dilute tin- and aluminum-base alloys. It should be noted that, especially at high values of $a$, the *eutectic* referred to here has an average composition rather different from the eutectic composition of the phase diagram; this is discussed further in Chap. 4.

Except for the qualitative confirmation afforded by the foregoing experiments there are no data available on overall microsegregation in cellular solidification to compare with theory. Such data should be readily obtainable using techniques employed in measuring dendritic microsegregation, particularly the electron microprobe.

## CELL SPACING

One basic assumption of the preceding analysis has been that cells are spaced closely enough so that constitutional supercooling in intercellular regions is negligible. Using available experimental data on intercellular spacing in crystal growth, it is possible to show that this assumption is reasonable, at least for two-dimensional cells, as sketched in Fig. 3-21. Assuming the concentration difference in the liquid between cells is small, Rohatgi and Adams[42] have shown that the change of liquid composition with time $\partial C_L/\partial t$ is to a good approximation independent of distance $y$ transverse to the heat-flow direction. The fact that the cells are thickening in the $y$ direction can also be neglected, and the resulting simplified differential equation describing solute redistribution in the liquid within the volume element is

$$D_L \frac{\partial^2 C_L}{\partial y'^2} = \frac{\partial C_L}{\partial t} = \text{const} \qquad (3\text{-}21)$$

where $y'$ is distance from the liquid-solid interface in the volume element (Fig. 3-21). For small composition differences within the volume element, $\partial C_L/\partial t$ is as if the liquid were uniform. Multiplying both sides of Eq. (3-19) by $\partial x/\partial t = R$, the result is

$$\frac{\partial C_L}{\partial t} = -\frac{GR}{m_L} \qquad (3\text{-}22)$$

where $G$ is the temperature gradient within the liquid-solid zone and $R$ is the isotherm velocity. Substituting Eq. (3-22) in (3-21) and integrating from $\partial C_L/\partial y' = 0$ at $y' = l$ (see Fig. 3-21) gives

$$\left(\frac{\partial C_L}{\partial y'}\right)_{y'=0} = -\frac{GRl}{m_L D_L} \qquad (3\text{-}23)$$

The maximum concentration difference in the liquid at any given moment is $\Delta C_L(\text{max})$, as shown in Fig. 3-21. This is calculated by integrating Eq. (3-23):

$$\Delta C_L(\text{max}) = -\frac{GRl^2}{2m_L D_L} \qquad (3\text{-}24)$$

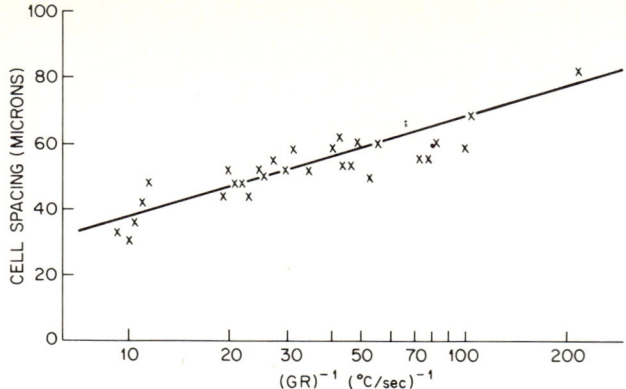

**FIGURE 3-24**
Cell spacing in cellularly solidified Sn–Pb alloys containing from 0.05 to 0.5 at% Pb. (*Data from Plaskett and Winegard.*[32])

Now, since temperature gradient in the direction perpendicular to heat flow is negligible, the gradient of constitutional supercooling in the liquid in intercellular regions is determined by multiplying Eq. (3-23) by $m_L$. The total maximum supercooling is obtained by multiplying Eq. (3-24) by $m_L$.

Several authors have measured intercellular spacings as a function of growth conditions, and Fig. 3-24 presents data of Plaskett and Winegard[32] for Sn–Pb alloys. The data are plotted as a function of $(GR)^{-1}$ and are for regular cells, not the two-dimensional cells assumed in deriving Eqs. (3-23) and (3-24). Nonetheless these equations can be used for an order-of-magnitude calculation of the supercooling. Substitution of a reasonable value for the diffusion coefficient shows that the gradient of constitutional supercooling is very small indeed, being on the order of 1°C/cm. Thus, the maximum undercooling within the intercellular spaces is about 0.001°C. Data on cell spacings by other investigators yield qualitatively similar results.[12,21,32,43]

The foregoing evidence suggests that cell spacing in cellular solidification adjusts itself to reduce constitutional supercooling to a very low value, probably to a value such that it is of the same order as the melting-point differences along the liquid-solid interface which result from the effect of radius of curvature. The mechanism for accomplishing this adjustment is presumably a branching mechanism, as sketched in Fig. 3-25. If the gradient of constitutional supercooling is above a small minimum value in the general vicinity of the cell tip, a perturbation should form and grow to reduce the local spacing, as sketched. From the simple equation (3-24), one would expect the cell spacing to depend on the product $GR$ and to be independent of composition. This is in general agreement with the data of Fig. 3-24, which cover a range

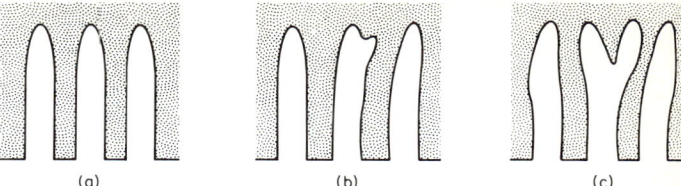

FIGURE 3-25
Reduction of cell spacing by a branching mechanism. Spacing of cell tips at (c) is 20 percent less than at (a).

of compositions from 0.05 to 0.5 at % Pb. Additional data of Plaskett and Winegard for other alloy systems (Sn–Bi and Sn–Sb) fall on the same curve with little additional scatter. Coulthard and Elliott[43] found that their data plotted linearly against $(GR)^{-1}$; however, they found an effect of solute concentration on cell spacing.

The theory of cellular solidification presented above is a considerable simplification of an extremely complex problem. Solid-state diffusion has been neglected (this is discussed in the following section). Details of solute diffusion at the dendrite tip have been highly simplified. Effect of radius of curvature on the melting point has been neglected. For the radii typically encountered in crystal growth (approximately 50 μm), the effect of this curvature on tip temperature is very small, but an exact description of cellular growth must include this effect as well as account for the multi-dimensional nature of the liquid diffusion. On the other hand, the simple theory accounts for most of the experimental observations to date, and the more detailed theoretical treatments available have involved approximations of limited validity.[44,45] There remains a need for more extensive and closely controlled experiments, especially experiments in which convection and solid-state diffusion are absent or quantitatively described.

## SOLID-STATE DIFFUSION

In some alloy systems, diffusion of solute in the solid occurs to sufficient extent that it modifies appreciably the solute distribution predicted above; Fig. 3-26 shows an example of such a phenomenon. Solute content at the cell center increases steadily during and after solidification; solute content at cell *nodes* (e.g., the shaded areas of the sketch at the upper right of Fig. 3-23) decreases steadily after solidification. Also, the rapid quench is seen to have entrapped enriched liquid that otherwise would have been depleted by diffusion through the liquid in the growth direction [Eq. (3-20)] and diffusion into the growing solid. Figure 3-27 shows another example from the work of Morris and Winegard on Pb–Sb alloys.[47] The structure is just back of a

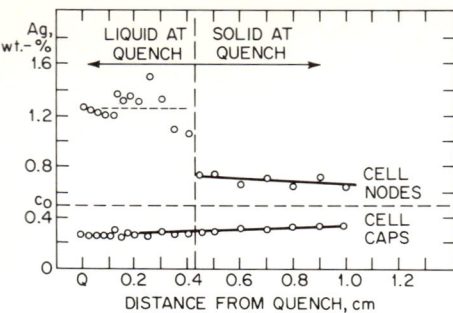

FIGURE 3-26
Variation in the solute concentration in nodes and centers of regular cells in Al–0.5 wt% Ag. Growth velocity 0.1 cm/min$^{-1}$. (*From Morris and Winegard.*[47])

rapidly quenched interface and shows sharp, highly segregated nodes. Figure 3-11*b* shows a comparable structure, but not quenched. Note how solid-state diffusion during cooling has blurred the segregation.

In Chap. 5 we describe an analysis of solid-state diffusion during dendritic solidification that is also applicable to cellular solidification. The essential conclusion of this work is that this diffusion becomes significant only when $D_s t_f k/l^2 \gtrsim 1$, where

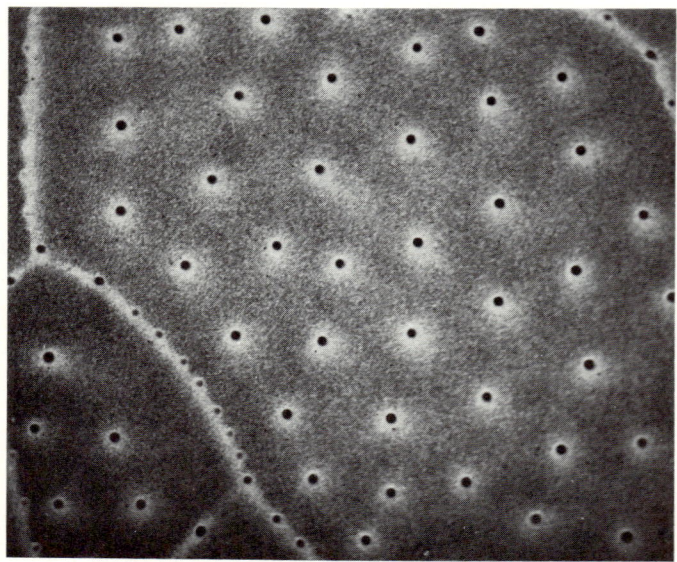

FIGURE 3-27
Nodes on a cross section taken near a quenched interface. The continuous boundaries coincide with striations (low angle boundaries); $C_0 = 100$ ppm Sb. (Magnification ×48.) (*From Morris and Winegard.*[21])

$D_S$ is the solid-state-diffusion coefficient and $t_f$ is the time from the beginning to the end of solidification of the volume element (i.e., it is the *local solidification time*). The appropriate dimensionless parameter governing diffusion after solidification is $(1/l^2) \int_{T_R}^{T_E} D_S t$, where $T_E$ and $T_R$ are the nonequilibrium solidus temperature and room temperature, respectively.

## TERNARY ALLOYS

Multicomponent single-phase alloys, like binary alloys, can be solidified with a plane front provided thermal gradient is sufficiently high and growth rate sufficiently low. Both the constitutional supercooling and interface stability theories can be applied to such alloys, and Coates et al.[48] have done so for ternary alloys. We summarize here the constitutional supercooling approach. It is exactly analogous to the approach for binary alloys in that the condition for stability is that the actual temperature gradient at the interface be equal to or greater than the liquidus temperature gradient. The essential complications in this case are that the liquidus is now a surface and solute diffusive interactions may occur.

Consider a ternary alloy that would be single-phase on solidification, containing solute elements $m$ and $n$ in solvent $q$ (Fig. 3-28). At steady-state solidification, assuming a plane front, no convection, and equilibrium at the interface, solute distribution in the liquid of elements $m$ and $n$ will exist also as shown in Fig. 3-28. Note that the diffusion boundary layers (as measured for example by the *characteristic diffusion distances*) are not generally equal for the two solutes since their diffusion coefficients are not generally equal. Using the liquidus surface of the phase diagram, we can now use the curves of Fig. 3-28 to calculate the variation of liquidus temperature in front of the growing interface. The remainder of the solution is as in the case of binary alloys.

The liquidus surface of the ternary phase diagram is given by an equation of the form

$$T_L = T_L(C_{Lm}, C_{Ln}) \qquad (3\text{-}25)$$

where $T_L$ is liquidus temperature, $C_{Lm}$ is concentration of element $m$ in the liquid, and $C_{Ln}$ is concentration of element $n$ in the liquid. Taking the total derivative with respect to distance from the liquid-solid interface $x'$ yields the slope of the liquidus curve at the interface $x' = 0$:

$$\left(\frac{dT_L}{dx'}\right)_{x'=0} = p\left(\frac{dC_{Lm}}{dx'}\right)_{x'=0} + s\left(\frac{dC_{Ln}}{dx'}\right)_{x'=0} \qquad (3\text{-}26)$$

where $p$ is the slope of the liquidus surface at constant $C_{Ln}$ and $s$ is the slope at constant $C_{Lm}$. Conservation of solutes at the advancing interface at steady state gives two

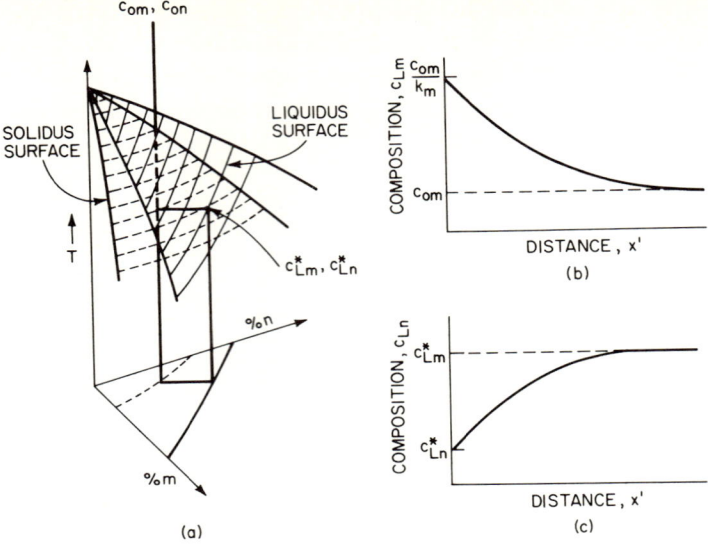

**FIGURE 3-28**
Portion of ternary phase diagram, and solute distributions in front of a single-phase ternary alloy solidifying with plane front.

equations analogous to Eq. (3-1). These are, for steady-state solidification,

$$D_{mm}\left(\frac{dC_{Lm}}{dx'}\right) + D_{mn}\left(\frac{dC_{Ln}}{dx'}\right) = -R(C_{Lm}^* - C_{0m}) \quad (3\text{-}27a)$$

$$D_{nm}\left(\frac{dC_{Lm}}{dx'}\right) + D_{nn}\left(\frac{dC_{Ln}}{dx'}\right) = -R(C_{Ln}^* - C_{0n}) \quad (3\text{-}27b)$$

where $D_{mm}$ and $D_{nn}$ are the on-diagonal diffusion coefficients of the two solutes in the liquid, and $D_{mn}$ and $D_{nm}$ are the off-diagonal coefficients resulting from interaction of the diffusing species. The constitutional supercooling criterion for stability is that the actual temperature gradient in the liquid at the interface $G_L$ be equal to or greater than $(dT_L/dx')_{x'=0}$. Equations (3-26) and (3-27), combined with this condition, give the criterion for ternary alloys. For the special case where interaction of diffusing species can be ignored ($D_{mn} \simeq D_{nm} \simeq 0$), they reduce to

$$\frac{G_L}{R} \geq -\frac{p(C_{Lm}^* - C_{0m})}{D_{mm}} - \frac{s(C_{Ln}^* - C_{0n})}{D_{nn}} \quad (3\text{-}28)$$

For the single-phase alloys considered here, this equation can be written

$$\frac{G_L}{R} \geq -\frac{pC_{0m}(1 - k_m)}{k_m D_{mm}} - \frac{sC_{0n}(1 - k_n)}{k_n D_{nn}} \quad (3\text{-}29)$$

where $k_m$ and $k_n$ are the equilibrium partition ratios of the solute elements $m$ and $n$, respectively, defined as in the case of binary alloys. Note the similarity of Eqs. (3-29) and the equivalent expression for binary alloys, Eq. (3-5). Equation (3-28), of course, is more difficult to apply since determination of the partition ratios from the phase diagram requires that tie lines of ternary phase be known and these have been determined only for a few systems. However, if the liquidus and solidus surfaces are planes

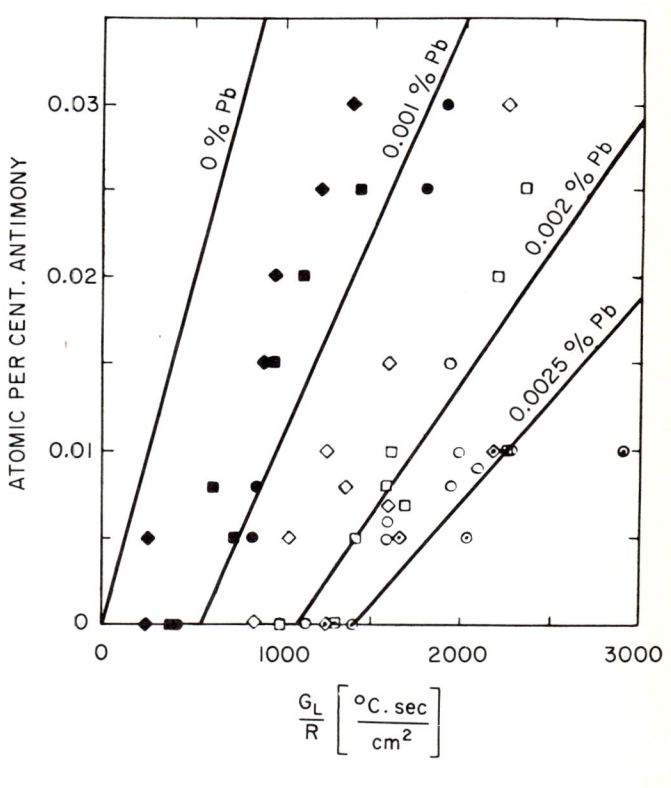

FIGURE 3-29
Conditions for interface stability in the Sn–Pb–Sb system. (*From Cole and Winegard.*[49])

joined by tie lines that generate constant partition ratios, $k_m$, $k_n$, $p$, and $s$ are all constants which can be determined from the two appropriate binary phase diagrams.

Figure 3-29 shows results of a study of Cole and Winegard[12] on interface stability in dilute Sn–Pb–Sb alloys. In qualitative agreement with Eq. (3-29), increasing amounts of either Pb or Sb require higher $G_L/R$ to maintain stability. Approximate quantitative agreement can also be obtained by assuming the partition ratios are as in the binary system and choosing the two diffusion coefficients to best fit the data. On the other hand, the authors suggest that some solute interaction is indicated by the data, and it may be that the off-diagonal diffusion coefficients $D_{mn}$ and $D_{nm}$ are significant or the tie lines are such that partition ratios depend on composition.

## REFERENCES

*1* PAPAPETROU, A.: *Z. Krist.*, **A92**:89 (1935).
*2* RUTTER, J. W., and CHALMERS, B.: *Can. J. Phys.*, **31**:15 (1953).
*3* TILLER, W. A., JACKSON, K. A., RUTTER, J. W., and CHALMERS, B.: *Acta Met.*, **1**:428 (1953).
*4* MULLINS, W. W., and SEKERKA, R. F.: *J. Appl. Phys.*, **34**:323 (1963); **35**:444 (1964). .
*5* SEKERKA, R. F.: *J. Appl. Phys.*, **36**:264 (1965).
*6* SEKERKA, R. F.: *J. Phys. Chem. Solids*, **28**:983 (1967).
*7* SEKERKA, R. F.: *J. Crystal Growth*, **3**:71 (1968).
*8* HUCKE, E., ADAMS, C., FLEMINGS, M. C., and TAYLOR, H. F.: in "Physical Chemistry of Process Metallurgy," pt. II, p. 815, Interscience Publishers, Inc., New York, 1961.
*9* HURLE, D. T. J.: *Solid-State Electron.*, **3**:37 (1961).
*10* JACKSON, K. A., and HUNT, J. D.: *Acta Met.*, **13**:1212 (1965).
*11* WALTON, D., TILLER, W. A., RUTTER, J. W., and WINEGARD, W. C.: *Trans. AIME*, **203**:1023 (1955).
*12* COLE, G. S., and WINEGARD, W. C.: *J. Inst. Met.*, **92**:322 (1963).
*13* PLASKETT, T. S., and WINEGARD, W. C.: *Can. J. Phys.*, **37**:1555 (1959).
*14* HUNT, M. D., SPITTLE, J. A., and SMITH, R. W.: The Solidification of Metals, p. 57, *Iron and Steel Inst.* Publ. No. 110, 1968.
*15* BARDSLEY, W., CALLAN, J. M., CHEDZEY, H. A., and HURLE, D. T. J.: *Solid-State Electron.*, **3**:142 (1961).
*16* COULTHARD, J. O., and ELLIOTT, R.: "The Solidification of Metals," p. 61, *Iron and Steel Inst.* Publ. No. 110, 1968.
*17* HARDY, S. C., and CORIELL, S. R.: *J. Crystal Growth*, **3,4**:569 (1968).
*18* SCHAEFER, R. J., and GLICKSMAN, M. E.: *Met. Trans.*, **1**:1973 (1970).
*19* SHARP, R. M., and HELLAWELL, A.: *J. Crystal Growth*, **6**:253 (1970).
*20* SHARP, R. M., and HELLAWELL, A.: *J. Crystal Growth*, **6**:334 (1970).
*21* MORRIS, L. R., and WINEGARD, W. C.: *J. Crystal Growth*, **5**:361 (1969).
*22* RUTTER, J. W.: in "Liquid Metals and Solidification," p. 243, American Society for Metals, Cleveland, Ohio, 1958.
*23* CHADWICK, G. A.: in M. Zief and W. R. Wilcox (eds.), "Fractional Solidification," p. 113, Marcel Dekker, Inc., New York, 1967.

24 BILONI, H., BOLLING, G. F., and COLE, G. S.: *Trans. Met. Soc. AIME*, **236**:930 (1966).
25 ATWATER, H., and CHALMERS, B.: *Can. J. Phys.*, **35**:208 (1957).
26 CHADWICK, G. A.: *Acta Met.*, **10**:1 (1962).
27 WEINBERG, F.: *Trans. AIME*, **224**:628 (1962).
28 BARDSLEY, W., BOULTON, J. S., and HURLE, D. T. J.: *Solid-State Electron.*, **5**:395 (1962).
29 BARDSLEY, W., MULLIN, J. B., and HURLE, D. T. J.: in "The Solidification of Metals," p. 93, *Iron and Steel Inst.* Publ. No. 110, London, 1968.
30 BARDSLEY, W., COCKAYNE, B., GREEN, G. W., and HURLE, D. T. J.: *Solid-State Electron.*, **6**:389 (1963).
31 COCKAYNE, B.: *J. Crystal Growth*, **4**:1–5 (1968).
32 PLASKETT, T. S., and WINEGARD, W. C.: *Can. J. Phys.*, **38**:1077 (1960).
33 FLEMINGS, M. C.: in J. Burke et al. (eds.), "Surfaces and Interface II," p. 313, Syracuse University Press, Syracuse, N.Y., 1968.
34 MORRIS, L. R., and WINEGARD, W. C.: *J. Crystal Growth*, **6**:61 (1969).
35 JACKSON, K. A., HUNT, J. D., UHLMANN, D., and SEWARD III, T. P.: *Trans. AIME*, **236**:149 (1966).
36 BILONI, H., BOLLING, G. F., and DOMIAN, H. A.: *Trans. AIME*, **233**:1926 (1965).
37 HOLMES, E. L., RUTTER, J. W., and WINEGARD, W. C.: *Can. J. Phys.*, **35**:1223 (1957).
38 DAVIES, G. J.: in "The Solidification of Metals," p. 66, *Iron and Steel Inst.* Publ. No. 110, 1968.
39 BILONI, H., BOLLING, G. F., and COLE, G. S.: *Trans. AIME*, **233**:251 (1965).
40 KRAMER, J. J., BOLLING, G. F., and TILLER, W. A.: *Trans. AIME*, **227**:374 (1963).
41 BOWER, T. F., BRODY, H. D., and FLEMINGS, M. C.: *Trans. AIME*, **236**:624 (1966), Appendix A.
42 ROHATGI, P. K., and ADAMS, C. M.: *Trans. AIME*, **239**:1737 (1967).
43 COULTHARD, J. O., and ELLIOTT, R.: *J. Inst. Metals*, **95**:21 (1967).
44 BOLLING, G. F., and TILLER, W. A., *J. Appl. Phys.*, **31**:2040 (1960).
45 DONAGHEY, L. F., and TILLER, W. A.: in "The Solidification of Metals," p. 87, *Iron and Steel Inst.* Publ. No. 110, 1968.
46 FLEMINGS, M. C., POIRIER, D. R., BARONE, R. V., and BRODY, H. D.: *J. Iron Steel Inst.*, **208**:371 (1970).
47 MORRIS, L. R., and WINEGARD, W. C.: *J. Inst. Metals*, **97**:220 (1969).
48 COATES, D. E., SUBRAMANIAN, S. V., and PURDY, G. R.: *Trans. AIME*, **242**:800 (1968).
49 COLE, G. S., and WINEGARD, W. C.: *J. Inst. Metals*, **92**:322 (1963).

## PROBLEMS

3-1 A Ge–Ga crystal is grown by normal freezing. Initial melt composition is 10 ppm Ga. Growth rate is $8 \times 10^{-3}$ cm/s. Assume $k = 0.1$, $m_L = -4°C/\%$, $D_L = 5 \times 10^{-5}$ cm²/s.
  (a) If convection is completely absent, what thermal gradient is required to maintain a plane front when the ingot is 50 percent solidified?
  (b) If convection is sufficiently vigorous that $k' \simeq k$, what thermal gradient is required to maintain a plane front when the ingot is 50 percent solidified?
  (c) If $\delta = 0.005$ cm, what thermal gradient is required to maintain a plane front when the ingot is 50 percent solidified?

**3-2** An Al–1% Cu alloy is grown by normal freezing at $3 \times 10^{-4}$ cm/s with convection completely suppressed. (The phase diagram for this alloy is as shown schematically in Fig. 2-3, with $C_E = 33\%$ Cu, $C_{sM} = 5.65\%$ Cu, $T_M = 660°C$, $T_E = 548°C$, and constant $k$ and $m_L$; $D_L = 3 \times 10^{-5}$ cm²/s.)
  (a) What will be the temperature of the planar liquid-solid interface at steady state?
  (b) What thermal gradient will be required to maintain the plane front according to the constitutional supercooling criterion?

**3-3** What thermal gradient would be required to maintain the plane front in Prob. 3-2, according to the Mullins and Sekerka interface stability criterion? (Assume the thermal conductivity of the solid is half that of the liquid; $\sigma = 1.2 \times 10^{-6}$ cal/cm². The volumetric heat of fusion can be calculated from Appendix B.)

**3-4** An Al–1% Cu ingot is solidified with no convection at $3 \times 10^{-4}$ cm/s, with a thermal gradient of 300°C/cm. Solidification is cellular. (See the data in Prob. 3-2).
  (a) What is the approximate liquid composition at the cell tips? The solid composition?
  (b) What is the temperature of the cell tips?
  (c) What is the distance from the cell tips to the cell roots?
  (d) How far does characteristic distance of the diffusion boundary layer extend in front of the cell tips?

**3-5** What weight fraction eutectic will form in intercellular regions of the ingot of Fig. 3-4?

**3-6** Controlled plane front solidification has been proposed as a means of making homogeneous large ingots of alloys for subsequent working. Evaluate this suggestion critically for the commercial age-hardenable alloy Al–4.5% Cu.

**3-7** When cells grow into a convecting fluid, they grow preferentially *upstream*. Explain qualitatively.

**3-8** Show schematically what you would expect to be the directions of fluid flow in the intercellular regions of Fig. 3-21 and in the bulk liquid in front of the cell tips. Driving forces for flow are density differences in the liquid and solidification shrinkage.

**3-9** What is the diffusion coefficient of Ga in Ge that you would estimate from the data of Fig. 3-5?

**3-10** Imagine a solid phase growing with a plane front in a supercooled pure liquid. Assume equilibrium at the liquid-solid interface and plot actual temperature distribution in the liquid in front of the growing solid. Plot also the liquidus temperature distribution in front of the interface. Show that there is a gradient of thermal supercooling in front of the interface that should result in interface breakdown just as does constitutional supercooling.

**3-11** What diffusion coefficients will give approximate agreement between Eq. (3-29) and the data of Fig. 3-29? Graphically show the extent of agreement. Assume partition ratios and liquidus slopes are constants. From the two binary phase diagrams, these are as follows: the partition ratios are 0.13 and 1.6 for Pb and Sb, respectively, in Sn; the liquidus slopes are $-36°C/$at for Pb and $+2.0°C/$at for Sb in Sn.

# 4
# PLANE FRONT SOLIDIFICATION OF POLYPHASE ALLOYS

## INTRODUCTION

The techniques of single-crystal growth described in the preceding chapters can be applied as well to alloys which are polyphase after equilibrium solidification. Since about 1960, much research has been done on plane front solidification of eutectic alloys. Many eutectics (when properly grown) possess fine, uniform, rodlike or plate-like structures. Other eutectics are more complex, with their structures described as *broken lamellar*, *spiral*, or other. Figure 4-1 shows an example of a rodlike structure; lamellar and other structures are shown in later figures. These finely mixed composites have suggested to many investigators a variety of possible applications, for example, as high-temperature materials or superconductors.[1-3] Recently, success has been obtained in producing identical composite structures in alloys of noneutectic composition, thus permitting control to be exercised over the volume fraction of constituents present. Closely similar structures have also been obtained in alloys of monotectic composition.

It will be seen in this chapter that crystal growth of polyphase alloys is closely analogous to that of single-phase alloys. Thermal gradient and growth rate must be controlled to maintain a stable plane front interface, and fluctuations in growth rate

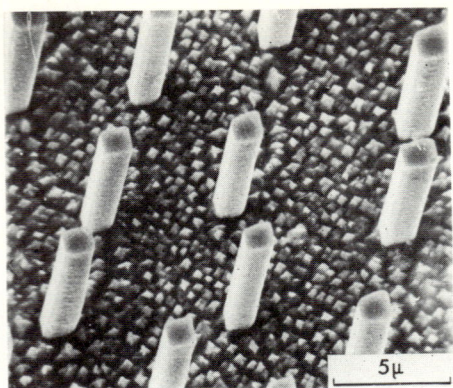

**FIGURE 4-1**
Eutectic rods of tantalum-carbide in a cobalt base matrix. Matrix etched away so rods protrude. (*From H. Bibring et al.*[39])

lead to compositional (as well as structural) variations along the growth direction. In polyphase alloys an added degree of complication arises from the fact that solute must diffuse transverse to the growing interface as well as in the growth direction.

## LAMELLAR EUTECTIC GROWTH

The mechanism of eutectic growth in nonfaceted alloys has been treated by several authors,[4,5] following closely the earlier work on eutectoid decomposition.[6-8] A somewhat simplified version of the treatment of Jackson and Hunt[4] is given below for lamellar (platelike) eutectic growth. Consider an alloy of exactly eutectic composition growing with a plane front except that the individual lamellae have slightly curved interfaces, as sketched in Fig. 4-2b. As the α phase grows, it rejects $B$ atoms into the liquid. Similarly, the growing β phase rejects $A$ atoms. Thus, there is a slight buildup of $B$ atoms in front of the α lamellae and depletion in front of the β lamellae, as shown in Fig. 4-3a. If equilibrium pertains at the liquid-solid interface, there must now be an undercooling in front of the lamellae which depends on the amount of solute buildup or depletion and the slope of the liquidus curve $T_{L\infty}$. This undercooling is

$$\Delta T_D = T_E - T_{L\infty} = m_L(C_E - C_L^*) \qquad (4\text{-}1)$$

where $\Delta T_D$ is undercooling due to solute diffusion, $m_L$ is liquidus slope, $C_E$ is eutectic composition, and $C_L^*$ is composition of the liquid at the location $y$ on the liquid-solid interface. $\Delta T_D$ is marked on the phase diagram in Fig. 4-2a for a specific location $y_1$ in front of an α lamella. It is shown schematically in Fig. 4-3b versus $y$. Note that it can be either positive or negative.

The actual interface temperature $T^*$ must be essentially constant in nonfaceted eutectics, as sketched in Fig. 4-3b. This is because the curved surfaces of lamellae in

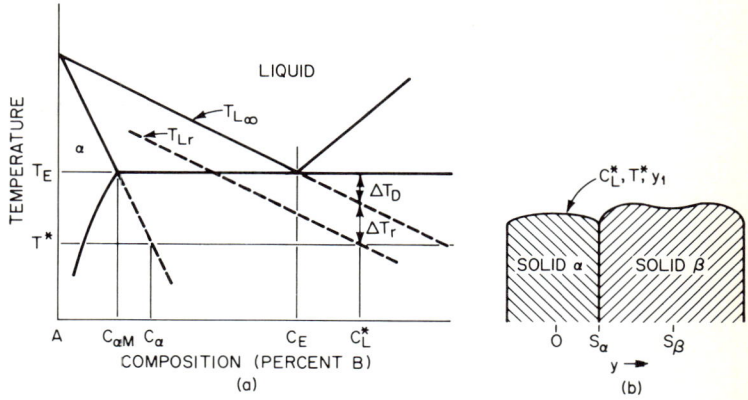

**FIGURE 4-2**
Undercoolings and interface liquid composition $C_L^*$ of a eutectic alloy at a point on the interface in front of the $\alpha$ phase. (*a*) Phase diagram; (*b*) interface showing point $y_1$ for which construction of (*a*) applies.

these eutectics usually deviate from a plane by less than a few microns. Hence, even if a temperature gradient exists in the growth direction, the variation in temperature from the leading to the lagging part of a lamella is negligible. The interface is maintained isothermal at $T^*$ by the lamellae adjusting their radii of curvature locally. Then total undercooling at the interface $\Delta T$ is constant and is given by

$$\Delta T = T_E - T^* = \Delta T_r + \Delta T_D \qquad (4\text{-}2)$$

where $\Delta T_r$ is undercooling due to radius of curvature.

This undercooling is given by Eq. (8-10), which is written here as

$$\Delta T_r = T_{L\infty} - T_{Lr} = \frac{\sigma T_E}{\rho_s H r} \qquad (4\text{-}3)$$

where $\Delta T_r$ is undercooling due to radius of curvature. In accordance with convention in solidification problems, this is taken as being positive where liquidus temperature is depressed. $T_{L\infty}$ is the equilibrium liquidus temperature of a surface of infinite radius of curvature, and $T_{Lr}$ is the liquidus temperature of a singly curved surface of radius $r$; $\sigma$ is liquid-solid surface energy, $T_E$ is eutectic temperature, $\rho_s$ is solid density, and $H$ is heat of fusion, a positive quantity. Figure 4-2 shows this depression of the liquidus temperature at a location $y_1$ on the interface of the $\alpha$ phase of a growing eutectic. The interface shape shown in Fig. 4-3*c* is that which produces the required values of $\Delta T_r$ along the interface. Note that at the center of the lamella, $\Delta T_r$ is negative, and so the radius of curvature here is also negative.

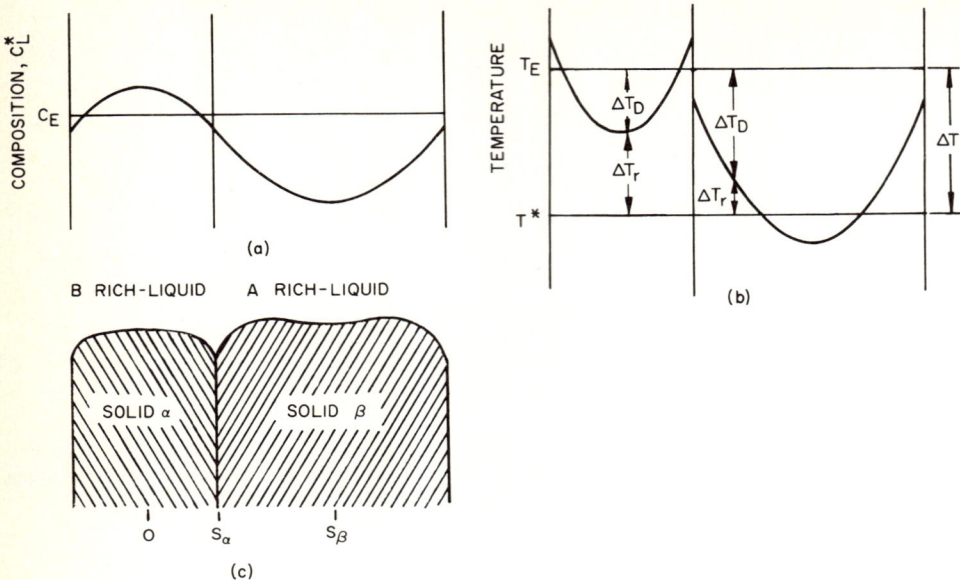

**FIGURE 4-3**
Lamellar curvature in eutectic growth. (a) Solute concentration in the liquid at the interface; (b) interfacial undercoolings $\Delta T_r$, $\Delta T_D$, and $\Delta T$; (c) predicted shape of the lamellar liquid-solid interface. (*From Hunt and Jackson.*[9])

Quantitative treatment of eutectic growth begins with a description of the solute-diffusion field in front of the growing lamellae using, for steady-state solidification, the differential equation

$$\nabla^2 C_L + \frac{R}{D_L}\frac{\partial C_L}{\partial x'} = 0 \qquad (4\text{-}4)$$

where $C_L$ = liquid composition at $x'$, $y$

$x'$ = distance in growth direction measured from the liquid-solid interface

$y$ = distance transverse to growth direction and to lamellae

$R$ = growth rate

$D_L$ = liquid-diffusion coefficient

Analytic solution of this equation is made possible by approximating the liquid-solid surface as being flat. The bulk alloy composition is assumed to be exactly $C_E$, and so one boundary condition is that $C_L = C_E$ at $x' = \infty$. Also, symmetry requires that $\partial C_L/\partial y = 0$ at $y = 0$ and at $y = S_\alpha + S_\beta$, where $y = 0$ at the center of an $\alpha$ lamella and

$S_\alpha + S_\beta$ at the center of a $\beta$ lamella, Fig. 4-2. It is further assumed that the undercooling $\Delta T$ is small so that $C_L^* \simeq C_E$ and the compositions of the solid phases forming are of the equilibrium composition at the eutectic temperature [for example, $C_\alpha = C_{\alpha M}$ (Fig. 4-2)]. The assumption of $C_L^* \simeq C_E$ can be alternatively stated: The amplitude of the composition distribution in the liquid transverse to the growing interface is small. Now, conservation of matter at the interface requires

$$\left(\frac{\partial C_L}{\partial x'}\right)_{x'=0} = -\frac{R(C_E - C_{\alpha M})}{D_L} \quad \text{for } 0 \le y \le S_\alpha \quad (4\text{-}5a)$$

$$\left(\frac{\partial C_L}{\partial x'}\right)_{x'=0} = -\frac{R(C_E - C_{\beta M})}{D_L} \quad \text{for } S_\alpha \le x \le S_\alpha + S_\beta \quad (4\text{-}5b)$$

The solution to Eq. (4-4) with the foregoing boundary conditions is

$$C_L - C_E = \sum_{n=1}^{\infty} B_n \cos \frac{2n\pi y}{\lambda} e^{-(2n\pi x'/\lambda)} \quad (4\text{-}6)$$

where
$$B_n = \frac{\lambda R(C_{\beta M} - C_{\alpha M})}{(n\pi)^2 D_L} \sin \frac{2n\pi S_\alpha}{\lambda} \quad (4\text{-}7a)$$

$$\lambda = 2(S_\alpha + S_\beta) \quad (4\text{-}7b)$$

Figure 4-4 shows schematically results of calculations for a typical example of eutectic growth. Maximum composition deviation from the eutectic is small, i.e., usually less than 0.1 percent. Moreover, even this small deviation is rapidly damped out at small distances in front of the interface. For example, at a distance $x'$ equivalent to $S_\alpha + S_\beta$ the liquid compositional differences essentially disappear. The distance $S_\alpha + S_\beta$ is half the *lamellar spacing*, which is the sum of the widths of an $\alpha$ and $\beta$ lamellae. The fineness of lamellar structures is usually described in terms of this *interlamellar spacing*. Typically, these are under 10 $\mu$m, decreasing with increasing growth rate.

To determine quantitatively curves such as those of Fig. 4-4 and the curve of $\Delta T_D$ in Fig. 4-3b, we need specify only lamellar spacing and growth rate. Somewhat more information is needed, however, to specify $\Delta T_r$. It must be such that $T^*$ is constant, as shown in Fig. 4-3b, but this statement alone does not uniquely determine $T^*$ (or $\Delta T_r$). This is done by specifying the angles between tangents to the lamellar surfaces at the lamellar edges and the growth direction ($\theta_\alpha$ and $\theta_\beta$, Fig. 4-5). Assuming equilibrium at the lamellar boundary juncture with the liquid, these angles are as required by the force equilibrium

$$\sigma_{\alpha L} \cos \theta_\alpha + \sigma_\beta \cos \theta_\beta = \sigma_{\alpha\beta} \quad (4\text{-}8)$$

where $\sigma_{\alpha L}$ and $\sigma_{\beta L}$ are the liquid-solid surface energies of the $\alpha$ and $\beta$ phases, respectively, and $\sigma_{\alpha\beta}$ is the interlamellar surface energy, as shown in Fig. 4-5. This expression

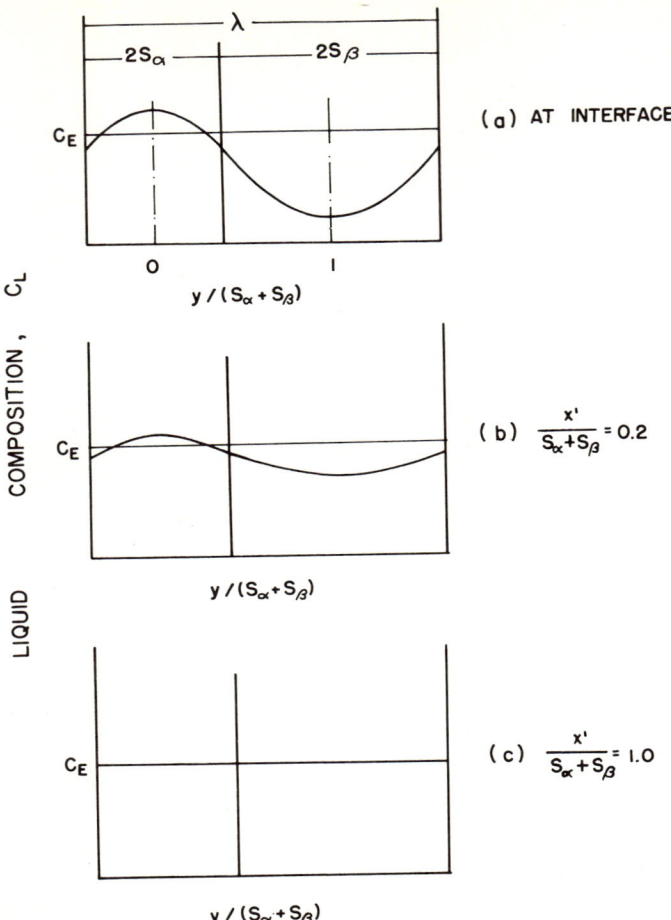

**FIGURE 4-4**
Typical calculated values of liquid composition at various distances $x'$ in front of a growing eutectic interface.

is equivalent to Eq. (8-27). With the condition of Eq. (4-8) and a given lamellar spacing, the curve of $\Delta T_r$ in Fig. 4-3 can now be uniquely specified using the procedure described by Jackson and Hunt.[4] The interface shape is then also uniquely specified. Figure 4-6 shows some interface shapes calculated by Hunt and Jackson,[9] superimposed on experimentally observed interfaces. They employed the procedure outlined above to obtain these shapes, except that they used measured angles at the lamellar boundary groove rather than Eq. (4-8).

In the foregoing calculations of interface undercooling and interface shape, it was necessary to specify both interface velocity $R$ and lamellar spacing $\lambda = 2S_\alpha + 2S_\beta$.

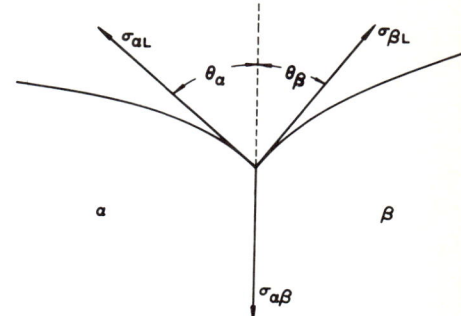

FIGURE 4-5
Equilibrium at the $\alpha\beta$ liquid juncture.

Alternatively, the problem might have been solved by specifying undercooling and growth rate. Manipulation of Eqs. (4-1) to (4-3) and (4-6) shows that $\lambda$, $\Delta T$, and $R$ are related by an expression of the form

$$\Delta T = AR\lambda + \frac{B}{\lambda} \qquad (4\text{-}9)$$

where $A$ and $B$ are constants depending on the particular alloy system. A schematic solution to this equation is shown in Fig. 4-7. For a fixed interface velocity, all the preceding equations can be satisfied by a range of $\Delta T$ and $\lambda$, as given by the curve

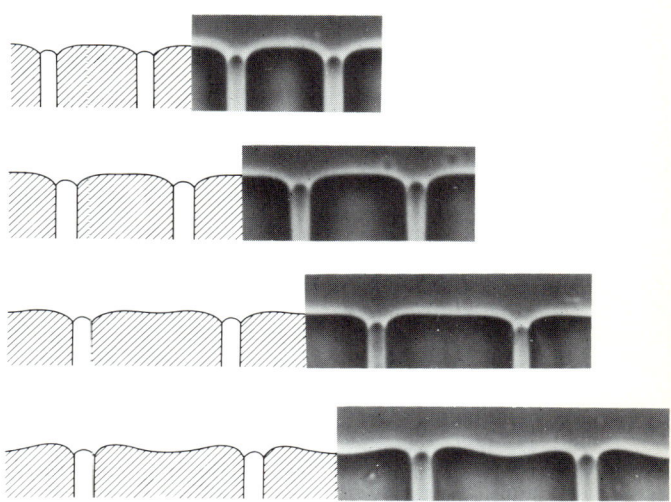

FIGURE 4-6
Comparison of calculated and observed shapes for a transparent organic eutectic. (*From Hunt and Jackson.*[9])

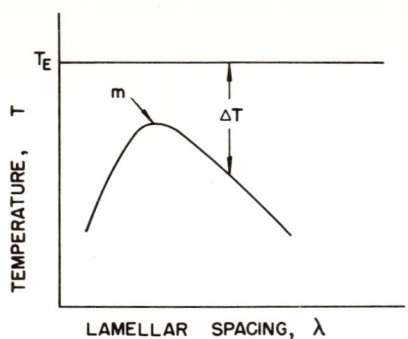

**FIGURE 4-7**
Change of interface temperature with lamellar spacing at constant growth rate.

shown. Experimental data, however, show no such ambiguity. For each $R$, one spacing is obtained, as shown by data on many systems; an example is given in Fig. 4-8. How does nature decide which spacing to adopt? The simplest possibility is that growth is preferred at the *extremum*, i.e., minimum undercooling for a given velocity, or, equivalently, maximum velocity for a given undercooling; this is represented by point $m$ on the curve of Fig. 4-7. With this condition, maximization of Eq. (4-9) gives a complete solution; that is,

$$\lambda^2 R = \frac{B}{A} \quad (4\text{-}10a)$$

or

$$\frac{\Delta T^2}{R} = 4AB \quad (4\text{-}10b)$$

Various investigators have assumed that growth does occur at this extremum and have derived equations of the form of Eqs. (4-10). In addition to simplicity of derivation, these equations have the advantage that they predict qualitatively the results obtained. Experiments on eutectic spacing invariably show $\lambda^2 R$ is a constant as predicted and illustrated by the experimental results of Fig. 4-8. The somewhat more limited data on interface temperature show that $\Delta T^2/R$ is a constant in agreement with Eq. (4-10b) and as shown by the example in Fig. 4-9. Moreover, the limited quantitative comparison of theory with experiment that has been attempted[4] shows this optimization is in reasonable quantitative agreement with results on metallic eutectics, although there is so much uncertainty in the values of the constants used in calculation that this limited agreement by no means proves that growth always (or ever) occurs exactly at the extremum. On the contrary, in work on thin specimens in transparent materials, Hunt and Jackson[9] have shown that after establishing a lamellar spacing at one growth rate, spacing can be maintained even though velocity is subsequently significantly increased. Growth could not have been at the extremum both before and after the velocity increase.

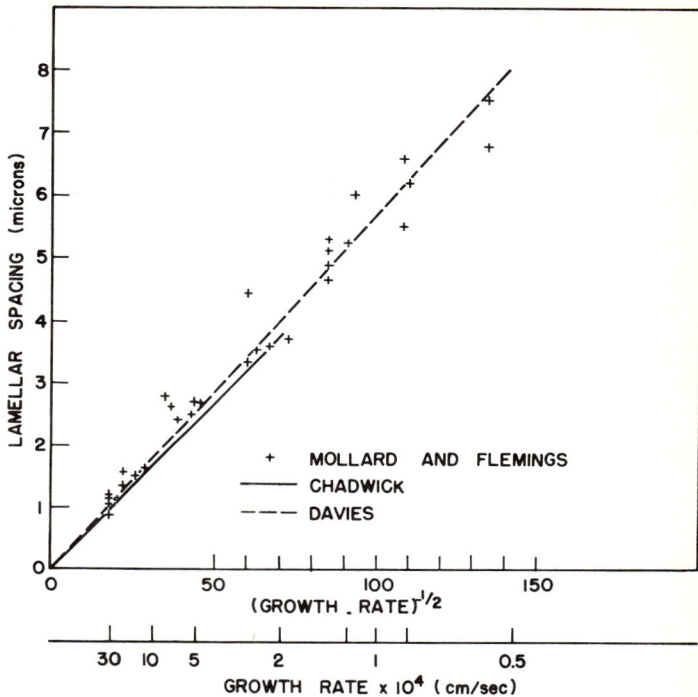

FIGURE 4-8
Lamellar spacing as a function of growth rate; tin–lead composites. (Results of Chadwick[31] and Davies[32] are for eutectic composition. Points from Mollard and Flemings[14] are for off-eutectic alloys of Fig. 4-16.)

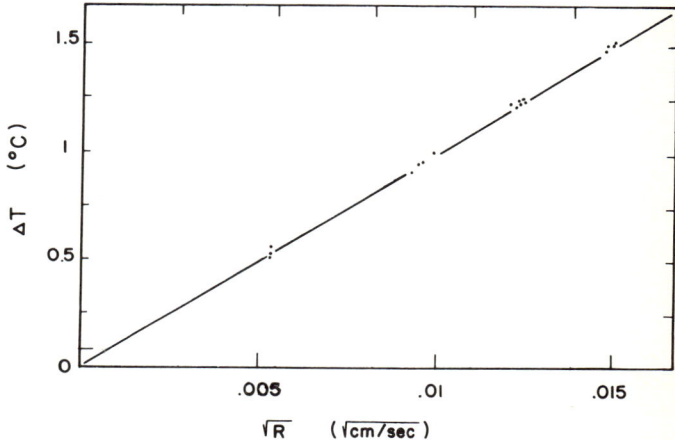

FIGURE 4-9
Interface undercooling for tin–lead eutectic. (*From Hunt and Chilton.*[33])

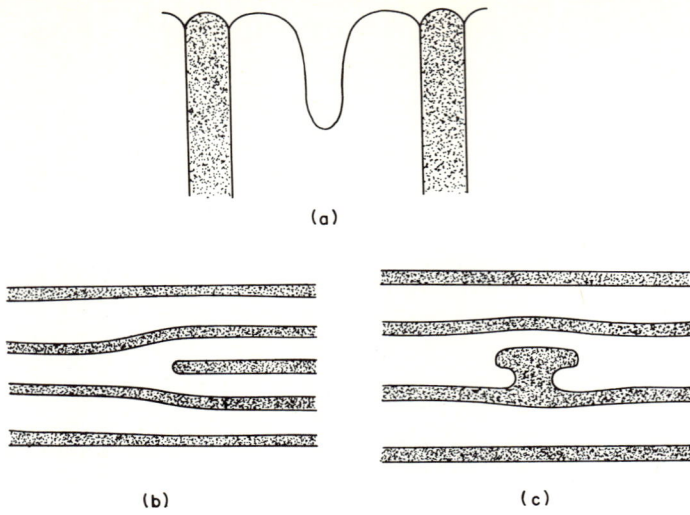

**FIGURE 4-10**
Schematic illustrations of mechanisms whereby plate spacing adjusts to increasing growth rate. (a) New lamellae form in a pocket of the wide phase; (b) extra lamellae move left; (c) a plate branches and new lamellae move left and right. In (a) growth direction is upward in the plane of the paper; in (b) and (c) it is perpendicular to the plane of the paper. (*After Jackson and Hunt*[4] *and Hunt and Hurle.*[11])

Although there is not complete agreement among investigators yet as to how nature chooses a lamellar spacing or undercooling at a given growth rate, it is clear that certain limits can be specified as to how far she can stray from the extremum condition. As pointed out by Cahn and discussed by Jackson and Hunt,[4] spacings smaller than that at the extremum are inherently unstable and thus the extremum spacing is expected to be the minimum observed. Spacings somewhat larger than the extremum are not obviously unstable, but when they become very much larger (a factor of 2 or more), interface curvature can no longer maintain an isothermal interface. Then a *pocket* develops in one phase and drops progressively back from the interface until growth of the other phase ultimately occurs in it (Fig. 4-10a). When this occurs, the local spacing is abruptly reduced by a factor of 2. This abrupt change in lamellar spacing is observed in thin-section transparent organics when growth rate is gradually increased and in bulk specimens when growth rate is abruptly increased by a large amount.

However, some additional mechanism or mechanisms must be operating to adjust lamellar spacing in bulk specimens because experiments on bulk specimens (as

**FIGURE 4-11**
Transverse microsection of directionally solidified Al–CuAl$_2$ eutectic. (*From Chadwick.*[34])

distinct from those on thin sections) show that lamellar spacing does adjust to relatively small changes in growth rate. One possible mechanism for lamellar structures is that described by Hunt and Jackson,[9] which develops an idea previously proposed by Jackson and Chalmers.[10] The proposed mechanism involves lateral movement during growth of structural defects called *terminations*. Figure 4-10b is a schematic example of such a termination as it would appear (ideally) looking at the liquid-solid interface. The average spacing to the left of the termination is larger than that to the right, and if the termination moves leftward during growth, overall average spacing is decreased; if it moves to the right, average spacing is increased. Now, suppose the lamellar spacing to the left of the termination is larger than that of the minimum $m$ (Fig. 4-7); then it will be growing at a larger undercooling than that portion to the right of the termination. A slight depression will form in the interface to the left of the termination, and the termination will move leftward. The result is that overall undercooling of the interface will be reduced. Presumably, when this mechanism operates, it should keep the lamellar spacing close to the minimum.

Lamellar terminations may be present in the structure or may be formed by plate branching, as in Fig. 4-10c.[11] Here, a local perturbation causes an incipient lamella to form as the interface progresses, with surrounding lamellae moving slightly so that locally there is only a small change in spacing between any two regions of the second phase. If the growth conditions are such that the finer spacing has the lower undercooling, this incipient lamella then spreads in the same way as the lamellar termination discussed above. Presumably the shape instability would occur preferentially in regions where the eutectic structure was least regular, as at eutectic grain boundaries or near regions of lamellar faults. The term *fault* is used here to describe the surface which separates two relatively perfect regions of eutectic. The two regions, however, are usually mismatched or misoriented and may contain a different number of lamellae. The faults may or may not contain lamellar terminations. Figure 4-11, a transverse microstructure of Al–CuAl$_2$ eutectic, shows examples of these features. Fault lines (traces of the fault surface) run vertically across the photograph, and the reader should be able to discern several terminations and a new lamella being formed by branching. Lamellar faults have been studied in detail by Hogan and coworkers,[12] but their origin is not yet fully understood. They appear to be an inherent structural feature of most unidirectionally solidified eutectics; an exception is the Sn–Cd eutectic.[13]

## ROD EUTECTIC GROWTH

The foregoing discussion has dealt only with lamellar eutectics, but quantitative treatment of rodlike eutectic growth has been done along exactly similar lines.[4] The solute redistribution [Eq. (4-4)] is solved as before but for the different boundary condition that the interface between the two solid phases is of circular cross section. The equation describing undercooling due to the radius of curvature is similar except that it must account for the double radius of curvature of the phases at the interface. The remaining equations, Eqs. (4-2) and (4-8), are identical, and the qualitative arguments leading to the quantitative relation between rod spacing $\lambda$, total undercooling $\Delta T$, and interface velocity $R$ are also identical. The form of the final expression is the same as for lamellae, with a minimum existing in the curve of $\Delta T$ versus $\lambda$ at some finite value of $\lambda$; and experimental results give the same functional dependence of phase spacing on growth rate. An example for Al–Al$_3$Ni eutectic is shown in Fig. 4-12. In these rodlike eutectics, spacing is probably maintained close to that of the extremum condition by a branching mechanism analogous to that of the plate branching.

Jackson and Hunt[4] show that the rod morphology should grow with lower undercooling than the lamellar morphology (and hence should be the stabler growth form) when the volume fraction of one phase is sufficiently small. With isotropic surface energies, the volume fraction is $1/\pi$. This result follows directly from the fact that total $\alpha\beta$ surface area (and hence free energy associated with surface) is less for the rodlike morphology when the minor phase is present in volume fraction less than $1/\pi$; it is less for the lamellar morphology when volume fraction is greater than $1/\pi$. The prediction of rodlike morphology at a low volume fraction is in qualitative agreement with observations on a large number of eutectics.[12] In direct confirmation, Mollard and Flemings[14] varied the volume fraction of the two phases in a lead–tin eutectic using a method to be described later in this chapter. They found that the structure was lamellar near the eutectic composition. When the volume fraction of the lead phase was reduced, the structure became fully rodlike at a volume fraction of about 0.18. The reason why the transition was complete only at 0.18 rather than

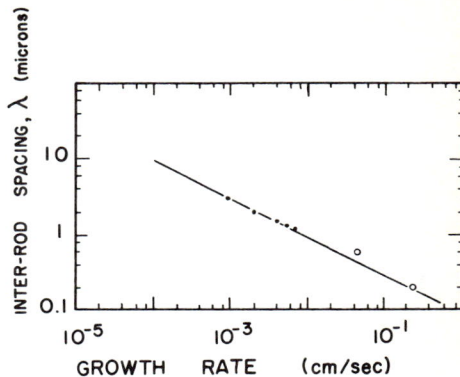

FIGURE 4-12
Interrod spacing of Al–Al$_3$Ni eutectic. (Solid points are from Lemkey et al.,[35] open points from Livingston et al.[36])

$1/\pi$ can be explained by assuming that the surface energy between solid phases is anisotropic. In this event, if lamellae are in their stable low-energy orientation, then a volume fraction smaller than $1/\pi$ is required before rods become the stable configuration.

Recent studies have shown that many, perhaps most, rodlike eutectic structures actually are not of circular cross section but are faceted with a few or many planes.[12] This does not alter the qualitative conclusion that rods (faceted or circular) are expected in phases which are present in a small volume fraction.

## FACETED–NONFACETED EUTECTIC GROWTH

A number of commercially important eutectic alloys, including cast iron and aluminum–silicon, solidify such that one phase has a faceted liquid-solid interface. Many other eutectics of potential commercial importance also solidify in this way, including many metal–metal oxide and metal–metal carbide eutectics.

In some materials, even though one phase is a strong facet-former, growth proceeds essentially as in the nonfaceting alloys described previously. This occurs when the facet, if it forms, does not reach the $\alpha\beta$ groove. The InSb–NiSb rodlike eutectic is an example. InSb is a strong facet former and is the matrix phase sketched in Fig. 4-13a. Facets, if they form in the InSb, cannot reach the $\alpha\beta$ groove since this groove is a region of negative curvature. In other eutectics, the faceting phase is the minor component or forms as lamellae. In these cases, the facet can extend to the $\alpha\beta$ groove, as sketched in Fig. 4-13b, and solidification behavior is now altered.[11]

Figure 4-14 shows the structure of two similar eutectics, one a transparent organic (succinonitrile–borneol) and the other aluminum–silicon. In spite of a rel-

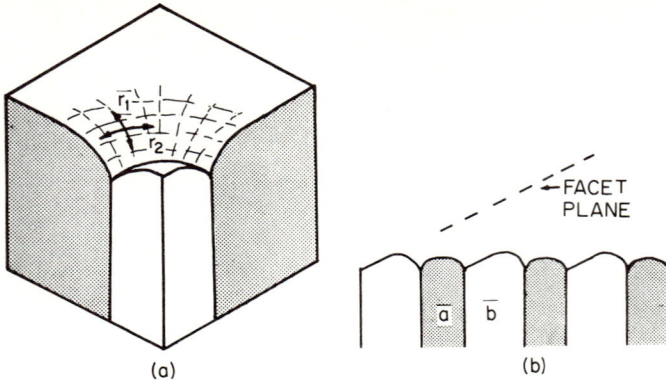

**FIGURE 4-13**
Schematic diagram of liquid-solid interfaces in eutectics. Facets cannot form in the $\alpha\beta$ interface groove in the major phase of (a); they can form in the minor phase or in lamellar eutectics as in (b). (*From Hunt and Hurle.*[11])

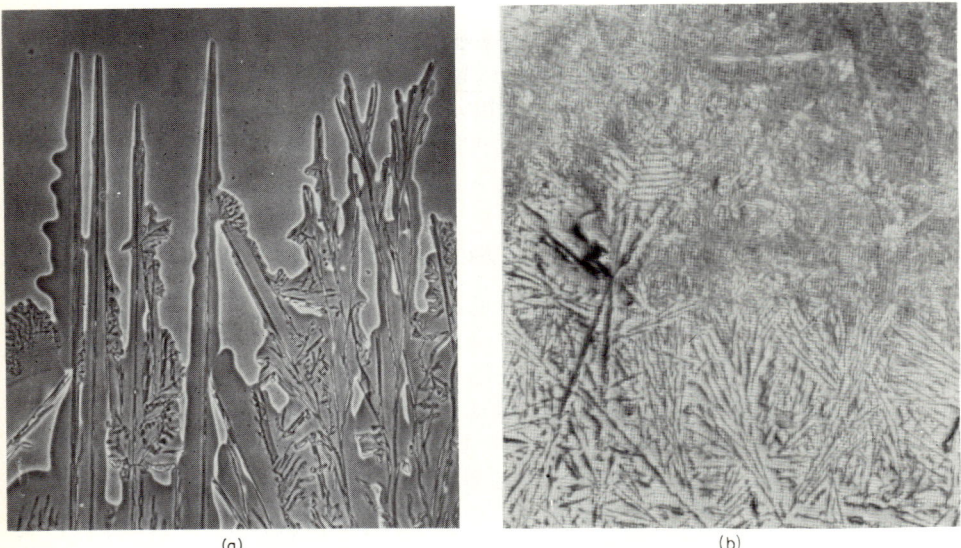

**FIGURE 4-14**
Irregular liquid-solid interface in a faceted-nonfaceted eutectic. (a) Transparent organic eutectic during growth, magnification ×200 (*from Hunt and Jackson*[9]); (b) Al-Si eutectic quenched during growth (*from Chadwick*[34]).

atively steep thermal gradient in both cases, the liquid-solid interface clearly is not planar on a scale the size of the phase spacing; therefore it is not isothermal. Furthermore, observations of the transparent eutectic show that locally no steady state is achieved. The faceted phase grows rapidly out from the interface with a platelike structure, and the nonfaceting phase immediately follows, apparently covering everything except the tip and leaving pools of liquid behind the tips of the solid. The two phases eventually form in these pools but at considerably lower temperature than that of the phase tips. Probably, other faceted–nonfaceted eutectics which form similar *irregular* structures solidify in like manner. No quantitative analysis of the solidification behavior of faceted–nonfaceted eutectics has been attempted. Such analysis would need to recognize two important differences between this type of solidification and that of the eutectics discussed earlier. Equilibrium can no longer be assumed at the liquid-solid interface, and, at least for the thermal gradients usually obtained, the interface can no longer be assumed to be isothermal. Even more difficult to describe quantitatively would be the faceted–faceted type of eutectic observed in some organic materials.

## INTERFACE STABILITY

The foregoing discussion has dealt primarily with pure alloys of exactly eutectic composition. However, it has been recognized for many years that structures fully eutectic in appearance are obtained with modest compositional deviations from the eutectic, and it has been shown recently that with suitable control of growth conditions very wide compositional variations can be tolerated while still maintaining a plane polyphase interface. The general problem, in its simplest view, is exactly analogous to that of solidification of single-phase alloys. If the interface can be maintained stable in the plane front configuration, eutectic or eutecticlike structures are obtained. When the interface breaks down, cellular and dendritic structures are obtained which are single-phase or polyphase, depending on whether the alloy is binary or multicomponent.[14,15]

Consider an alloy of composition $C_0$, off eutectic composition, but which solidifies with a two-phase structure at equilibrium, Fig. 4-15a. Assume for the moment that plane front can be maintained. An equation can now be written describing liquid composition at positions $x'$ and $y$ in front of the interface that is closely similar to Eq. (4-6). It is[4]

$$C_L - C_0 = (C_E - C_0)e^{-(R/D_L)x'} + \sum_{n=1}^{\infty} B_n \cos \frac{2n\pi y}{\lambda} e^{-(2n\pi x'/\lambda)} \qquad (4\text{-}11)$$

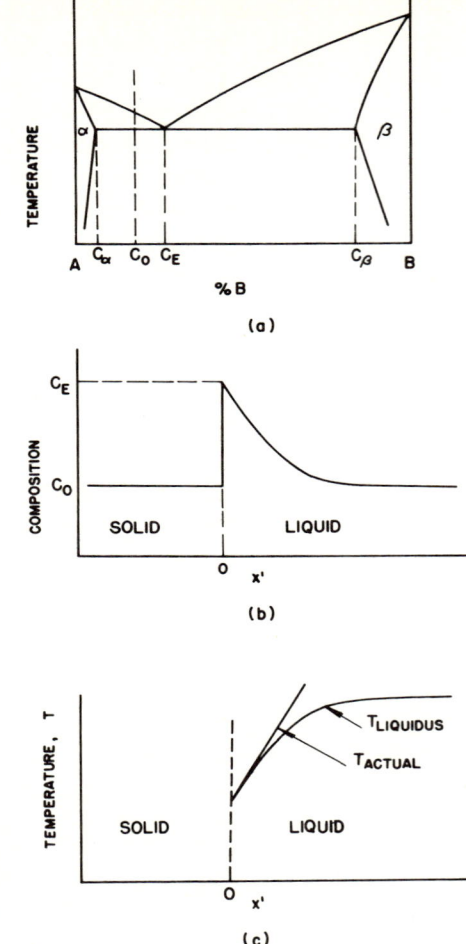

**FIGURE 4-15**
Constitutional supercooling criterion for stability of composites. (*a*) Phase diagram; (*b*) solute distribution at steady state in growth direction; (*c*) required condition for stability.

where $B_n$ is as given by Eq. (4-7a). (Equation 4-11) is identical to Eq. (4-6) except for the term involving $Rx'/D_L$. The amplitude of the variations in the $y$ direction (transverse to the interface) is given by the second term and is exactly as in alloys of eutectic composition (e.g., Fig. 4-4). The mean liquid composition decreases exponentially with distance into the liquid as in single-phase alloys. Now, one may distinguish two different characteristic diffusion distances in the $x'$ direction. The first is the boundary layer within which solute is transported transverse to the interface; this is the order of $S_\alpha + S_\beta$, one-half the interlamellar spacing. Beyond this distance, transverse concentration gradients are negligible, as shown in Fig. 4-4. The second characteristic diffusion distance is that analogous to the one in single crystal growth, or $D_L/R$. Over a distance about $D_L/R$, the mean liquid solute content is appreciably above $C_0$.

For usual crystal-growing velocities, the thickness of the former boundary layer is much less than that of the latter; that is, $S_\alpha + S_\beta \ll D_L/R$. Hence, perturbations in the $y$ direction damp out much more quickly than those in the $z$ direction. When $S_\alpha + S_\beta \ll D_L/R$, Eq. (4-10) reduces simply to

$$\frac{C_L - C_0}{C_E - C_0} \simeq e^{-(R/D_L)x'} \qquad (4\text{-}12)$$

and the solution is exactly as in growth of single-phase alloys. Equation (4-12) is plotted schematically in Fig. 4-15b. Now, *constitutional supercooling* will develop in front of the interface unless thermal gradient is sufficiently steep, as sketched in Fig. 4-15c. By following steps identical to those used in developing the constitutional supercooling criterion for single-phase alloys, one obtains the similar requirement for stability in two-phase alloys. In the absence of convection, this is

$$\frac{G_L}{R} \geq -\frac{m_L(C_E - C_0)}{D_L} \qquad (4\text{-}13)$$

where $m_L$ is slope of the liquidus line and $G_L$ is temperature gradient in the liquid at the liquid-solid interface. Equation (4-13) is the two-phase analog to Eq. (3-4).

Figure 4-16 is a plot of experimental results of Mollard and Flemings[14] on tin–lead alloys testing the relation of Eq. (4-13). Eutecticlike, or *composite*, structures are obtained with adequately high values of $G_L/R$ even in compositions substantially removed from the eutectic. Rough agreement is obtained with the simple theory above, except close to the eutectic where only low values of $G_L/R$ are sufficient for stability. This discrepancy plus some observations of Cline and Livingston[16] that the interface can be surprisingly stable at very high interface velocities are discussed later. Figure 4-17 shows two typical structures from the work of Mollard and Flemings. One near eutectic composition shows the typical lamellar structure; the other, with only about one-half the lead content of the first, is rodlike as expected from the discussion given earlier.

When the plane interface of binary alloys such as the foregoing breaks down, single-phase cells, or dendrites, form. An example is shown in Fig. 4-18 for an Al-rich Al–Cu alloy. Although there is presumably some local perturbation of the two-phase interface at the initial stage of breakdown, only a single-phase cell, or dendrite, can move very far ahead of the isothermal interface.

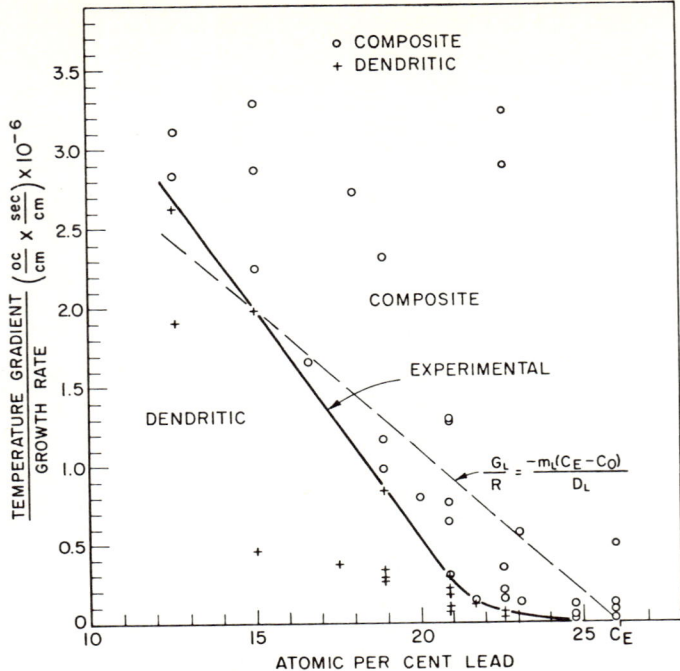

**FIGURE 4-16**
Plane front stability in directionally solidified tin–lead alloys. (*From Mollard and Flemings.*[14])

Solute redistribution occurring when cells or dendrites form is as given by the approximate treatment in Chap. 3, with the solute concentration in the liquid decreasing with distance from the eutectic front as shown schematically in Fig. 3-21a. The liquid composition at the eutectic front is just that of the eutectic, but the composition of the two-phase solid forming there is not. Its exact average composition $\bar{C}_S$ depends on $G/R$, where $G$ is the temperature gradient in the liquid-solid region. This may be seen by writing a mass balance at the eutectic front for steady-state solidification similar to Eq. (3-1):

$$\left(\frac{dC_L}{dx'}\right)_{x'=0} = -\frac{R}{D_L}(C_E - \bar{C}_S) \qquad (4\text{-}14)$$

The two-phase solid forming is of composition $\bar{C}_S$ and is different from $C_E$ by an amount which balances the solute flux down the gradient $dC_L/dx'$. Now, from Eq.

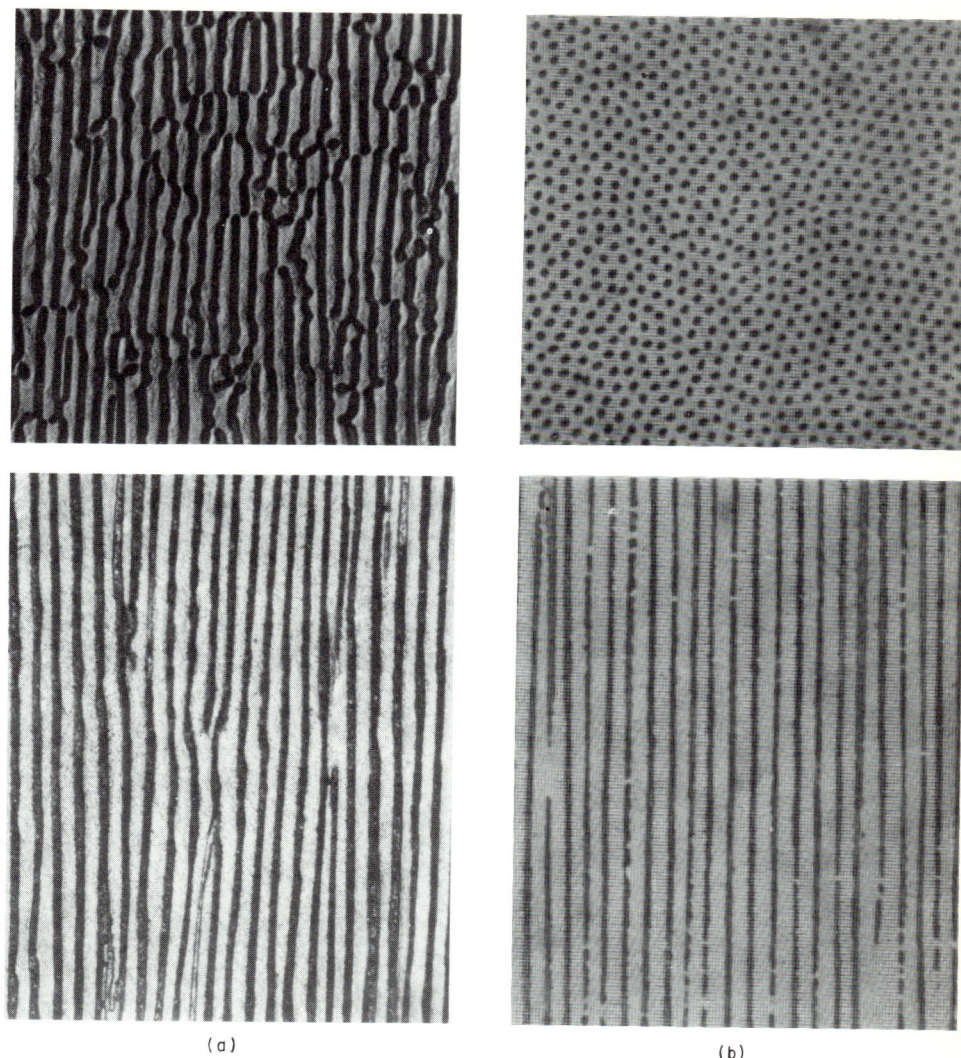

(a)            (b)

FIGURE 4-17
Transverse and longitudinal microstructures of tin–lead alloys solidified with plane front as *composites* (magnification ×390). (*a*) Lamellar structure (approximate eutectic composition, 25 at% Pb); (*b*) rodlike structure (12.6 at% Pb). (*From Mollard and Flemings.*[14])

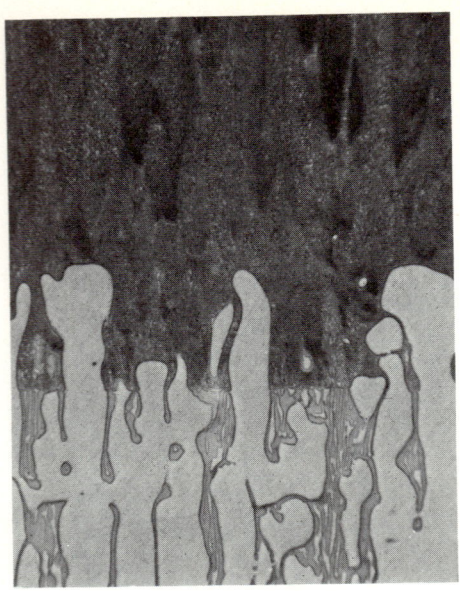

FIGURE 4-18
Quenched interface of directionally solidified Al–2.6% Cu alloy. $G/R$ slightly less than that required for stability of plane front. (*Courtesy W. Brown.*)

(3-19), $(dC_L/dx')_{x'=0}$ is just $G/m_L$, and so Eq. (4-14) becomes

$$\bar{C}_S = C_E + \frac{D_L G}{m_L R} \quad (4\text{-}15)$$

Thus, $\bar{C}_S = C_E$ at low values of $G/R$. As $G/R$ increases, it approaches $C_0$ as a limit. At this limit, single-phase dendrites are no longer stable and Eq. (4-14) becomes identical to the stability criterion [Eq. (4-13)].

## PERTURBATION ANALYSIS FOR STABILITY

The simple constitutional supercooling criterion for stability is useful for predicting approximately the effects of alloy and growth variables on structure. However, agreement with experiment is only approximate, and deviation is especially marked near the eutectic in binary alloys. Here, results such as those of Fig. 4-16 often show that much lower values of $G_L/R$ are sufficient for stability than predicted by a simple linear dependency on $G_L/R$. Cline and Livingston[16] have shown further that compositions within a few percent of the eutectic (in tin–lead) can be solidified at exceedingly high growth rates without encountering interface breakdown. They obtained fully lamellar tin–lead structures several percent from the eutectic at growth rates as high as $\frac{1}{2}$ cm/s. This is shown by the graph in Fig. 4-19 and by the photograph in the

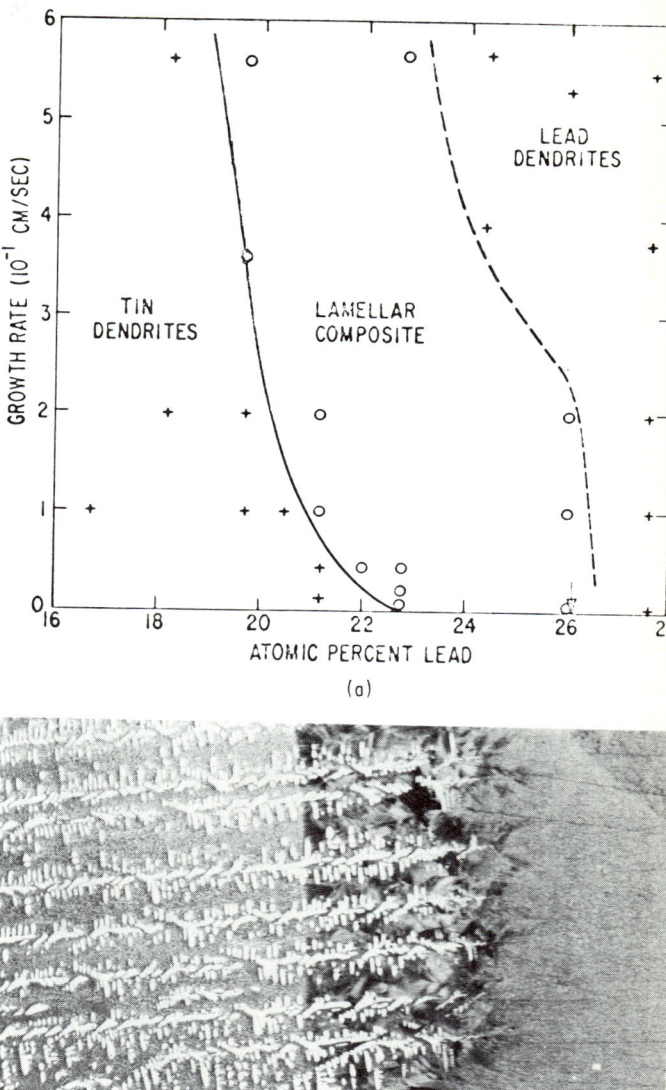

FIGURE 4-19
Growth rate–composition plot for Sn–Pb alloys showing regions over which various microstructures are observed. Photomicrograph shows the transition from dendritic to composite structure obtained in Sn–21 at% Pb by abruptly increasing the growth rate. (*From Cline and Livingston.*[16])

same figure showing a structure that was dendritic at relatively low solidification velocity which became composite when that velocity was increased. The experiments of this figure were conducted with temperature gradients in the neighborhood of 100°C/cm, and so values of $G_L/R$ were very low indeed. Structures obtained at the high growth rates employed by Cline and Livingston are very fine, with lamellar spacings as low as 0.1 μm. Small spacings are also obtained in rodlike composites at high growth rates. Some data for Al–Al$_3$Ni at high growth rates are given in Fig. 4-12.

Quantitative description of these results clearly requires a different stability theory than the simple constitutional supercooling criterion. Several have been attempted, and these have been reviewed by Sekerka[17] and by Hunt et al.[18] Each of the theories presented to date contains assumptions or approximations that make the validity of the calculated results uncertain, and the area remains one where more study is needed.

## CRYSTALLOGRAPHIC FEATURES OF DIRECTIONALLY SOLIDIFIED EUTECTICS

An interesting aspect of directionally solidified eutectics is that, at least after several centimeters of steady-state growth, a tendency toward a preferred crystallographic-orientation relationship is observed between the phases in all lamellar or fibrous systems. Furthermore, the phases themselves develop a preferred crystallographic relation with the growth axes.[12] For example, in the tin–lead eutectic, the preferred orientations are expressed by the relations[14]

$$\text{Lamellae} \quad //(0\bar{1}1)\text{Sn}//(1\bar{1}\bar{1})\text{Pb}$$
$$\text{Growth direction} \quad //[211]\text{Sn}//[211]\text{Pb}$$

Preferred orientation is developed in eutectics in a number of ways. In some instances, epitaxial effects in nucleation of the second phase on the first are important. In other cases, the important effects are change of the crystallographic direction of the phases during growth or encroachment of favorably oriented eutectic *grains* of other grains. Hopkins and Kraft[19] studied the growth of lead–tin eutectics using a tin single crystal to nucleate a single *eutectic grain*. In their seeding experiments, they found that if the tin seed crystal was oriented for growth according to the description above, the lamellar structure developed immediately in its equilibrium orientation. They then misoriented the seed so the $(0\bar{1}1)$ plane of the Sn remained parallel to the growth direction but the [211] direction did not. In this case, the lamellar structure developed immediately, and, as growth proceeded, the crystal orientation of the Sn phase "rotated" until it reached its equilibrium orientation with the [211] direction parallel to the growth axis. The mechanism of crystal rotation was not studied. When the (0$\bar{1}$1) plane of the Sn single crystal was not parallel to the growth direction, the lamellar structure did not form immediately. Instead, a wavy *semidegenerate* structure was obtained.

Jaffrey and Chadwick[20] showed that similar orientation changes occur in rodlike as well as platelike eutectics. They used a *boat* whose cross section abruptly increased at some point along the growth direction so that the eutectic growth would occur around a corner. In the rodlike Al–Al$_3$Ni eutectic, the Al$_3$Ni fibers followed the contour of the section increase, at least when this was of radius greater than about 1 in. The crystallographic orientation of the rod changed as well during growth, and so the Al$_3$Ni maintained its preferred growth direction. Apparently, this change in crystallographic orientation involved frequent branching with formation of associated subboundaries. The crystallographic orientation of the aluminum phase was found not to change as growth occurred around the corner. This is the way the aluminum would have behaved had it been a single crystal.

When eutectic growth is occurring not from a single nucleus, there is the additional mechanism of development of preferred orientation by grain encroachment. The eutectic *grains* which are oriented most favorably in the beginning tend to grow at the expense of their neighbors and eventually eliminate those most poorly oriented. This is expected since the most favorably oriented grains should grow with the least undercooling and hence lead very slightly their less fortunate neighbors. Experiments of Hopkins and Kraft[19] on samples containing more than one eutectic grain showed that preferred orientation developed both through crystallographic rotation, as discussed above, and then by encroachment of favorably oriented grains on their neighbors.

## NON-STEADY-STATE GROWTH

When a two-phase *composite* alloy is growing that is not of exactly eutectic composition, a boundary layer exists in front of the plane front interface, as sketched in Fig. 4-20. The existence of this boundary layer means that fluctuations in growth rate result in fluctuations of composition in much the same way as in single-phase alloys. Increases in velocity decrease the width of the boundary layer and increase the average composition of the solid; decreases in interface velocity do the reverse. Neglecting the small variations in solute content of the liquid in the $y$ direction, solute redistribution along the growth axis is treated as in the single-phase alloys discussed in Chap. 2. The only difference is that the boundary condition at the liquid-solid interface is fixed during growth; it is constant at $C_E$ as long as two phases are forming. Figure 4-21 is a set of calculated curves showing effect of sudden increase in growth rate on solid

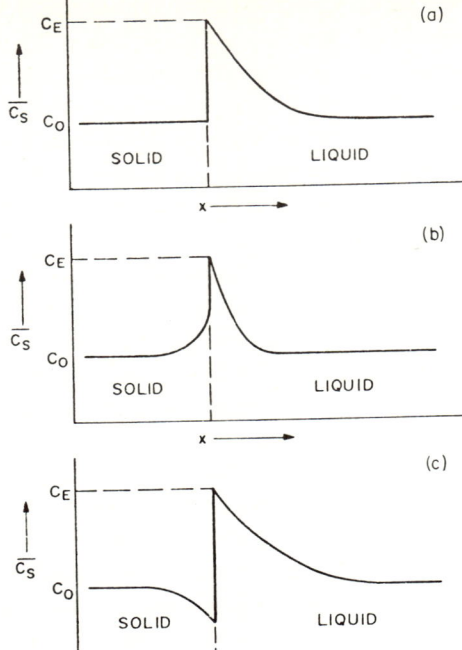

FIGURE 4-20
Schematic representation of concentration changes due to changes in growth velocity in plane front growth of two-phase alloys. (a) Steady state; (b) velocity increase; (c) velocity decrease. (From Mollard and Flemings.[14])

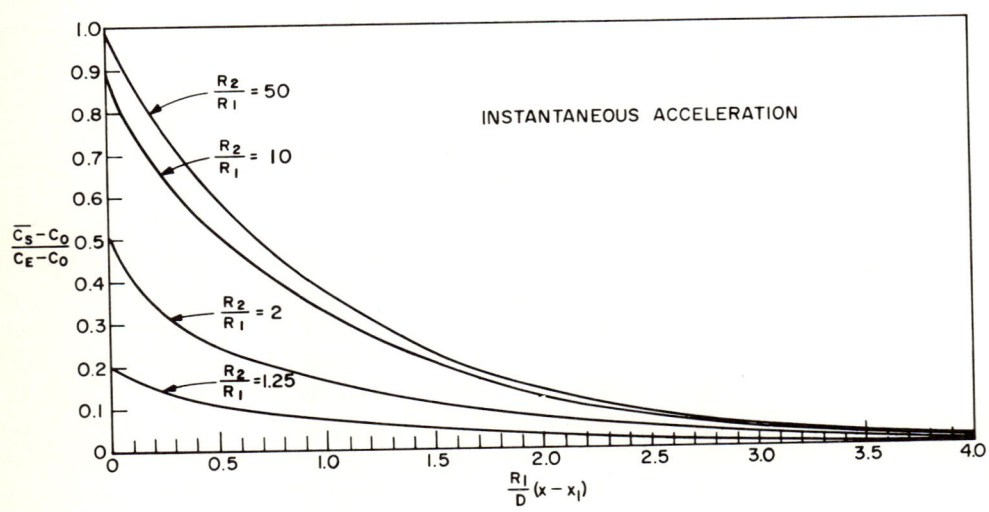

FIGURE 4-21
Solute redistribution due to an infinitely fast increase in interface velocity (from $R_1$ to $R_2$). (From Mollard and Flemings.[14])

composition. These curves are similar to those of a single-phase alloy undergoing a rapid increase in growth rate (Fig. 2-7). The major difference is that they show a maximum exactly at the time of the velocity increase because of the fixed boundary condition in the liquid at the interface.

Experimental results confirm the foregoing, and Fig. 4-22 shows one such set of results. Of course, concomitant with the increase in velocity is also an abrupt decrease in lamellar spacing. Decreases in velocity result in reduced solid composition $\bar{C}_S$ and associated increases in interlamellar spacing. When the interface velocity is slowed sufficiently so that $\bar{C}_S$ drops below the maximum solubility of the primary phase, the structure abruptly becomes single phase and this is the origin of the single-phase *bands* often seen in eutectic growth.

In this section and in preceding sections of this chapter, it has been assumed that convection is negligible. No direct effect of convection on solute diffusion is expected in eutectic alloys since the boundary layer for transverse diffusion is much smaller than is affected by usual convection. The situation is different when the alloys are of off-eutectic composition, i.e., when a boundary layer exists in the growth direction whose dimension is the order of $D_L/R$. Here, an analysis directly analogous to that presented for single-phase alloys shows that convection lowers average composition $\bar{C}_S$ of the composite forming and that with sufficient convection only a single-phase structure can form until the remnant liquid is sufficiently enriched to $C_E$. Changes in either growth rate or thickness of convective boundary layer $\delta$ will result in local variations of composition. Changes in growth rate will change both lamellar spacing and composition, while changes in $\delta$, to a first approximation, will change only composition.

## OTHER TWO-PHASE STRUCTURES

Cooperative growth (e.g., plane front solidification) of polyphase alloys is not limited to alloys containing a eutectic. Livingston and Cline[21] have shown that monotectics can be similarly solidified. Figure 4-23 is an example of rods of Pb in Cu. Growth of this structure occurs in essentially the same manner as does the growth of rodlike eutectics. At low growth rates, the lead appears as droplets rather than rods presumably because relative surface energies favor this form. It should also be possible to solidify alloys with plane fronts that undergo a peritectic reaction on solidification. For the reaction $L + \alpha \to \beta$, a composite two-phase structure $\alpha + \beta$ would result if the initial alloy composition $C_0$ were such that it is two-phase after equilibrium solidification.

Another type of two-phase structure results when foreign particles are present in the liquid. These particles may be incorporated into the growing solid, or they may

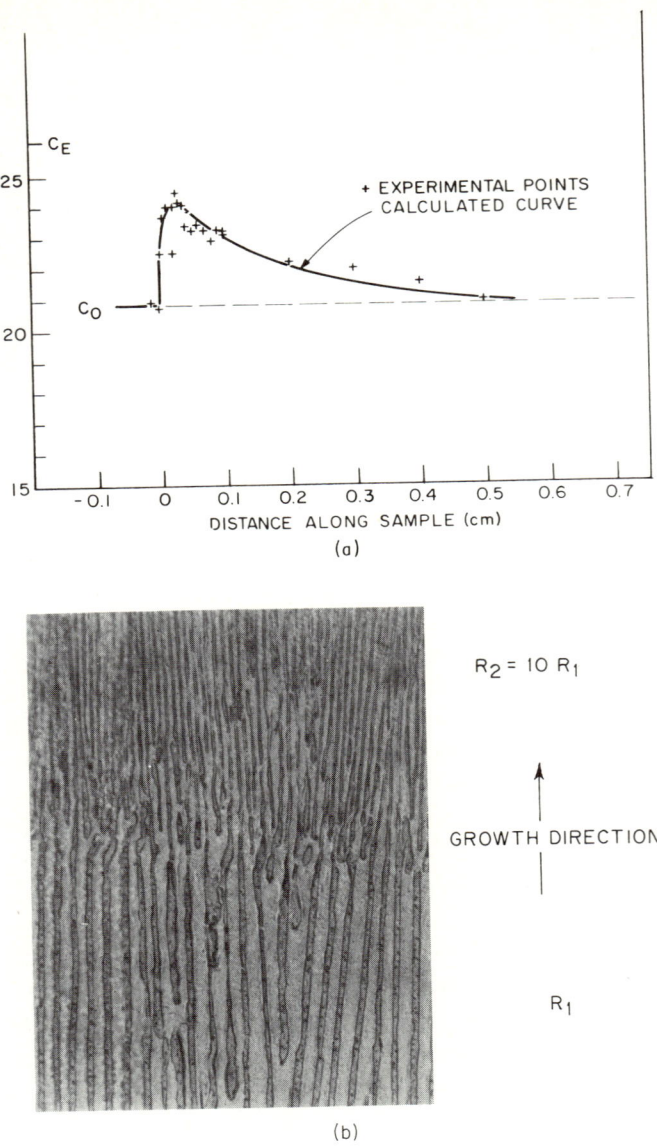

**FIGURE 4-22**
Composition and structural changes in Sn–Pb alloy resulting from an increase in growth velocity of approximately 10:1. (*From Mollard and Flemings.*[14])

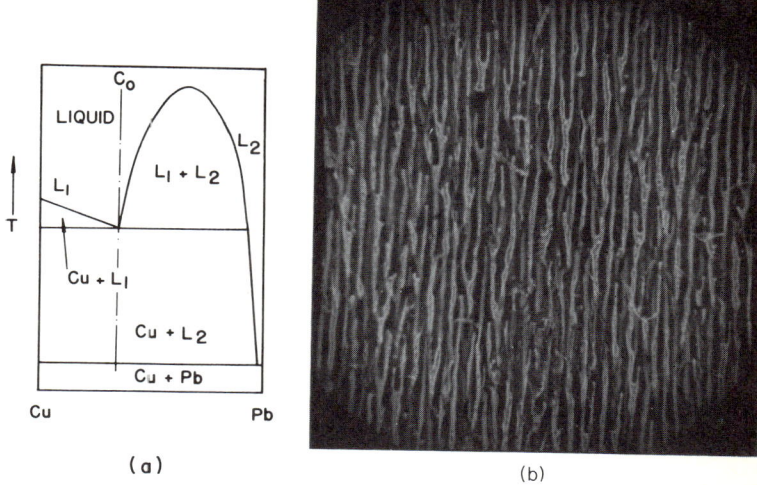

**FIGURE 4-23**
(*a*) Schematic diagram of Cu–Pb system showing monotectic; (*b*) electron micrograph of directionally solidified structure. (*From Livingston and Cline.*[21])

be "pushed" by the solid front as it moves. The pushing constitutes a method of crystal purification in addition to the diffusional mechanism discussed in Chap. 2. When the particles are entrapped, they form local heterogeneities in the crystal. It has been suggested by Jackson[22] that incorporation of the small heterogeneous nuclei that are present in most liquid metals is a source of dislocations in melt-grown crystals.

Particles present in liquid melts are not always entrapped. Uhlmann et al.[23] have studied "pushing" of particles by liquid-solid interfaces in several different organic materials, including salol, thymol, and water. For each particular type of particle, a critical velocity was observed, below which the particles were pushed by the interface and above which they were trapped. For particles under about 15 $\mu$m in size, the critical velocity was in the range of 0.2 to 20 $\mu$m/s, depending on particle material but independent of particle size. In Chap. 6 examples will be given of pushing of particles in metallic systems, for example, $SiO_2$ inclusions in an iron alloy. The mechanism of pushing can be viewed qualitatively by considering a system with $\sigma_{SP} > \sigma_{SL} + \sigma_{LP}$, where $\sigma_{SP}$ is the surface energy between the solid and the particle, $\sigma_{SL}$ is solid-liquid surface energy, and $\sigma_{LP}$ is the surface energy between the particle and liquid. As the particle and liquid-solid interface approach one another, the sum of the surface energies per unit area for the two surfaces in near contact increases continuously from

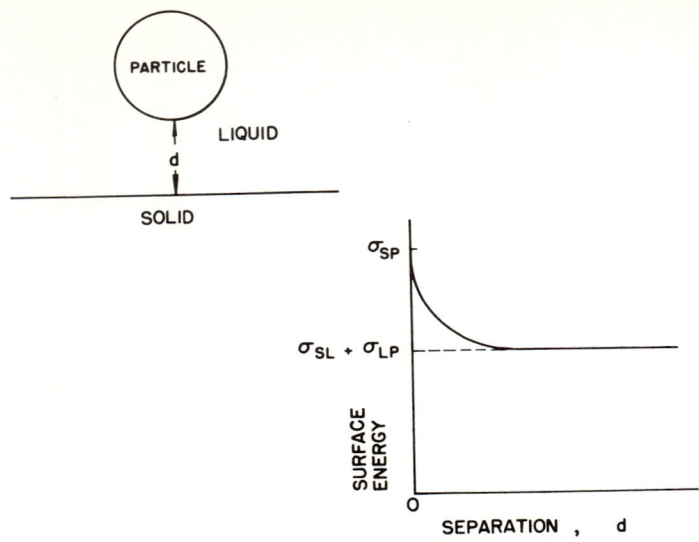

**FIGURE 4-24**
Pushing of a particle by a liquid-solid interface. (*Left*) Particle at distance $d$ from the interface; (*right*) surface free energy for the solid-liquid and liquid-particle interfaces versus $d$. (*After Uhlmann, Chalmers, and Jackson.*[23])

$\sigma_{SL} + \sigma_{LP}$ to $\sigma_{SP}$ at zero separation, Fig. 4-24. The increase in surface energy and hence in free energy with decreasing separation constitutes a driving force tending to keep a finite separation and therefore "push" the particle. In order to maintain the separation, fluid must reach the region of the liquid-solid interface underneath the particle. It does so by diffusion and fluid flow, processes driven by the surface-energy effect outlined above.

## TERNARY ALLOYS

For illustration of the principles of polyphase crystal growth in ternary alloys, consider the simple ternary eutectic system sketched in Fig. 4-25. Isothermal sections of this diagram are shown in Fig. 4-26, and a vertical section in Fig. 4-27. The vertical section extends from pure component $q$ through the ternary eutectic point. If interface kinetics pose no great barrier to growth, any alloy in this system can be grown with a plane front. The number of phases present in the solidified crystal is the number of phases present at the equilibrium solidus of the alloy. For example, alloy 1 (Fig. 4-27) is single-phase, alloy 2 is two-phase, and alloy 3 is three-phase after solidification. In

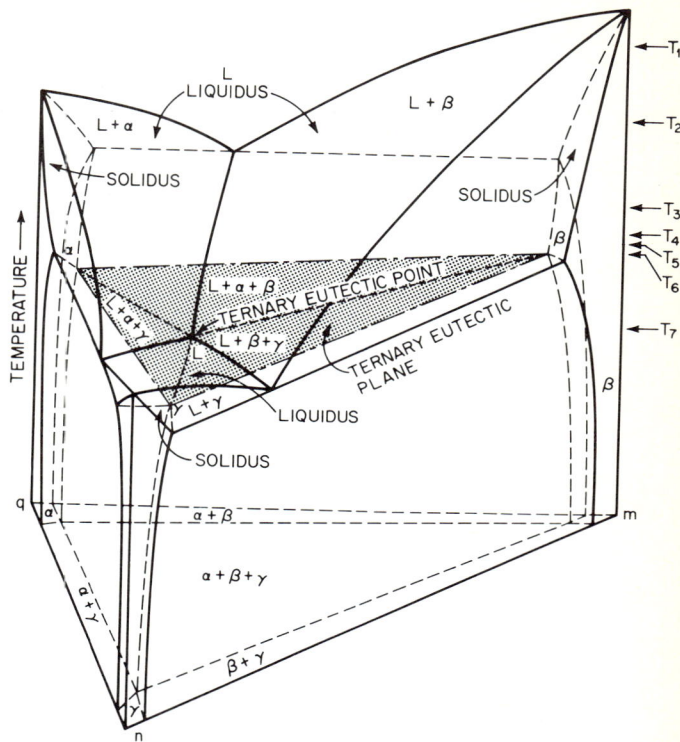

**FIGURE 4-25**
Phase diagram for a three-component system containing a ternary eutectic. (*After Rhines.*[37])

Chap. 3, we examined conditions for stability of the single-phase alloy. Here we look at stability of the two- and three-phase alloys using a simple constitutional supercooling criterion.

First of all, for either one-, two-, or three-phase alloys, the solute distributions in the growth direction in front of the growing interface are given by Fig. 3-28. When applied to polyphase alloys, Fig. 3-28 is analogous to Fig. 4-15*b* in that it neglects the small variations in solute content near the interface perpendicular to the growth direction. The arguments used in Chap. 3 to develop the diffusion and stability equations [Eqs. (3-27) and (3-28)] for single-phase alloys apply equally well to polyphase alloys once the assumption is made of negligible variation of solute perpendicular to the growth direction. For the case of three-phase alloys, the interface composition is approximately eutectic ($C_{Em}, C_{En}$). The stability is therefore written

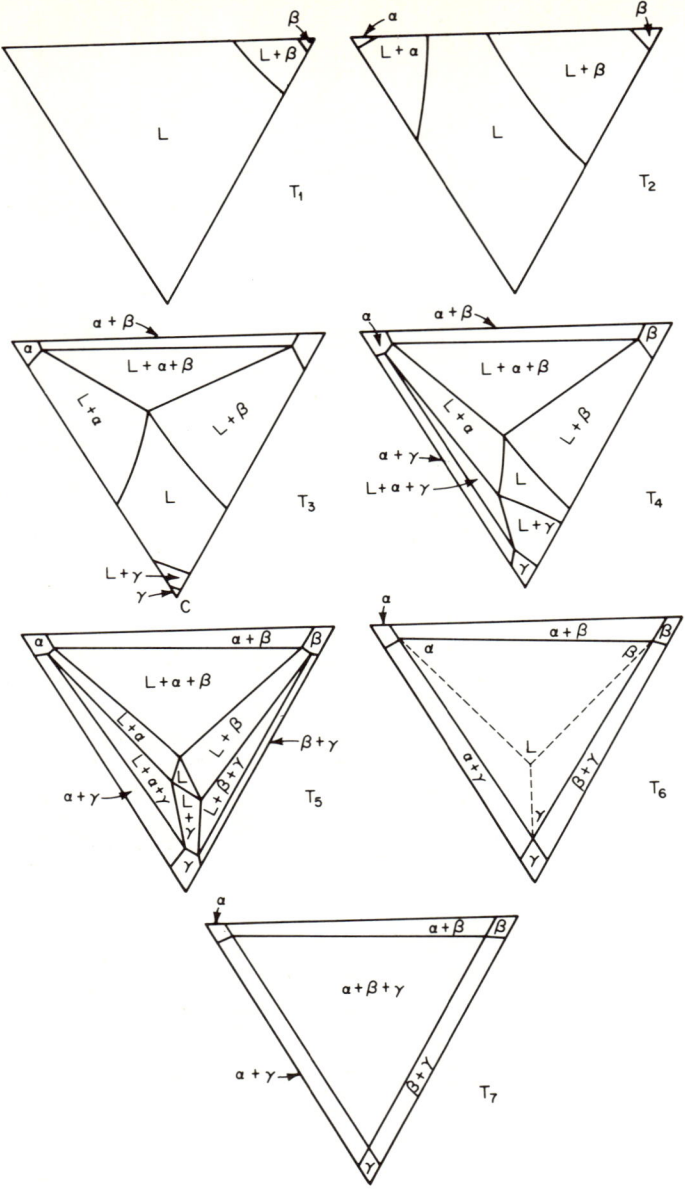

**FIGURE 4-26**
Isothermal sections of the phase diagram of Fig. 4-25. (*After Rhines.*[37])

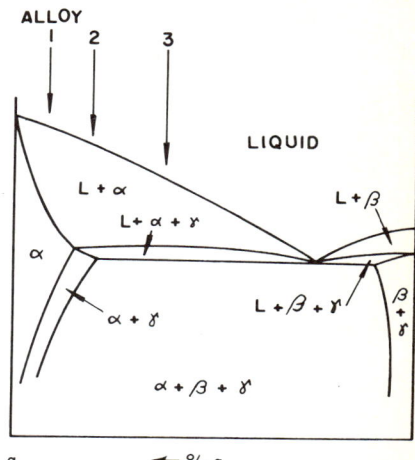

**FIGURE 4-27**
Vertical section through the phase diagram of Fig. 4-25. Section extends from pure component $q$ through the ternary eutectic.

directly for this case by simply substituting the eutectic compositions for the interface composition in Eq. (3-28):

$$\frac{G_L}{R} = -\frac{p(C_{Em} - C_{0m})}{D_{mm}} - \frac{s(C_{En} - C_{0n})}{D_{nn}} \qquad (4\text{-}16)$$

where $C_{Em}$ and $C_{Cn}$ are solute contents of elements $m$ and $n$, respectively, at the ternary eutectic point (other symbols are defined in Chap. 3).

Rinaldi, Sharp, and Flemings[27] have compared Eq. (4-16) with experiments on the stability of three-phase Al–Cu–Ni alloys. Comparison is good (Fig. 4-28) in spite of the many simplifications of this constitutional supercooling analysis. Figure 4-29 shows a typical structure from this work; the nickel-rich phase is present in smallest quantity and is rodlike. The two major phases, α–aluminum–solid solution and $CuAl_2$ are lamellar. When the value of $G_L/R$ is not high enough to maintain stability, cells, or dendrites of one or two phases protrude from the three-phase interface as shown in Fig. 4-30.

Alloys of close to ternary eutectic composition can be grown with plane front with little difficulty provided interface kinetic limitation to growth is small. A number of workers have grown these materials and observed a wide array of interesting morphologies. For example, in the Sn–Pb–Cd eutectic, the three phases are lamellar, that is, in ABCBA arrangement, where A, B, and C refer to the tin-, lead-, and cadmium-rich phases, respectively.[26] Thus, lead-rich lamellae are found on both sides of the cadmium-rich lamellae, and there are twice as many of these as of the tin or cadmium lamellae. Cooksey and Hellawell[28] describe a number of eutectic structures of this type, others with structures where one phase is fibrous and the

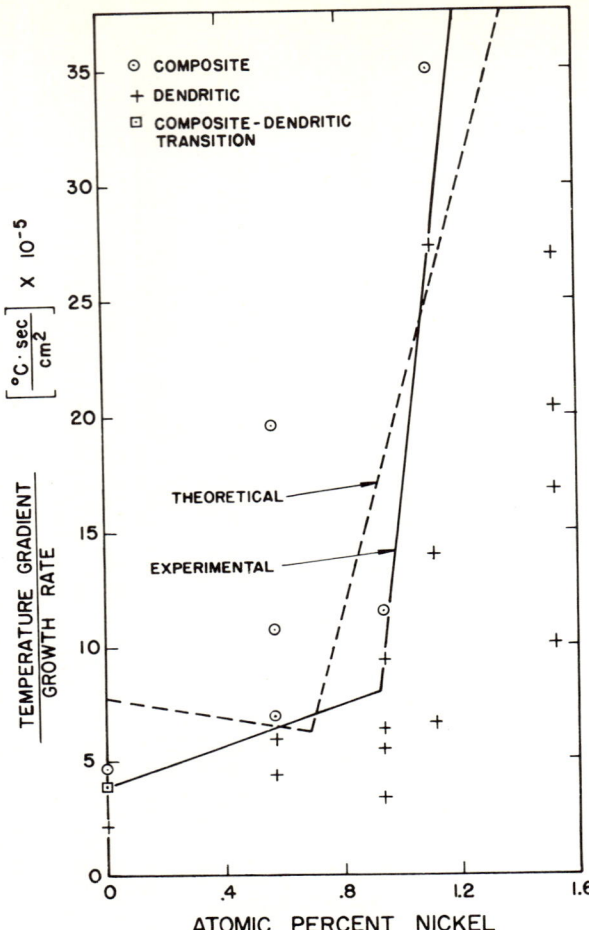

**FIGURE 4-28**
Interface stability of Al-rich ternary Al–Cu–Ni alloys. Cu content is constant at 14.4 atomic percent. (*From Rinaldi, Sharp, and Flemings.*[27])

remaining two are lamellar (as in Fig. 4-29), and still others in which phases are irregular.

Ternary alloys that are two-phase after equilibrium solidification also freeze with a plane front when the stability criterion of Eq. (3-28) is met. In the example of alloy 2 in Fig. 4-27, the two phases forming are $\alpha + \gamma$. The equilibrium solidus temperature of this alloy is section $T_5$ of Fig. 4-26; this is also the temperature of the liquid-solid interface in plane-front solidification. The compositions of the $\alpha$, $\beta$, and liquid phases at the interface are given by the corners of the tie triangle in this isothermal section.

TERNARY ALLOYS 125

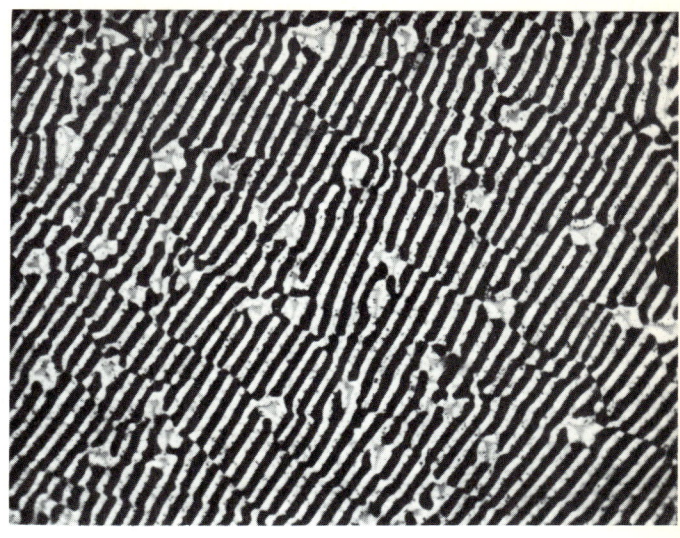

FIGURE 4-29
Transverse section from an Al–Cu–Ni composite crystal. Phases are $CuAl_2$ (dark), $Cu_3NiAl_6$ (light gray), and $\alpha$ Al (white). (*From Rinaldi, Sharp, and Flemings.*[27])

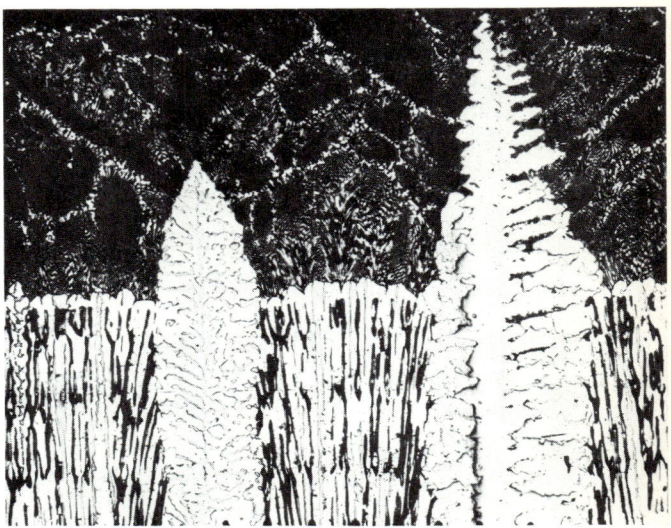

FIGURE 4-30
Quenched interface of ternary Al–Cu–Ni alloy grown with nonplanar interface. Leading phase is $Cu_3NiAl_6$. Two-phase dendrites are $Cu_3NiAl_6 + \alpha$. (*From Rinaldi, Sharp, and Flemings.*[27])

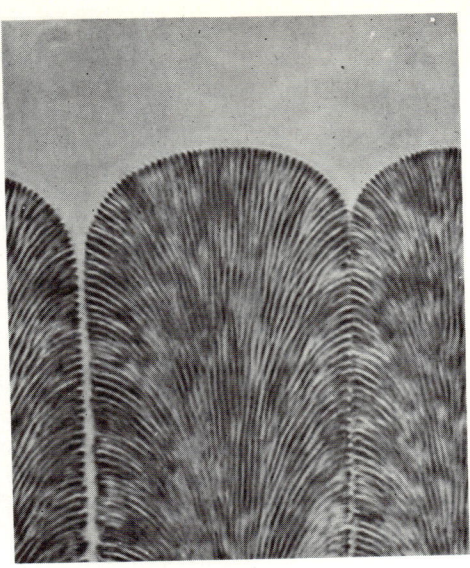

FIGURE 4-31
Eutectic cells in a transparent organic alloy. (*From Jackson.*[22])

Thus, to solve the stability equation for two-phase ternary alloys requires more information from the phase diagram than for three-phase alloys.

A simplification of Eq. (3-28) is possible for the special case of ternary alloys which lie near a binary eutectic composition, e.g., any one of the three alloys shown in Fig. 4-25. This case is of practical importance in understanding the effects of impurities on solidification of binary eutectics. It is also of practical importance if one wishes to add a solid solution-alloying element to one or another of the phases of a binary eutectic. The simplification results by assuming that the diffusion coefficients of the two solutes are equal and that the ternary alloy composition lies on the eutectic trough. The equation is

$$\frac{G_L}{R} \geq \frac{m_{em}C_{0m}(1 - \bar{k}_m)}{\bar{k}_m D_L} \qquad (4\text{-}17)$$

where $m_{em}$ is the slope of the liquidus trough with respect to element $m$, and $\bar{k}_m$ is the *average partition ratio* of element $m$ (the average composition of the solid divided by the composition of the liquid at the interface); $D_L$ is diffusion coefficient of both solutes. This equation has been compared to experiments by Gruzleski and Winegard[29] and by Bullock et al.[30]

At least for small additions to binary eutectics, two-phase cells usually form when $G_L/R$ is less than that required for plane front. These are analogous to those that form in constitutionally supercooled single-phase alloys. An example is shown in Fig. 4-31 for a transparent organic material and in Fig. 4-32 for a tin–cadmium

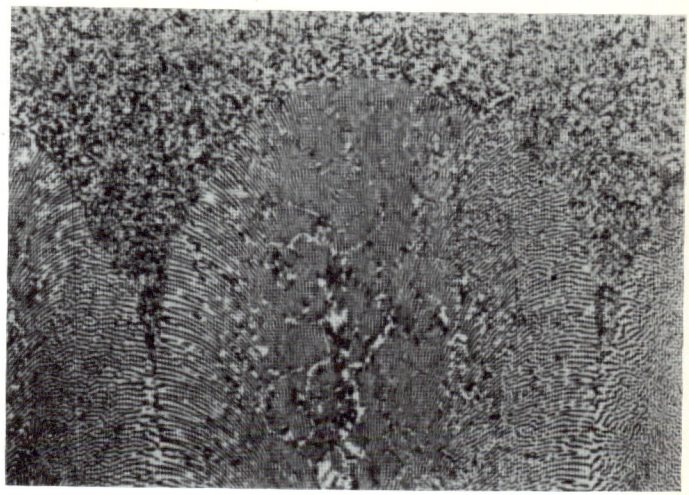

FIGURE 4-32
Cells in tin–cadmium eutectic. (Magnification ×170.) (*From Gruzleski and Winegard.*[13])


alloy. At sufficiently low thermal gradient, these two-phase cells can assume a more complex morphology—as in the case of their single-phase counterparts, branches form on the cells. The resulting structure is then best described as "two-phase dendrites." A different type of more complex morphology can occur in cellular faceted–nonfaceted alloys. Here, the cell boundaries tend to be faceted in the orientation of the *facet plane* sketched in Fig. 4-13. The nonfaceting phase (rods or lamellae) grows perpendicular to the facets of the cells. The result is a *complex-regular* eutectic structure[11] such as that seen in $Zn–Mg_2Zn_{11}$ (Fig. 4-33).

## EFFECT OF CONVECTION

In the growth of alloys of exactly eutectic composition, we expect no direct effect of convection on structure or composition. This is because the solute redistribution takes place within a boundary layer that is very much smaller than the momentum boundary layer resulting from the flow. In effect, even with very vigorous convection, the solute redistribution in eutectic growth takes place in a quiescent liquid. The solute boundary layer in eutectic growth extends outward only the order of an interlamellar spacing, as shown in Fig. 4-4. Convection, of course, can have the indirect effect of thermally perturbing the interface. The thermal perturbations result in bands of increased or decreased interlamellar (or interrod) spacing.

**FIGURE 4-33**
The complex regular structure formed in the Zn–Mg$_2$Zn eutectic when growth is cellular. Overall growth direction normal to the plane of the paper. (Magnification × 560.) (*From Hunt and Hurle.*[11])

Convection much more readily affects the structure and composition of polyphase alloys that are not of exactly eutectic or monotectic composition. Here, a solute boundary layer, as sketched in Fig. 4-15, extends outward the order of $D_L/R$ in the absence of convection. This distance is usually much greater than the interlamellar spacing and is large enough so that solute redistribution in it can be significantly affected by flow. As an example, consider the alloy of composition $C_0$ shown schematically in Fig. 4-34a. If this alloy were solidified with plane front and without convection, the final ingot would be a two-phase solid of uniform composition $\bar{C}_S$ equal to $C_0$.

Assume now that the alloy is solidified in a boat with convection so that a solute boundary layer $\delta$ is present at the liquid-solid interface as sketched in Fig. 4-34b. Convection is such that the solid forming $\bar{C}_S^*$ is greater than $C_{SM}$ and so is two-phase. The liquid at the interface is in equilibrium with these two solids and so is $C_E$. The bulk liquid is of uniform composition $C_b \geq C_0$. The width of the boundary layer $\delta$ in Fig. 4-34b is of negligible size compared with the boat length. Average solute composition of the solid forming is given by Eq. (2-10), which can be rewritten using the symbols of Fig. 4-34b as

$$\frac{C_b - \bar{C}_S^*}{C_E - \bar{C}_S^*} = e^{-(R\delta/D_L)} = a \qquad (4\text{-}18)$$

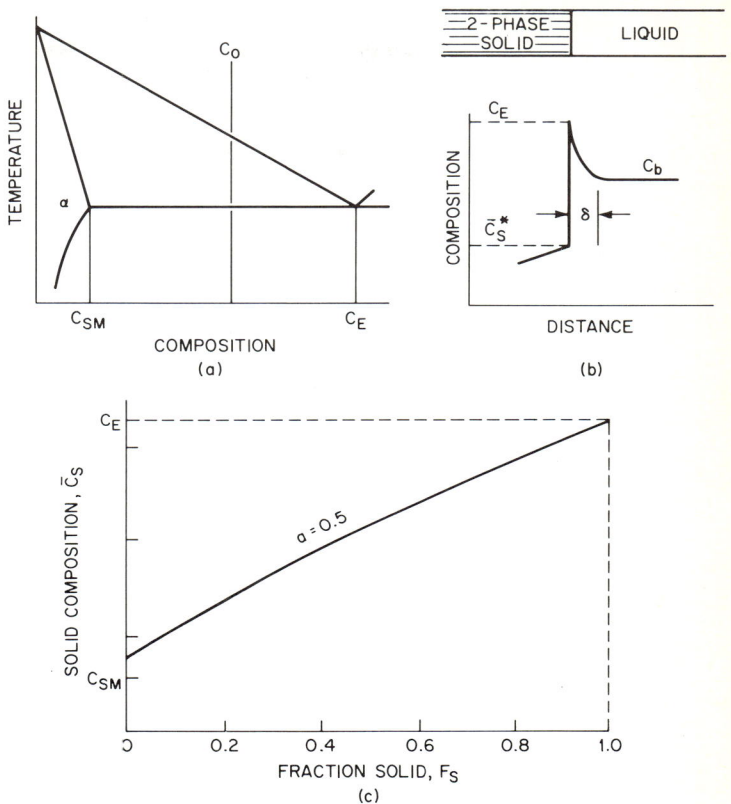

**FIGURE 4-34**
Solidification of a two-phase alloy with convection. (*a*) Phase diagram; (*b*) solute buildup in boundary layer $\delta$; (*c*) final solute distribution for convection such that $a = 0.5$.

where $a$ is a constant for constant growth rate and convection. A differential mass balance equivalent to Eq. (2-3) can now be written, assuming that excess solute in the small boundary layer is negligible:

$$(C_b - \bar{C}_S^*)\, df_S = (1 - f_S)\, dC_L \quad (4\text{-}19)$$

Now Eqs. (4-18) and (4-19) can be combined and integrated from $C_b = C_0$ at $f_S = 0$. The result, expressed in terms of $\bar{C}_S^*$ and $f_S$, is

$$\frac{C_E - \bar{C}_S^*}{C_E - C_0} = \frac{1}{1-a}(1 - f_S)^{a/(1-a)} \quad (4\text{-}20)$$

Figure 4-34c shows a schematic example of the final solid solute distribution calculated from Eq. (4-20), assuming a value of $a$ equal to 0.5. The composition of the two-phase solid forming varies continuously up to a maximum of $C_E$.

With sufficiently vigorous convection, the first solid to form is less than $C_{SM}$ and hence is single-phase. Here, Eqs. (2-11) and (2-12) apply. Equation (4-20) is then applicable when $\bar{C}_S^* > C_{SM}$. Note that the $C_0$ of Eq. (4-20) in this case becomes the bulk liquid composition at the moment $\bar{C}_S^* = C_{SM}$, and $f_S$ is fraction of the liquid remaining after single-phase solidification. The curve for $k' = 0.7$ in Fig. 2-9 shows the general form of the result obtained, and Prob. 4-16 (at the end of this chapter) also serves to illustrate the point.

The effect of stirring on stability of plane front in two-phase growth is readily seen by calculating the constitutional supercooling criterion comparable to Eq. (4-13) but assuming convection. The result is

$$\frac{G_L}{R} \geq -\frac{m_L(C_E - \bar{C}_S^*)}{D_L} \qquad (4\text{-}21)$$

and $\bar{C}_S^*$ is given by Eq. (4-18). Comparison of Eqs. (4-21) and (4-13) shows that for a given solid composition forming, the required $G_L/R$ is the same in the presence or absence of convection; this is also the case for single-phase alloys, as shown by Eq. (3-3). However, contrary to the case for single-phase alloys, increasing convection in an alloy of given bulk composition $C_b$ increases required $G_L/R$ for stability.


## REFERENCES

1 LEMKEY, F. D., and SALKIND, N. J.: in H. S. Peiser (ed.), "Crystal Growth," p. 171, Pergamon, Oxford (1967).
2 GALASSO, F. S.: *J. Metals*, **19**:17 (1967).
3 ALBERS, W., and VERBERKT, J.: *J. Materials Sci.* (to be published).
4 JACKSON, K. A., and HUNT, J. D.: *Trans. AIME*, **236**:1129 (1966).
5 TILLER, W. A.: "Liquid Metals and Solidification," p. 276, American Society for Metals, Cleveland, Ohio, 1958.
6 ZENER, C.: *Trans. AIME*, **167**:550 (1946).
7 BRANDT, W. H.: *J. Appl. Phys.*, **16**:139 (1945); *Trans. AIME*, **167**:405 (1946).
8 HILLERT, M.: *Jernkontorets Ann.*, **144**:520 (1960).
9 HUNT, J. D., and JACKSON, K. A.: *Trans. AIME*, **236**:843 (1966).
10 CHALMERS, B.: "Principles of Solidification," John Wiley & Sons, Inc., New York, 1964.
11 HUNT, J. D., and HURLE, D. T. J.: *Trans. AIME*, **242**:1043 (1968).
12 HOGAN, L. M., KRAFT, R. W., and LEMKEY, F. D.: in H. Herman (ed.), "Advances in Materials Research," **5**:83 (1971).

*13* GRUZLESKI, J. E., and WINEGARD, W. C.: *J. Inst. Metals*, **96**:301 (1968).
*14* MOLLARD, F. R., and FLEMINGS, M. C.: *Trans. AIME*, **239**:1534 (1967).
*15* MOLLARD, F. R., and FLEMINGS, M. C.: *Trans. AIME*, **239**:1526 (1967).
*16* CLINE, H. E., and LIVINGSTON, J. D.: *Trans. Met. Soc. AIME*, **245**:1987 (1969).
*17* SEKERKA, R. F.: *Bull. Soc. Franc. Mineral. Crist.*, **92**:540 (1969).
*18* HUNT, J. D., HURLE, D. T. J., JACKSON, K. A., and JAKEMAN, E.: *Met. Trans.*, **1**:318 (1970).
*19* HOPKINS, R. H., and KRAFT, R. W.: *Trans. Met. Soc. AIME*, **242**:1627 (1968).
*20* JAFFREY, D., and CHADWICK, G. A.: "The Nucleation, Growth Morphology, and Thermal Stability of Sn–Zn and Al–Al$_3$Ni Eutectic Alloys" (to be published).
*21* LIVINGSTON, J. D., and CLINE, H. E.: *Trans. AIME*, **245**:351 (1969).
*22* JACKSON, K. A.: in "Solidification," American Society for Metals, Metals Park, Ohio (1972).
*23* UHLMANN, D. R., CHALMERS, B., and JACKSON, K. A.: *J. Appl. Phys.*, **35**:2986 (1964).
*24* KIRKALDY, J. S.: *Can. J. Phys.*, **36**:907 (1958).
*25* COATES, D. E., and KIRKALDY, J. S.: *J. Crystal Growth*, **3**, No. 4:549 (1968).
*26* KERR, H. W., PLUMTREE, A., and WINEGARD, W. C.: *J. Inst. Metals*, **93**:63 (1964–1965).
*27* RINALDI, M. D., SHARP, R. M., and FLEMINGS, M. C.: *Met. Trans.*, **3**:3139 (1972).
*28* COOKSEY, D. J. S., and HELLAWELL, A.: *J. Inst. Metals*, **95**:183 (1967).
*29* GRUZLESKI, J. E., and WINEGARD, W. C.: *J. Inst. Metals*, **96**:304 (1968).
*30* BULLOCK, J. B., SIMPSON, C. J., EADY, J. A., and WINEGARD, W. C.: *J. Inst. Metals*, **99**:212 (1971).
*31* CHADWICK, G. A.: *J. Inst. Metals*, **92**:18 (1963–1964).
*32* DAVIES, V. L.: *J. Inst. Metals*, **93**:10 (1964–1965).
*33* HUNT, J. D., and CHILTON, J. R.: *J. Inst. Metals*, **92**:21 (1963–1964).
*34* CHADWICK, G. A.: in T. J. Hughel (ed.), "Liquids: Structure, Properties, Solid Interactions," p. 326, Elsevier Publishing Company, Amsterdam, 1965.
*35* LEMKEY, F. D., HERTZBERG, R. W., and FORD, J. A.: *Trans. AIME*, **233**:334 (1965).
*36* LIVINGSTON, J. D., CLINE, H. E., KECK, E. F., and RUSSELL, R. R.: "High Speed Solidification of Several Eutectic Alloys," General Electric Co., Schenectady, N.Y., *G.E. Rept.* 69-C-328, 1969.
*37* RHINES, F. N.: "Phase Diagrams in Metallurgy," McGraw-Hill Book Company, New York, 1956.
*38* CHADWICK, G. A.: "The Solidification of Metals," Iron and Steel Institute Publ. No. 110, p. 138, 1968.
*39* BIBRING, H., SEIBEL, G., and RABINOVITCH, M.: *Mem. Sci. Rev. Met.*, **49**(5):341 (1972).

## PROBLEMS

*4-1* A tin–lead eutectic alloy ($C_0 = C_E = 26.1$ at% Pb) is directionally solidified at $R = 10^{-4}$ cm/s. Use the experimentally determined value of $\lambda$ (Fig. 4-8) to calculate $S_\alpha$ and $S_\beta$. What is the liquid composition at the center of the $\alpha$ lamella? At the center of the β lamella? Assume $D_L = 6.7 \times 10^{-6}$ cm²/s and the densities of the tin-rich and lead-rich phases are 7.3 and 11.5 g/cm³.

4-2 At what distance in front of the eutectic interface does the deviation of the liquid composition from $C_E$ in Prob. 4-1 become negligible? Show that this is small compared with $D_L/R$.

4-3 A tin–lead alloy of 15 at% Pb is directionally solidified at $R = 10^{-4}$ cm/s. What thermal gradient is required according to the constitutional supercooling criterion for stability? Assume $D_L = 6.7 \times 10^{-6}$ cm²/s, $m_L = -1.2$°C/at% Pb.

4-4 Using sketches show that if lamellar spacing in a growing eutectic is smaller than that at the extremum, the spacing is inherently unstable.

4-5 Using sketches show how rod branching can change rod spacing to adjust to relatively small changes in growth rate.

4-6 Explain why facets cannot form in a region of a growing crystal that has negative curvature.

4-7 Derive the constitutional supercooling criterion for binary two-phase alloys [Eq. (4-13)].

4-8 What mechanism or mechanisms do you think permit rodlike phases in a eutectic to change crystallographic orientation during growth so that they grow in their preferred growth direction?

4-9 An Al–10 wt% Cu alloy ingot is "normally" solidified with plane front.
(a) Schematically show composition versus distance, assuming vigorous convection.
(b) Repeat for no convection.
(c) Schematically show the structures of the two ingots.
(d) What fraction of each of the two ingots is two-phase? Assume $D_L = 5 \times 10^{-5}$ cm²/s.

4-10 Derive a constitutional supercooling criterion for stability of a planar two-phase interface in a binary alloy growing with a convective boundary layer $\delta$.

4-11 Derive a constitutional supercooling criterion for stability of a plane front in an alloy that undergoes a peritectic reaction and is two-phase after equilibrium solidification.

4-12 Suppose you could halve the growth rate $R$ of the eutectic of Fig. 4-4 without changing the lamellar spacing. Redraw schematically all of the curves shown, and draw the new interface shape. Repeat if you were to double the growth rate.

4-13 A three-phase crystal is grown from a ternary alloy at steady state in a large convecting melt. Melt composition is $C_{0m}$, $C_{0n}$, and stirring is such that solute boundary layer thickness is $\delta$. Derive an equation relating the average effective partition ratio of element $m$, $\bar{k}'_m$ to growth conditions. $\bar{k}'_m = \bar{C}_{sm}/C_{0m}$, where $\bar{C}_{sm}$ is average concentration of solute in the solid.

4-14 For a given final solid composition $\bar{C}_{sm}$, $\bar{C}_{sn}$, will stirring a melt make it possible to grow three-phase composites with plane front at lower $G_L/R$ than when convection is absent?

4-15 Show that Eq. (4-17) can be derived directly from Eq. (3-28) providing diffusion coefficients of the two solutes are equal and alloy composition lies on the eutectic trough.

*4-16* The schematic diagram of Fig. 4-34 was calculated for an Al–20% Cu alloy, with $C_0 = 20\%$ Cu and convection such that $a = 0.5$. If convection is increased such that $a = 0.55$, the first solid to form is single-phase. Plot the final solute distribution along the length of the crystal solidified with $a = 0.55$. What fraction of crystal is single-phase?

*4-17* Derive Eq. (4-21). Explain physically why increasing convection increases the value of $G_L/R$ required for stability.

# 5

# SOLIDIFICATION OF CASTINGS AND INGOTS

## GRAIN STRUCTURE

The classical picture of the grain structure of a casting or ingot is shown schematically in Fig. 5-1, with the outer *chill zone*, the intermediate *columnar zone*, and the central *equiaxed zone*. All three of these zones are sometimes seen in castings and ingots of real materials, especially in plain carbon or low-alloy steel. More often, however, one or another of the zones is absent. In stainless steels, the structure is often fully columnar, with no central equiaxed zone and little or no chill zone. In well-grain-refined aluminum alloys, on the other hand, the structure is fully equiaxed. The presence and extent of these zones in ingot structures are now known to depend both on nucleation and on *crystal multiplication*, as discussed later in this chapter.[1,2]

The grain structure in Fig. 5-1 is the solidification grain structure. Particularly in relatively pure alloys, grain boundary movement can occur after solidification. Solid-state phase transformations also modify the final grain structure. Low-carbon steels are an interesting case in point. The dendrites which form first are $\delta$ ferrite, and these subsequently transform to austenite and still later to ferrite and pearlite. Each of these grain structures can be delineated in the fully solidified ingot by suitable metallographic procedures.

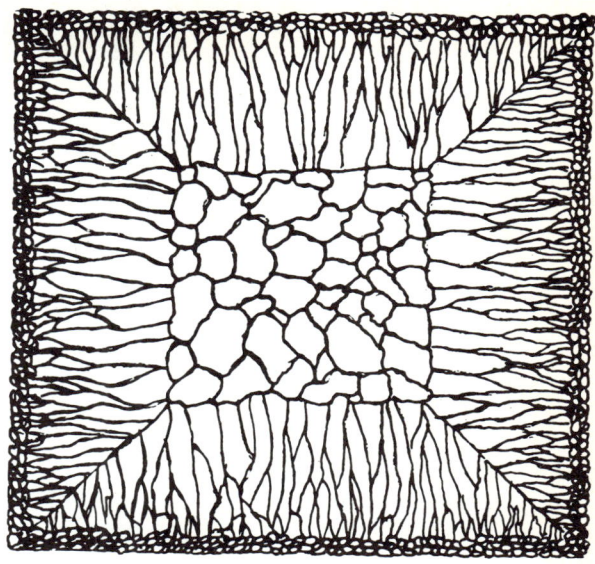

FIGURE 5-1
Sketch of ingot structure showing chill zone, columnar zone, equiaxed zone. (*From Bower and Flemings.*[1])

## COLUMNAR STRUCTURES

In Chap. 3, it was shown that at the cellular-dendrite transition, perturbations form on the cell stalk which develop into short, stubby dendrite arms. Commercial alloys which are very dilute or freeze over a narrow range, such as the Fe–10% Ni alloy of Fig. 3-19, show such structures. Most commercial alloys, however, exhibit a much more highly branched morphology similar to the transparent organic alloy shown in Fig. 5-2 (from the work of Jackson et al.[2]). Figure 5-3 is a drawing of a portion of a dendrite from one such material, low-alloy steel. The figure was drawn from a three-dimensional study of isoconcentration surfaces of a steel dendrite after complete solidification. The contour lines approximately represent successive portions of the liquid-solid interface. An important aspect of the sketch of Fig. 5-3 is that spaces between dendrite arms tend to fill in during the latter stages of solidification to form *plates*. These plates produce some rather surprising effects on polished surfaces, as shown in several figures which follow.

    Low-alloy steel has a particular advantage for the study of dendrites in that it is possible to heat-treat it so that the low-alloy central portions of the dendrite transform to pearlite while the outer portions are quenched to martensite. The boundary between the pearlite and martensite then represents an isoconcentration surface, and the heat treatment can be modified to delineate isoconcentration surfaces of higher or

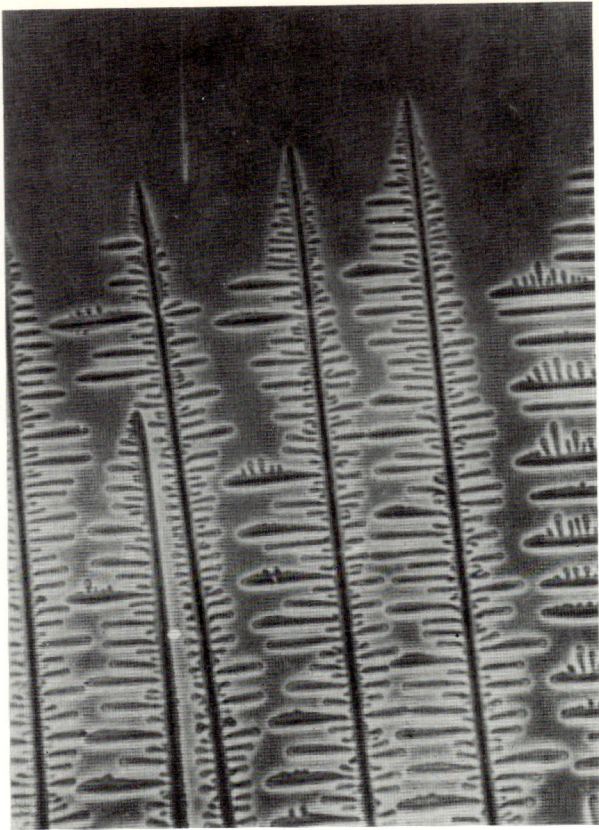

FIGURE 5-2
Columnar structure of a transparent organic alloy. (*From Jackson et al.*[2])

lower magnitude.[3] Figure 5-4 shows one such structure. At lower magnifications, the structure of this typical low-alloy steel appears as in Fig. 5-5. The dendrite arms are all at right angles ($\langle 100 \rangle$ direction), and the apparently noncrystallographic "ghost" patterns are simply the traces of $\{100\}$ plates intersecting the specimen surface at low angles. There is probably a grain boundary in each of the photographs of Fig. 5-5, but all of the arms in Fig. 5-4 are part of the same cast grain, or dendrite (using the terminology of Chap. 3). At lower magnification (Fig. 5-6) it is possible to delineate the original cast grains. This casting has 10 to 20 columnar grains across the diameter in the upper portion.

The *plates* observed in columnar structures often extend over many primary dendrite arms. In Fig. 5-4 note how well aligned the dendrite arms are along a direction about 20° from the horizontal. A somewhat higher isoconcentrate would

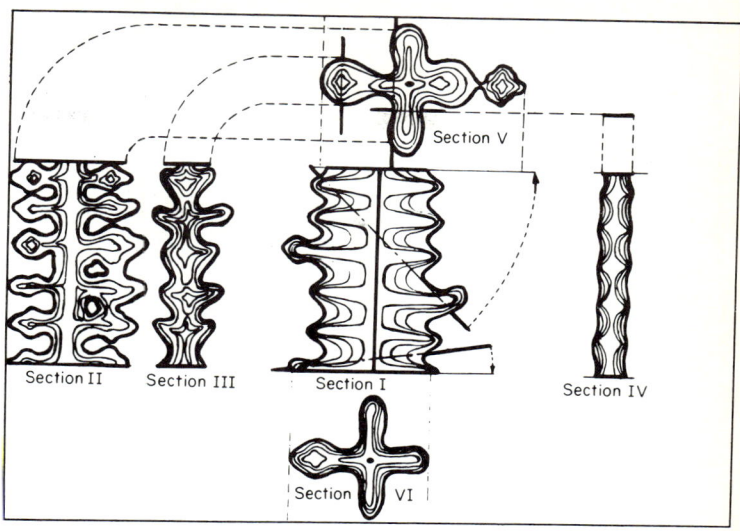

**FIGURE 5-3**
Isoconcentration surfaces of a columnar dendrite, low-alloy steel. (*From Kattamis and Flemings.*[3])

show plates extending across the entire photomicrograph in this direction. Extended plates in both dendritic and cellularly solidified cubic alloys are generally found to comprise the (100) plane that is most nearly parallel to the heat-flow direction.[4-6] The degree to which dendrites show a platelike structure in columnar solidification depends on alloy composition. Alloys which freeze with only a minor fraction of eutectic usually show pronounced plates; Al–4.5% Cu alloy is one example (Fig. 5-7). On the other hand, alloys in which a relatively large fraction of the solid forms as eutectic usually have much less well-developed plates.

Dendritic structures in noncubic metals are more complex geometrically than those discussed above because of nonorthogonality of dendrite arms; otherwise, these structures have many similarities to those discussed above. Figures 5-8 and 5-9 show dendrites of one such material, i.e., body-centered tetragonal tin dendrites in a tin–bismuth alloy. The columnar growth direction (growth direction of primary arms) in this alloy is [110]. Secondary arms grow in the [1$\bar{1}$1] and [$\bar{1}$1$\bar{1}$] directions; these are perpendicular to the [110] direction and are 138° apart. Secondary arms in ⟨111⟩ directions were found only on one side of the primary arm; that is, secondary arms did not grow in the [$\bar{1}$11], [1$\bar{1}$1], or [111] directions in dendrites in which they grew in the [1$\bar{1}\bar{1}$] and [$\bar{1}$1$\bar{1}$] directions. Presumably, this is because these arms, had they formed, would have been at relatively small angles to the primary arm or other secondary arms. Instead, the additional growth direction [112] is found at 37.6°

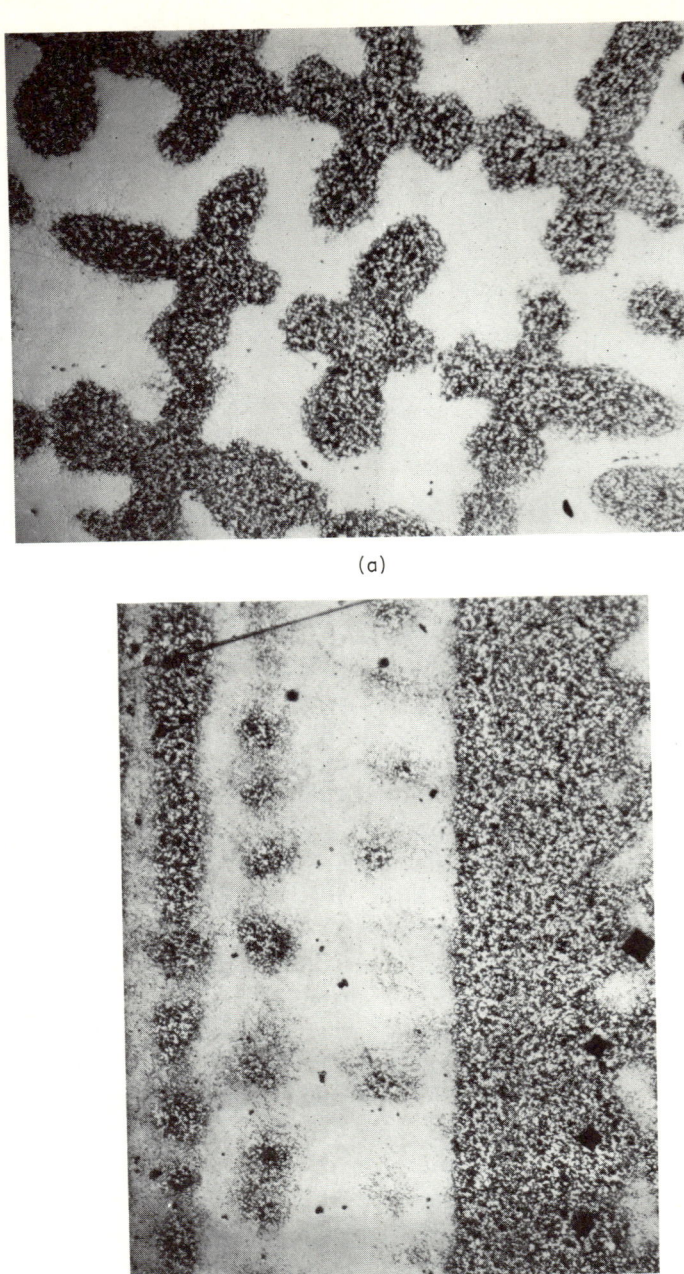

**FIGURE 5-4**
Columnar dendrite, low-alloy steel (magnification ×55). (*a*) Section transverse to heat flow; (*b*) section parallel to heat flow. (*From Kattamis and Flemings.*[4])

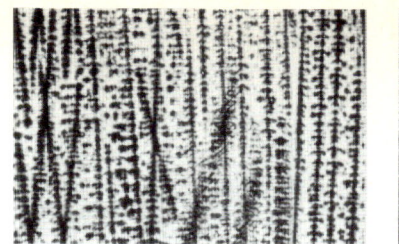
3 INCHES FROM CHILL

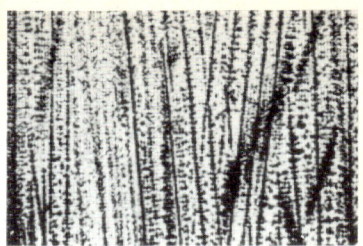
1¾ INCHES FROM CHILL

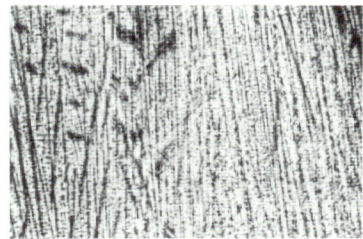
½ INCH FROM CHILL

FIGURE 5-5
Dendrite structures in a unidirectionally solidified low-alloy steel ingot. (Magnification ×11.)

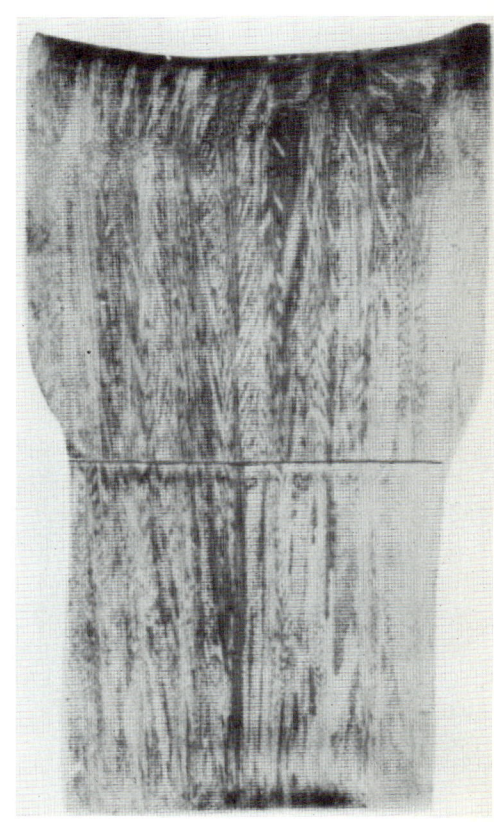

FIGURE 5-6
Structure of a unidirectionally solidified low-alloy steel ingot. (Magnification approximately ×½.)

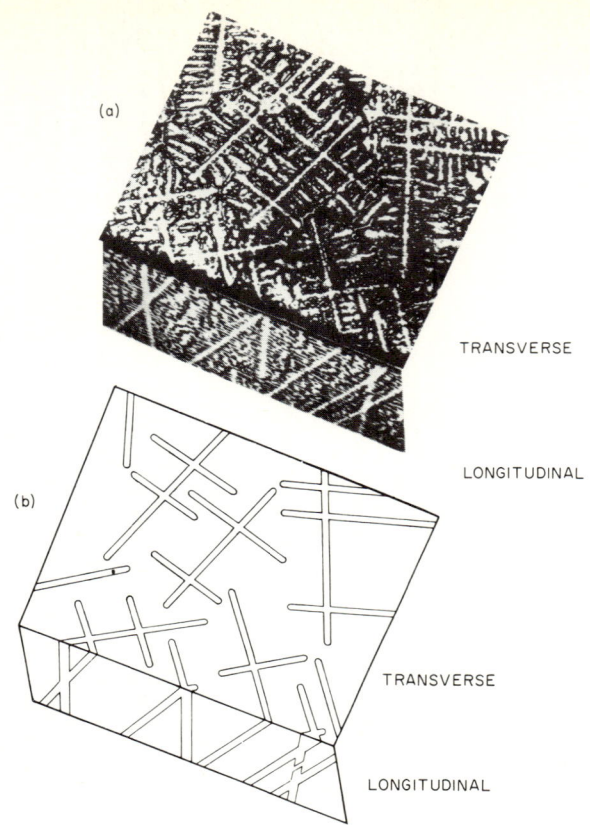

**FIGURE 5-7**
Unidirectionally solidified Al–4.5% Cu alloy. (*a*) Photomicrograph; (*b*) schematic drawing showing primary dendrite plates. (Magnification ×16.) (*From Bower et al.*[6])

**FIGURE 5-8**
Columnar tin-rich tin–bismuth alloy. (Magnification ×50.) (*From Ahearn and Flemings.*[7])

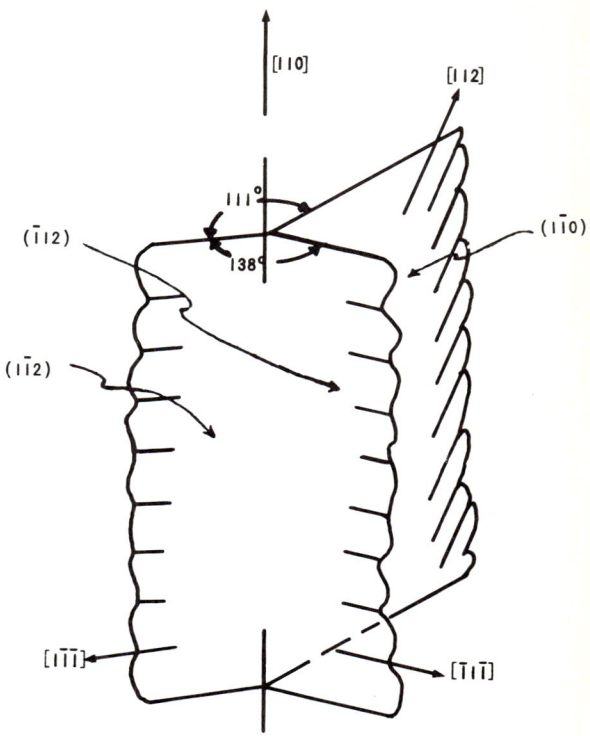

FIGURE 5-9
Growth features of columnar tin-rich tin–bismuth alloy. (*From Ahearn and Flemings.*[7])

from the [110] direction. Interstices between dendrite arms tend to fill in preferentially to form planes parallel to the heat-flow direction, as observed in dendrites of cubic materials.[7]

## MICROSEGREGATION IN COLUMNAR STRUCTURES

A convenient method of measuring microsegregation in columnar structures is by use of the electron microprobe; Fig. 5-10 shows results of a typical series of measurements across a dendrite arm in low-alloy steel. This alloy solidified as single-phase, and so there are no abrupt concentration changes from dendrite centers to interdendritic regions. When a second phase forms in the interdendritic spaces, very large concentration differences are often found, as in Al–4.5% Cu alloy (Fig. 5-11).

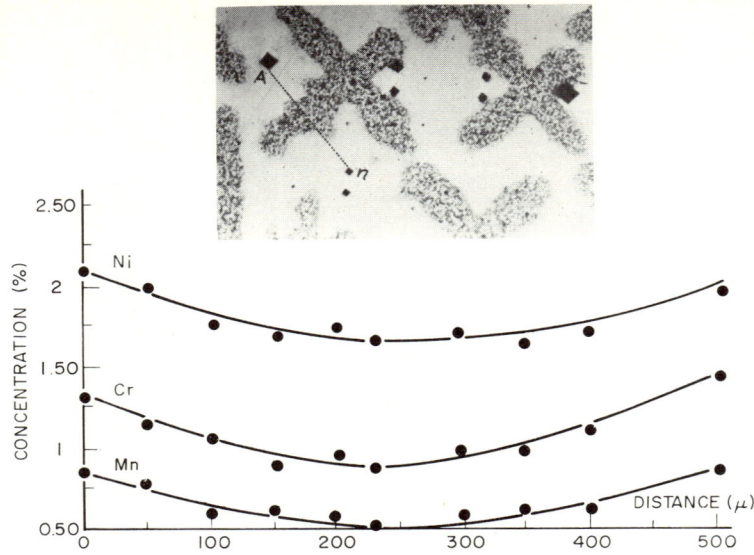

**FIGURE 5-10**
Solute distribution in low-alloy steel along path *A–n* as determined by electron microprobe. (Magnification × 55.) (*From Kattamis and Flemings.*[3])

The simplest first approach to a quantitative description of microsegregation in dendritic solidification is that outlined in Chap. 3 for cellular solidification. A small-volume element is chosen which is small enough to be treated as a differential element but large enough to contain a representative local fraction of solid. Such a suitable element would contain one or more primary arms with associated secondary and higher-order arms. Other assumptions of the previous analysis are then also made, including negligible local constitutional supercooling, no effect of radius of curvature on solidification behavior, and no solid diffusion. Dendritic solidification generally occurs at a sufficiently low ratio of thermal gradient to growth rate that the parameter $a$ of Eq. (3-20) is negligible[6,8] and the equation reduces to

$$C_s^* = kC_0(1 - f_s)^{k-1} \tag{5-1}$$

where $C_s^*$ is the composition of the isoconcentrate surrounding $f_s$ weight fraction of solid, both during and after solidification of the volume element.

Equation (5-1) is exactly that derived in Chap. 2 for crystal growth under conditions of no solid diffusion and uniform liquid composition. It applies here, not to the casting as a whole, but to a tiny region whose size is the order of a dendrite arm. Using the procedure outlined in Chap. 3 for cellular solidification, it can be shown

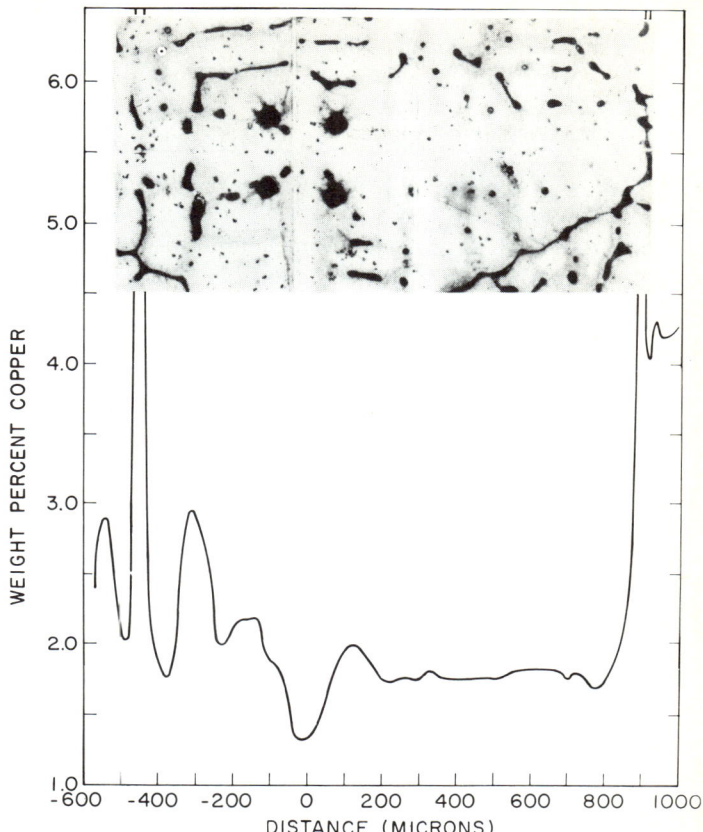

FIGURE 5-11
Solute distribution in Al–4.5% Cu alloy as determined by electron microprobe. (*From Bower et al.*[6])

that the assumption of very low undercooling in interdendritic regions must be a good one, and for commercial alloys that solidify over a wide temperature range most of the other assumptions are warranted as well. Discrepancies between quantitative predictions of this analysis and experiments performed to date arise primarily from two factors (discussed below). One factor is that some solid diffusion does occur in many alloys before and after solidification, and the other is that the radius of curvature effect does modify solidification in an interesting way by causing local melting in some parts of the volume element while solidification is proceeding in others.

Equation (5-1) predicts (for alloys of constant $k$) that some eutectic will form no matter how low the initial composition, and this is in general agreement with experience. In aluminum–copper alloys, eutectic is found when initial composition is well below $\frac{1}{2}\%$ Cu although the limit of solid solubility is in excess of 5% and the eutectic composition is 33%. The equation also predicts no effect of solidification rate on microsegregation, as measured by maximum composition, minimum composition, or amount of interdendritic eutectic. In practice, the effect of solidification rate on these variables is usually small. Quantitatively, the agreement of the simple equation with experiment is less good. Minimum composition predicted by equation (5-1) for Al–4.5% Cu is about 0.6%, whereas in practice this is usually found to be significantly greater. The amount of eutectic predicted is about 9%, whereas the amount actually observed is always less.[6,9]

It has been demonstrated convincingly that solid diffusion during and after solidification is primarily responsible for the foregoing numerical discrepancy. One way this has been shown is by abruptly quenching samples from various temperatures during and after solidification. Studies made this way show that the composition of the solid which forms first (at the dendrite tip) is near $kC_0$ and that the composition of this central portion of the dendrite then gradually increases during and after solidification.[6,10,11] The reason why microsegregation is so nearly constant over wide ranges of cooling rates is that the coarseness of the dendrite structure (as measured by the dendrite arm spacing $d$) varies with cooling rate. This variation is such that the extent of diffusion occurring during and after solidification is nearly constant.[8,12] This qualitative conclusion is independent of dendrite morphology provided the morphology is independent of cooling rate. Quantitative calculations, however, require specification of dendrite morphology. The simplest such expression is one obtained by assuming a platelike morphology and modifying the differential solute-redistribution equation [Eq. (2-2)] by an additional term to account for back diffusion at the liquid-solid interface. This is[8]

$$\begin{matrix} \text{Solute rejected} \\ \text{at interface} \end{matrix} = \begin{matrix} \text{solute increase} \\ \text{in liquid} \end{matrix} - \begin{matrix} \text{solute back diffusing} \\ \text{in solid} \end{matrix}$$

$$(C_L - C_s^*)\frac{dy^*}{dt} = (l - y^*)\frac{dC_L}{dt} - D_s\left(\frac{\partial C_s}{\partial y}\right)_{y=y^*} \quad (5\text{-}2)$$

where the volume element is as pictured in Fig. 3-21. Its length is $l$, and the position of the liquid-solid interface in the volume element is $y^*$. Hence, $y^*/l$ is fraction solidified.

Two assumptions now permit simple analytic solution of Eq. (5-2). The first of these is that diffusion in the solid is small so that solute gradient at the interface is not changed appreciably by the diffusion; that is,

$$\left(\frac{\partial C_s}{\partial y}\right)_{y=y^*} = \frac{dC_s^*}{dy^*} \quad (5\text{-}3)$$

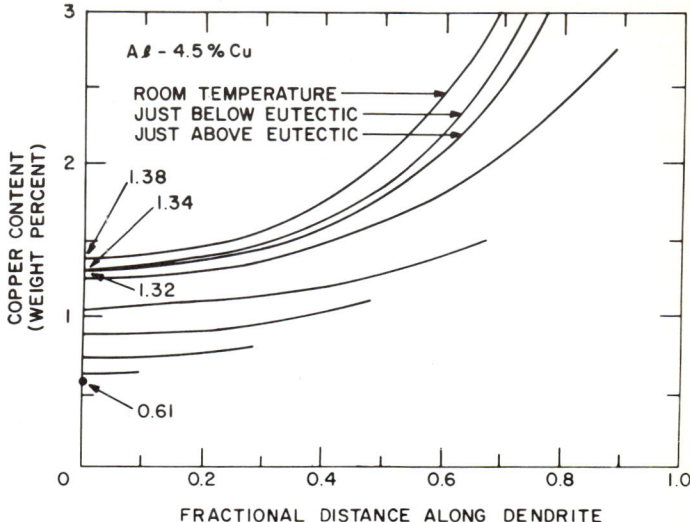

**FIGURE 5-12**
Effect of solid diffusion during solidification on composition distribution across dendrite arms calculated for $t_f/d^2 = 3.3 \times 10^8$. (*From Brody and Flemings.*[8])

The second assumption is that the rate of dendrite-arm thickening is constant so that

$$\frac{dy^*}{dt} = \frac{l}{t_f} \qquad (5\text{-}4)$$

where $t_f$ is the time from beginning to end of solidification, i.e., the *local solidification time*. Substituting Eqs. (5-3) and (5-4) and integrating as in previous similar examples yields

$$C_s^* = kC_0\left(1 - \frac{f_s}{1 + \alpha k}\right)^{k-1} \qquad (5\text{-}5)$$

where $C_s^*$ is the isoconcentrate enclosing $f_s$ weight fraction of solid during (but not after) solidification and $\alpha = D_s t_f/l^2$. Since $l$ is one-half the dendrite arm spacing $d$, $\alpha = 4D_s t_f/d^2$. Equation (5-5) has been used by Bower, Brody, and Flemings[6] to estimate the effect of solid diffusion on amount of eutectic in cast structures. The extent of this diffusion depends, not on solidification time alone, but on the dimensionless ratio $\alpha k$, and it becomes significant only for values of $\alpha k$ equal to or greater than about 0.1. Diffusion in the solid after solidification depends on a similar dimensionless ratio, as discussed in Chap. 3.

Detailed description of solute distribution within the dendrite during and after solidification requires solution of the diffusion equation by numerical techniques, and Fig. 5-12 shows results of one such set of calculations for a value of $t_f/d^2$ that gives results that are most closely in agreement with experiment. Again the solute redistribution depends only on the dimensionless ratio $\alpha$. It will be seen later in this chapter that the ratio $\alpha$ is relatively constant over wide ranges of cooling rate, thus leading to the relative constancy of segregation observed at different cooling rates.

In many important commercial alloys, the partition ratio is far from constant and no second phase precipitates during solidification. An example is iron–nickel binary alloys in which the partition ratio goes to unity as the liquidus minimum is approached during solidification. For these alloys segregation calculations are done by numerical integration of the solute-redistribution equations. Experimental and analytical results which have been obtained are qualitatively similar to those described above. In these alloys microsegregation is most conveniently described in terms of a *segregation ratio S*, which is the ratio of maximum to minimum compositions observed in the structure. Variations of this ratio with cooling rate, when found, are generally not large.[13-15]

The actual morphology of dendrites is, of course, much more complex than the simple platelike (or cylindrical) morphology generally used for calculations involving solid diffusion. Moreover, solute is not uniformly distributed in interdendritic areas but tends to be somewhat more concentrated between primary arms than between secondary arms. Schematic examples of dendrite morphology at two late stages of solidification are given in Fig. 5-13. The periphery of the portion of the dendrite drawn is the region between primary arms; there is somewhat more solute-rich liquid here than within the figure (between secondary arms). Note the similarity between the top surface of Fig. 5-13 and the photomicrograph of Fig. 5-11. The complex morphology of isoconcentrates makes it difficult to solve, with much precision, problems involving solid diffusion. Important problems in this category include the formation of microsegregation and the homogenization during subsequent heat treatment or thermomechanical processing. In the case of the microsegregation calculations which have been made,[6,16] quantitative agreement of experiment with theory has required assumption of a smaller dendrite spacing than is actually observed. This discrepancy is partly attributable to the simple dendrite geometries assumed, but it may also result in part from improper choice of diffusion coefficient[17] or from coarsening effects (to be discussed later).

## DENDRITE ARM SPACING

Dendrite morphology in usual casting and ingot-making processes remains largely unchanged over wide ranges of cooling rates. It simply becomes finer as heat is extracted at greater rate. Thus, the structure in Fig. 5-5 near the chill is not very different from the one that is far away, except in scale. One exception to this rule is

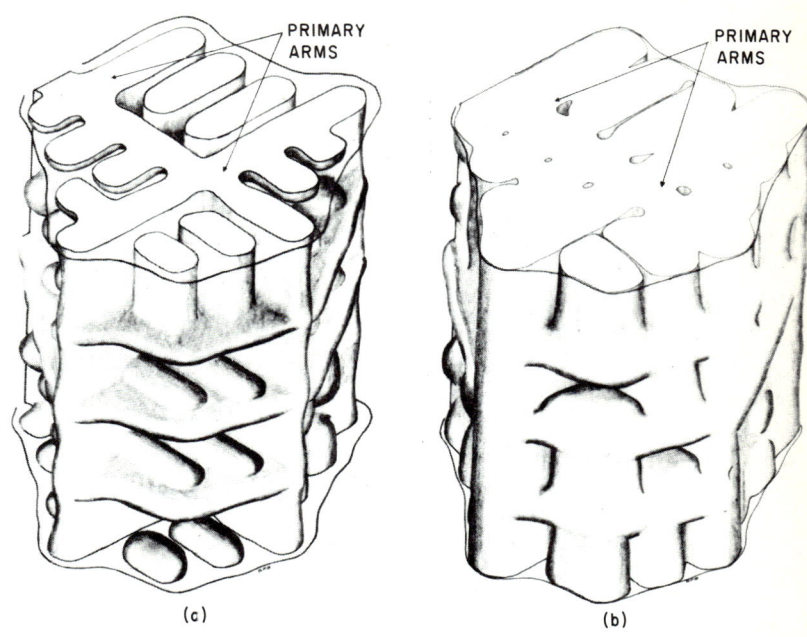

**FIGURE 5-13**
Schematic diagram of dendrite structure in Al–4.5% Cu alloy at (a) 50 percent solid and (b) 90 percent solid. (*From Singh et al.*[16])

that at very high cooling rates, when primary arm spacing becomes very small, secondary and tertiary arms may be absent; this effect is seen in Fig. 3-17.

A convenient and widely used measure of the effects of solidification conditions on dendrite structure is *dendrite arm spacing*, i.e., the spacing between primary, secondary, or higher-order branches. Generally, the spacings measured are the perpendicular distances between branches. However, especially on more poorly defined structures, the random intercept method has been employed. This latter method gives results somewhat higher than the former.

Studies on transparent alloys[2] show that columnar dendrites can adjust their primary spacing during growth without difficulty. If spacing is too close, one or another primary arm falls behind and is subsequently engulfed; this has happened to one of the primary arms in Fig. 5-2. If spacing is too large, a tertiary arm growing from a secondary arm catches up to the growing primary tips and becomes one of them, as sketched in Fig. 5-14. This mechanism of spacing adjustment is similar to that by which cells adjust their spacing (discussed in Chap. 3). The driving force is the constitutional supercooling in the region between the two primary dendrite arms.

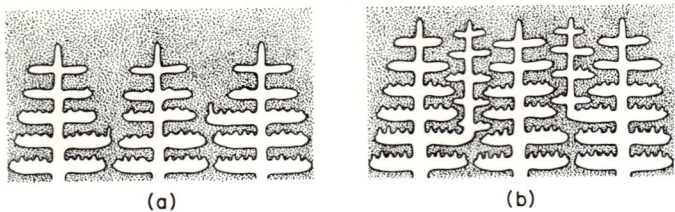

**FIGURE 5-14**
Formation of new primary arms by branching from secondaries.

Apparently the dendrite is able to branch sufficiently to reduce this supercooling to a very low value. By analogy with the closely similar cellular solidification, it would be expected that primary dendrite arm spacing depends on the product of thermal gradient and growth rate $GR$, as does cellular spacing, and results which have been reported correlate well with this parameter.[15,18,19] Figure 5-15a shows one example. Note the product $GR$ has the units of cooling rate (for example, degree centigrade per second), and experimental results are often expressed in this way.[6,15,18-22]

Secondary dendrite arm spacings also depend directly on cooling rate as has now been shown for a wide variety of alloys, both columnar and equiaxed.[6,15,20-22] Results are plotted either versus average cooling rate during solidification $GR$ or versus local solidification time $t_f$. The resulting plots are closely similar since

$$t_f = \frac{\Delta T_s}{GR} \qquad (5\text{-}6)$$

where $\Delta T_s$ is the nonequilibrium temperature range of solidification. Some data are shown in Figs. 5-15 and 5-16. Relationships found between dendrite arm spacing and thermal variables have the form

$$d = a t_f^n = b(GR)^{-n} \qquad (5\text{-}7)$$

where the exponent $n$ is in the range of $\frac{1}{3}$ to $\frac{1}{2}$ for secondary spacings and generally very close to $\frac{1}{2}$ for primary spacings. It has been shown that the final secondary dendrite arm spacing one sees and measures in a fully solidified casting is usually much coarser than the one that forms initially. When this coarsening proceeds to a sufficient extent, the spacing which originally forms must have little influence on that of the final casting[23] and one must look for some mechanism other than simple constitutional supercooling as that determining the final spacing.

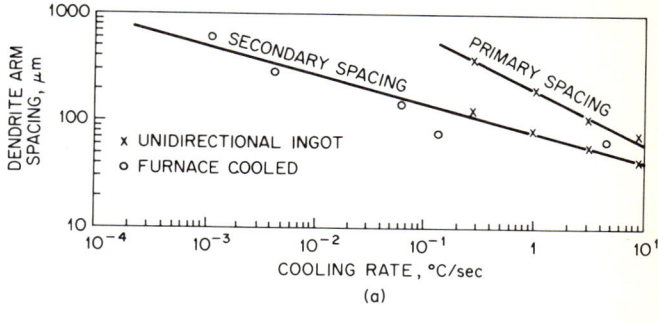

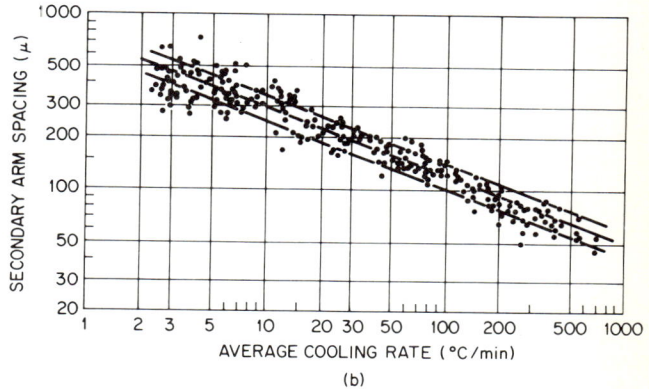

**FIGURE 5-15**
Some experimental data on dendrite arm spacings in ferrous alloys. (a) Fe–25% Ni alloy (*from Flemings et al.*[15]); (b) commercial steels containing from 0.1 to 0.9% C (*from Suzuki et al.*[22]).

The coarsening comes about because some of the arms which form initially become unstable later in solidification and melt while others continue to grow. Surface energy is the driving force for the remelting. This phenomenon is best observed directly in transparent organics; excellent examples are shown in Jackson's motion pictures on solidification of such material.[24] Figure 5-17 shows the process schematically. The less fortunate dendrite arms (those that have regions of smaller-than-average radius of curvature) grow more slowly than their neighbors or melt away. Constitutional supercooling, which was sufficient near the dendrite tip to form the arms, is reduced to such a low value back from the tips that the effect of the radius of curvature on melting point becomes relatively more important. The result is *remelting* of some of the arms.

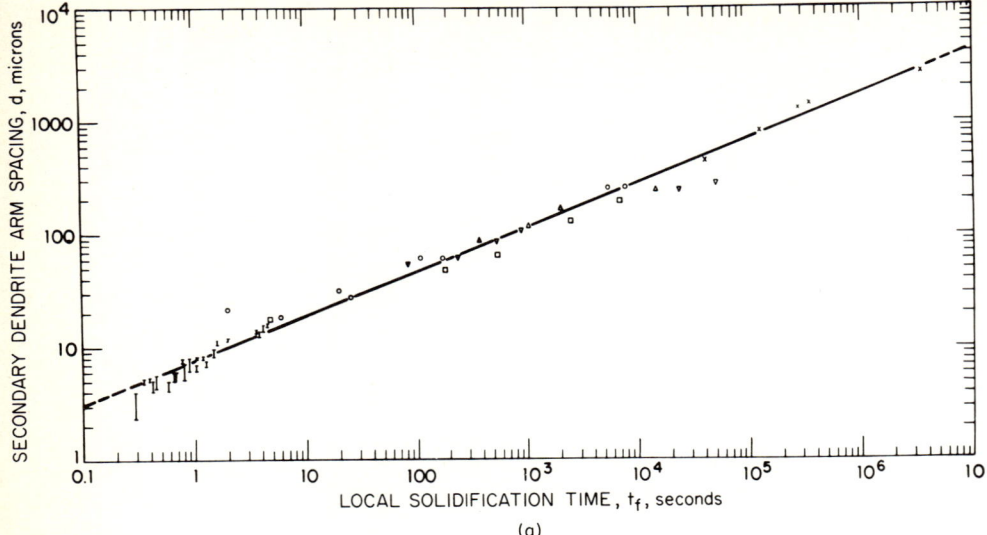
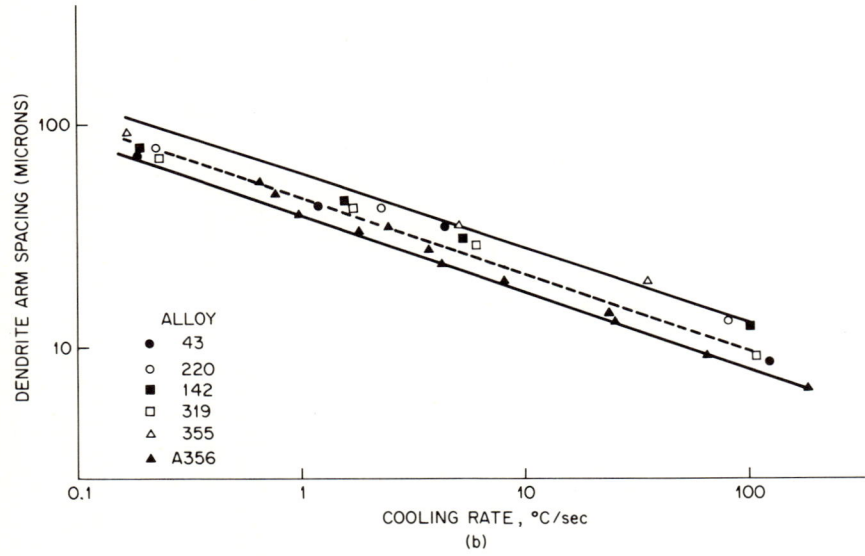

FIGURE 5-16
Experimental data on dendrite arm spacings in aluminum alloys. (a) Al–4.5% Cu alloy (*from Bower et al.*[6]); (b) commercial aluminum alloys (*from Spear and Gardner*[20]).

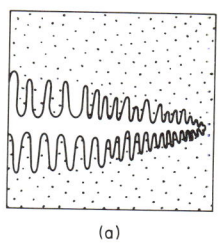 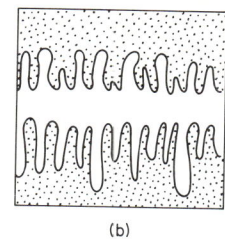 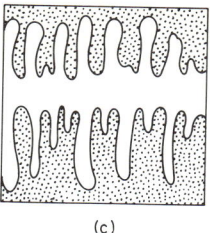 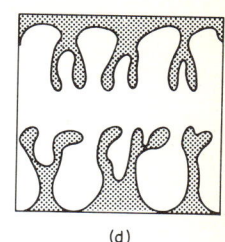

FIGURE 5-17
Schematic diagram of growth of a dendrite in an alloy. (*a*) through (*d*) show a fixed position at various stages of solidification. Note that many smaller arms disappear while larger ones grow.

No theoretical study has been made of the process of coarsening in dendritic solidification. However, work has been done on the closely similar but much simpler problem of isothermal coarsening. Three simple idealized models have been used in these treatments, as sketched in Fig. 5-18. The first of these has been considered by Kattamis et al.,[23] the second by Chernov,[25] Klia,[26] and Kattamis et al.,[23] and the third by Kahlweit.[27] These studies were antedated by the extensive work of Papapetrou.[28] In the first model, the radius of dendrite arms is considered to be constant (radius = $a$) except for one arm which is $r_0$, where $r_0 < a$. The melting point of the smaller arm is therefore less than that of the remaining arms, and in an isothermal melt it will disappear (by transport of material from the smaller to the larger arms). Consequently, the local dendrite arm spacing increases. In the second model, a dendrite arm is considered whose root is slightly smaller than its remainder; this arm tends to melt off by transport of solid from the necked region. In the third model, a

FIGURE 5-18
Dendrite coarsening models.

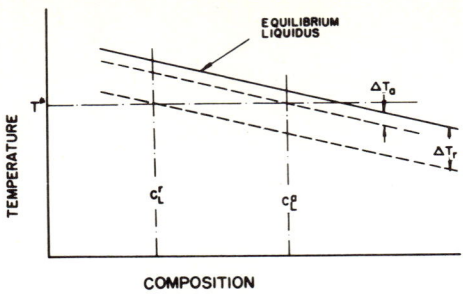
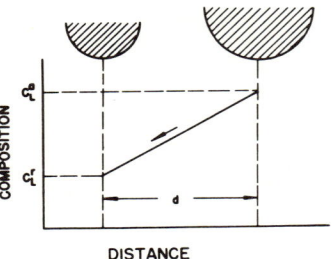

**FIGURE 5-19**
Dendrite coarsening. *Top:* Depression of equilibrium liquidus temperature by curvature effect; *bottom:* solute gradient in liquid.

single dendrite arm of radius $a$ is considered. The twofold radius of curvature at its tip causes it to "melt" off by transport of material back to the cylindrical surface.

The first of the three models sketched in Fig. 5-18 is easiest to describe quantitatively. The equilibrium temperature of the liquid-solid interface depends on local curvature, and for the singly curved cylindrical surfaces the depression of the liquidus temperature $\Delta T_r$ is given by Eq. (8-10), which we write here as

$$\Delta T_r = \frac{\sigma T_L}{r \rho_s H} \quad (5\text{-}8)$$

where $\sigma$ is liquid-solid surface energy, $T_L$ is equilibrium liquidus temperature, $r$ is local radius of curvature, $\rho_s$ is solid density, and $H$ is heat of fusion, a positive quantity. Thus, the liquidus temperature of arms of radius $r$ and $a$ is depressed by $\Delta T_r$ and $\Delta T_a$, respectively (Fig. 5-19). Assuming equilibrium at the two liquid-solid interfaces, a diffusion couple is now established in the liquid between the two dendrite arms such that

$$j \simeq -\rho_L D_L \frac{(C_L^a - C_L^r)}{d} \quad (5\text{-}9)$$

where $j$ is flux density, $D_L$ is the diffusion coefficient, $C_L^r$ and $C_L^a$ are weight fractions of solute in the liquid in equilibrium with arms of radius $r$ and $a$, respectively, and $d$

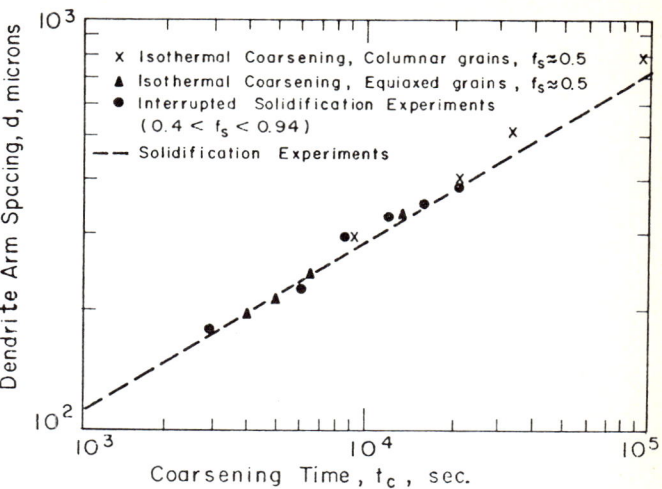

**FIGURE 5-20**
Coarsening of dendrite arms in Al–4.5% Cu alloy. (*From Kattamis et al.*[23])

is dendrite arm spacing. From Fig. 5-19, solute is seen to diffuse toward the smaller arm. This is equivalent to solvent diffusing away, and the smaller arm melts, or *dissolves*, at a rate $-dr/dt$, according to the relation

$$j \simeq -\rho_L C_L^r (1 - k) \frac{dr}{dt} \quad (5\text{-}10)$$

Equations (5-8) to (5-10) are now combined, recognizing that concentration differences are very small so that $C_L^r \simeq C_L^a \simeq C_L$ and the liquidus slope is $m_L$. Integrating from the start of coarsening ($t = 0$, $r = r_0$) to the time $t_c$ (at which the smaller arms disappear) gives a rather long expression, which, for intermediate fractions of solid, reduces to[23]

$$t_c \simeq -\frac{\rho_s H C_L (1 - k) m_L \, d^3}{\sigma D_L T_L} \quad (5\text{-}11)$$

Observations of Klia[26] on a transparent inorganic show that dendrite arms do, in fact, disappear in isothermal coarsening in times given approximately by Eq. (5-11). In metallic alloys, it is not possible to observe directly the disappearance of arms. However, one can observe dendrite structures before and after isothermal coarsening, and these observations confirm expectations that large numbers of dendrite arms disappear in times $t_c$, as given by Eq. (5-11). Figure 5-20 shows these experimental results.

Many experiments on both metallic and nonmetallic alloys have now shown that coarsening occurs during solidification as well as during isothermal coarsening. One way this has been shown on nontransparent materials is to solidify a sample part way and then rapidly quench it. These *interrupted solidification experiments* show that the longer solidification proceeds before quenching, the larger the dendrite arm spacing becomes. The data of Fig. 5-20 suggest that, for this alloy at least, coarsening proceeds at about the same rate during solidification as during isothermal coarsening. This is seen by the data for the interrupted solidification experiments and by the fact that the line relating all data in this figure is exactly the line in Fig. 5-14a which relates spacing to *local solidification time* in castings of the same alloy.

The coarsening process influences dendrite structures in other ways than simply by altering final dendrite arm spacing. One effect is to decrease microsegregation. This happens because some arms which form early in solidification (of low solute content) later dissolve and reprecipitate at higher solute content. Dendrite morphology must also be altered by coarsening. *Bulbous* dendrite arms, such as those in Fig. 5-21a, have a shape suggestive of a strong influence of coarsening. In the later stages of solidification when the dendrite framework is well formed, surface area can be most effectively reduced, not by forming larger cylinders or spheres from the arms already existing, but by filling in the spaces between cylinders to form plates. Thus, coarsening seems likely to be the important driving force leading to plate formation. Support is lent to this idea by structural studies performed, including especially the work of Subramanian et al.,[10] who showed the gradual development of the plate structure during growth while higher-order dendrite arms were disappearing.

## EQUIAXED GRAINS

It was once assumed that each crystal in a casting or ingot represented a new nucleation event. Perhaps this is true in castings in which a strong nucleating agent is present. However, it has now been amply demonstrated that *grain multiplication* is an important and general source of crystals in castings and ingots. Apparently, the grain multiplication is caused primarily by dendrite remelting.

One way in which the remelting can occur is by the coarsening mechanism, as shown in the second model of Fig. 5-18. If the arm that is separated is then carried away into slightly supercooled liquid, a new crystal is formed without an added nucleation event. Convection provides an admirable mechanism, not just for carrying away the dendrite arm, but also for dissipating superheat in the liquid so that the transported arm can grow when reaching it. Also, turbulent convection has the added effect of bringing heat pulses to the interface, as discussed in Chap. 7. These heat pulses accelerate the melting off of dendrites, as shown experimentally by Jackson et al.[29]

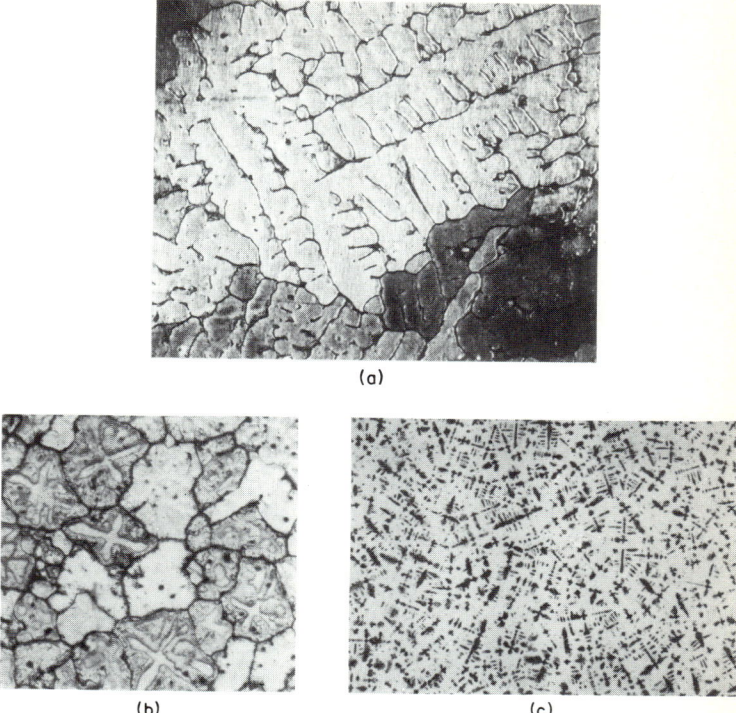

FIGURE 5-21
Some equiaxed dendritic structures. (*a*) Al–4.5% Cu alloy (magnification ×75); (*b*) high-strength grain refined aluminum alloy (magnification ×75); (*c*) Fe–25% Ni alloy (magnification ×12).

Many experiments have now shown the strong effect of convection on grain size of cast metals. When this convection is reduced, grain size is larger and columnar structures are much more readily obtained. Convection appears to play a dominant role in formation of the outer *chill zone* as well as in the columnar-equiaxed transition. When convection is absent, no outer chill zone is observed even though rates of heat extraction may be very high indeed.[1,29-31]

It has been suggested that the roots of secondary dendrite arms may have a slightly higher solute content than outer portions of the arms.[29] Thus, the melting point here would be lower, and thermal fluctuations would tend to cause melting just at this location. In any case, the roots of dendrite arms are often smaller in diameter than exterior portions, and even if melting in response to a thermal fluctuation were uniform, it would result in separation of the arm from the main stalk. It has

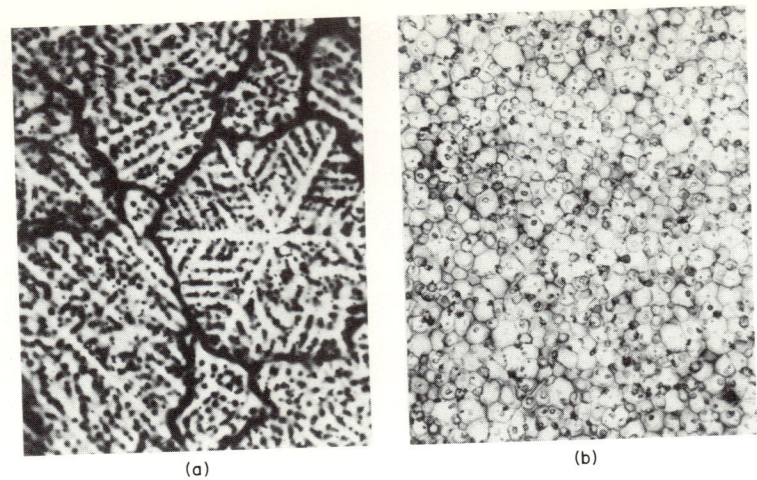

FIGURE 5-22
Magnesium–zinc alloy (magnification ×55). (a) Not grain refined; (b) grain refined by zirconium addition. (*From Kattamis et al.*[33])

also been suggested that mechanical fracture resulting from the stress caused by fluid flow might be enough to lead to grain multiplication,[32] but there is no experimental verification that this mechanism operates.

There have been few detailed structural studies of equiaxed dendrites, although many examples of etched structures are to be found in the foundry literature and in textbooks on metallography. Some examples of different types of structures are shown in Figs. 5-21 and 5-22. Generally, coarse grained equiaxed structures show dendrites that are broadly similar to the columnar dendrites described earlier, except that there is little, if any, preferential orientation in the heat-flow direction. Plates are often evident except in alloys which contain a relatively large amount of eutectic. As grain size is reduced until it begins to approach the dendrite arm spacing, the dendrite structure assumes a less crystallographic appearance and more second phase is found at grain boundaries than in interdendritic regions.

When, at the limit, grain refinement is so effective that a dendritic structure cannot form or at least cannot survive the initial stages of solidification, the final structure is as in Fig. 5-22*b*, with spherical morphology of isoconcentrates within each grain. The minimum grain size that can be obtained in these *nondendritic* alloys is determined by coarsening kinetics in the same way as is secondary dendrite arm spacing in dendritic alloys. Unfortunately, highly refined nondendritic structures such as that of Fig. 5-22*b* have been obtained in very few alloys, the only important commercial example being zirconium-refined magnesium alloys.[33] The limited data available show little difference in microsegregation between columnar and equiaxed grains, as would be expected from the simple analysis described earlier. When differences are found, there is usually more segregation in the equiaxed region. One possible reason for this is interdendritic fluid flow, as discussed in Chap. 7.

It is generally possible to distinguish only one or two major (primary) dendrite arms within an equiaxed dendrite, and so it is not often feasible to measure a primary spacing in these alloys. Secondary arm spacings are readily measured. Results which have been obtained show this spacing equal to that of columnar structures at equivalent cooling rates. Several of the plots of Figs. 5-15 and 5-16 show this in that data from equiaxed grains are plotted with data from columnar grains. Other studies show grain refinement has either no effect on secondary arm spacing or an effect so small it is probably within experimental error.[13,34] This lack of dependence of arm spacing on grain structure again suggests a strong dependence on coarsening of final spacing.

## FACETED DENDRITES AND PREFERRED GROWTH DIRECTIONS

Perhaps to an even greater extent than nonfaceted dendrites, faceted dendrites present a very wide array of morphologies, depending on alloy and solidification variables. As in nonfaceted alloys, these dendrites form because of instability arising from constitutional supercooling in front of the liquid-solid interface. The facets arise because of kinetic difficulties in forming new planes of atoms, as discussed in Chap. 9. The assumption of interface equilibrium used up to this point in description of all nonfaceted phases is not applicable to these structures.

Faceted dendrites are found in many nonmetallic materials, including such commercially important ones as semiconductors and crystalline oxides used in refractory and abrasive applications. Faceted dendrites are also found in *semimetals* such as bismuth and in metals when the element forming the primary phase is very dilute. Two examples of faceted dendrites are shown in Fig. 5-23. In impure commercial metal alloys, nonmetallic phases sometimes precipitate as faceted dendrites, and an example of this is given in Chap. 6.

The directions assumed by growing dendrite arms in faceted materials are always ones that are "capped" by relatively slow-growing (usually low-index) planes.[35] Figure 5-24a is a schematic example of a dendrite growing in the $\langle 100 \rangle$ direction with its tip capped by four slow-growing $\{111\}$ planes. The reader may wish to verify for himself (from geometric considerations alone) that if conditions are altered during growth so that the $\{100\}$ planes become the slowest growing, the $\{111\}$ planes will grow out, leaving $\{100\}$ facets and a new dendrite growing in a $\langle 111 \rangle$ direction.

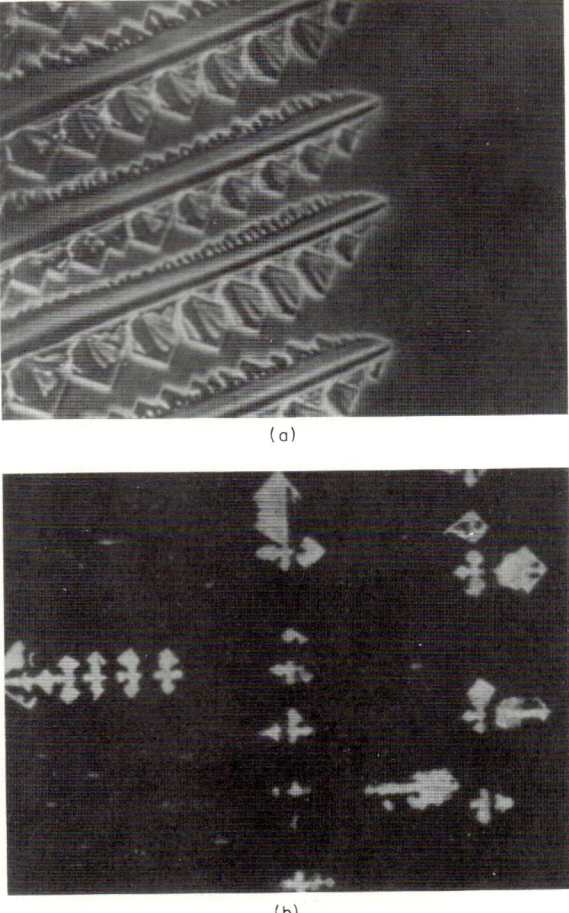

**FIGURE 5-23**
Faceted dendrites. (*a*) Transparent organic material (magnification ×50) (*from Jackson*[62]); (*b*) aluminum dendrites in a tin-rich Al–Sn alloy (magnification ×65) (*from Miller and Chadwick*[63]).

Urea, when added to an ammonium chloride–water alloy, causes the ammonium chloride dendrites to change growth direction in just this way. Many other examples are to be found in the literature.[36,37]

In metals which grow with nonfaceted dendrites, it is not so easy to decide why dendrites grow in the directions they do. Experimentally, it is found that the major dendrite direction in metals is that which would be expected if the dendrites grew with

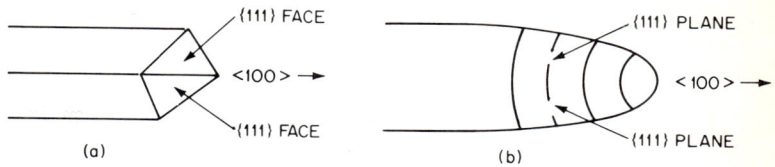

FIGURE 5-24
Schematic diagram of dendrite tip for growth of cubic materials in the $\langle 100 \rangle$ direction. (*a*) Faceted; (*b*) nonfaceted.

faceted caps, as sketched in Fig. 5-24*b*. The slowest-growing planes would be expected to be the closest-packed planes, and, as pointed out by Weinberg and Chalmers,[38] the major dendrite direction is generally the axis of a pyramid whose sides are the most closely packed planes with which a pyramid can be formed. These directions are $\langle 100 \rangle$ for the body- and face-centered cubic structure, $\langle 10\bar{1}0 \rangle$ for the hexagonal close-packed structure, and $\langle 110 \rangle$ for the body-centered tetragonal (tin). Of course, as seen earlier for tin, higher-order dendrite arms may form in other directions as well as these.

The driving forces leading to preferred dendrite directions are interface kinetics, anisotropic surface energy, or both, as discussed in Chap. 9. These driving forces are very small for metals since small driving forces can cause the dendrites to grow out of their crystallographic orientation. One example is the tin dendrite discussed earlier in this chapter, where overlapping diffusion fields apparently influence secondary arm directions. Another example is that quite modest fluid flow causes dendrites to change orientation.[39] A curious related effect is seen in the dendrites of aluminum alloys observed on chill surfaces. The dendrite arms appearing on the surface are orthogonally arranged and so give the appearance of marked preferred orientation, i.e., that {100} planes lie parallel to the chill face. However, no such orientation exists. The dendrite *arms* observed represent intersections of {100} planes with the chill surface (as would be expected), while the orthogonal arms represent projections of the $\langle 100 \rangle$ direction.[41]

Distinct from but determined by the dendrite directions discussed above is the direction of columnar growth, or *texture*, of a columnar region. In cubic metals, the orientation of initial grains is random, but as columnar growth begins from the chill zone those grains most favorably oriented crowd out their less fortunate neighbors so that a preferred texture rapidly develops, as shown schematically in Fig. 5-25. The most favorable orientation for growth is always the preferred dendrite direction, for example, $\langle 100 \rangle$ for cubic metals. Dendrites with this orientation should have a slightly higher tip temperature and should, therefore, lead their neighbors slightly. Thus, they can be expected to encroach gradually on the domain of their neighbors as

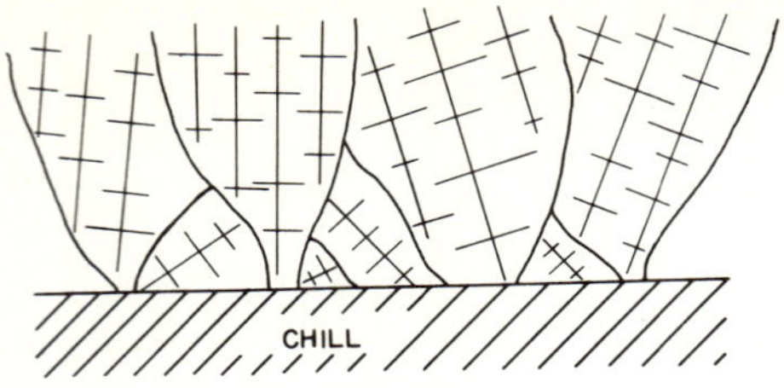

**FIGURE 5-25**
Development of a preferred texture at a chill face.

growth proceeds. An additional factor favoring selection of grains oriented in this way may be that such grains provide the most favorable path for heat flow through the liquid-solid region.

Zinc alloys present an interesting difference to the case of cubic metals discussed above in that a preferred orientation develops at the chill face and this orientation is different from the preferred orientation at the interior. The alloys are of hexagonal close-packed structure and prefer to grow in the basal plane (in the $\langle 10\bar{1}0 \rangle$ direction), and near the chill this plane is oriented parallel to the chill while at the interior it is preferentially located perpendicular to it.[40,41]

The process of texture formation has been utilized industrially to obtain castings with oriented, fully columnar structures. Crystal orientation in these structures is random about the growth axis unless the solidification front is caused to grow around a corner. Then a high degree of *double orientation* can be achieved. If conditions are such that only one grain survives a corner or series of corners, then the remainder of the casting will be a single dendrite of predetermined orientation. These various structures find use as magnetic materials, high-temperature materials, and materials with enhanced room-temperature properties.[42–46] As example, Fig. 5-26 shows a fully columnar and a single dendrite (*monocrystal*) turbine blade from the work of Versnyder and Shank.[44]

## TEMPERATURE AND FRACTION OF SOLID

In metal alloys and other alloys where the liquid-solid interface is close to equilibrium, liquid composition within any given interdendritic space (a *volume element*) is close to uniform and uniquely defined by the liquidus line of the phase diagram. Thus, for a phase diagram of constant liquidus slope $m_L$,

$$C_L = m_L(T^* - T_M) \qquad (5\text{-}12)$$

where $T^*$ is temperature of the volume element and $T_M$ is melting point of the pure solvent. Fractions of liquid and solid at this temperature are now also uniquely

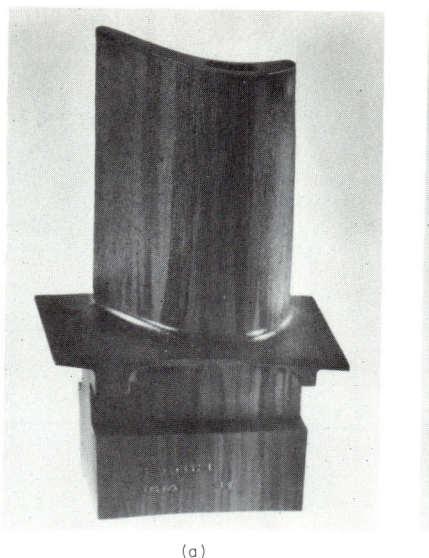 

FIGURE 5-26
Modern air-cooled gas turbine blades directionally solidified to obtain (a) columnar grain and (b) monocrystal. (*From Versnyder and Shank.*[44])

defined, depending on assumptions made regarding the solidification mode. For no solid diffusion, Eq. (5-12) is combined with Eq. (5-1) and with the relation $C_S^* = kC_L$, and so

$$f_L = \phi^{-1/1-k} \quad (5\text{-}13)$$

where $f_L$ is fraction of liquid, $\phi$ is dimensionless temperature during solidification $T_M - T^*/T_M - T_L$, and $T_L$ is liquidus temperature of the alloy. Similar equations are obtained for limited solid diffusion by combining Eqs. (5-5) and (5-12), and for very steep temperature gradient and slow growth rate by combining Eqs. (5-12) and (3-7). The form of Eq. (5-13) is such that in dilute alloys a preponderant portion of the solidification occurs very near the liquidus temperature, with the small remaining amount of liquid then solidifying over a relatively large temperature range. In alloys of richer composition, solidification of the primary phase is spread more evenly across the temperature range and the percentage of eutectic gradually increases as eutectic composition is approached. Figure 5-27a shows this for a series of aluminum–copper alloys. The solid diffusion that occurs during solidification of these alloys changes the shapes of the curves only slightly, as shown in Fig. 5-27b for Al–4.5% Cu alloy. More pronounced changes can be effected at high-temperature gradients and slow growth

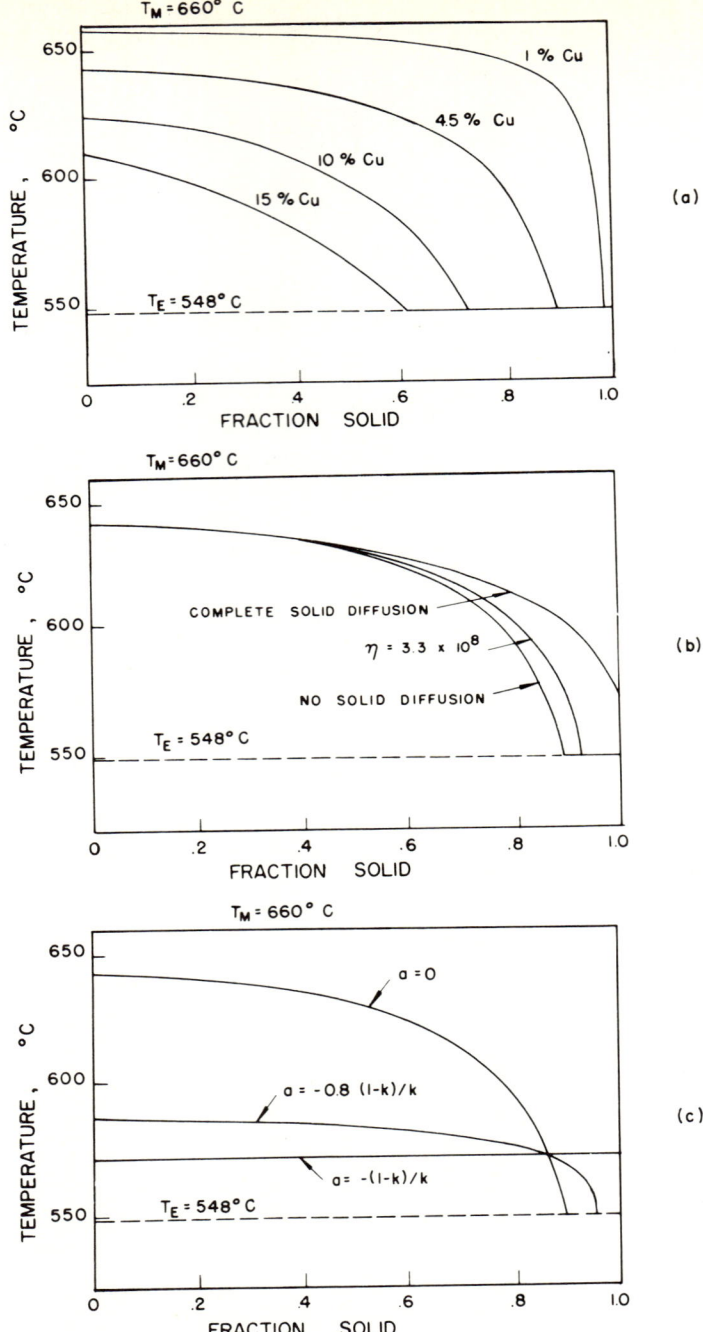

**FIGURE 5-27**
Curves of fraction solid versus temperature for aluminum–copper alloys. (*a*) Effect of alloy analysis: no solid diffusion and no long-range liquid diffusion; (*b*) effect of solid diffusion: Al–4.5% Cu alloy ($\eta = t_f/d^2$); (*c*) effect of long-range liquid diffusion: Al–4.5% Cu. (*a* is defined by Eq. 3-20.)

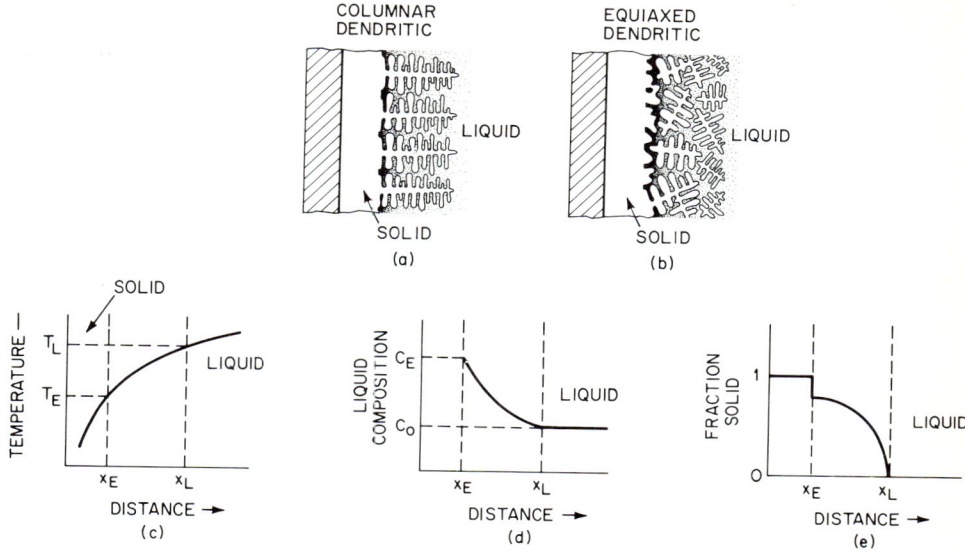

FIGURE 5-28
Model of the liquid-solid region.

rates (i.e., as conditions for cellular solidification are approached), as shown in Fig. 5-27c.

Curves such as those of Fig. 5-27 are useful in various ways for understanding cast structures. As one example, intuition and observation suggest that cells or stubby dendrites are preferred to dendrites with well-developed secondary and tertiary branches when the slope of these curves is low at low fractions of solid, that is, when $(df_s/dT)_{f_s=0}$ is large. From an engineering point of view, the major usefulness of curves such as those of Fig. 5-27 is that they permit describing the fraction of solid at a given point in a casting or ingot, knowing only the temperature at that point. Thus, thermal data can be directly transformed to a physical picture of the solidifying structure. This is done schematically in Fig. 5-28 for two cases of unidirectional solidification, equiaxed and columnar. In both sketches, the dendrites are very greatly magnified. Typically, there would be several orders of magnitude more dendrite arms across the mushy zone than are drawn here. For the same temperature distribution, the curves of liquid composition and fraction solid versus distance, given by Eqs. (5-12) and (5-13), respectively, are identical for the two types of grain structure. Very large differences in liquid composition exist over the liquid-solid *mushy* zone. Flow of this solute-rich liquid is an important source of macrosegregation. Since the temperature range of solidification is often quite large, the distance over which fluid

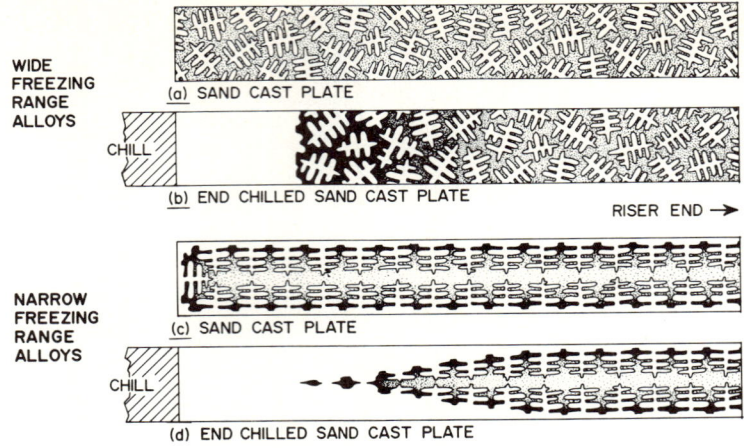

FIGURE 5-29
Cross sections of plate castings showing solidification in a sand mold. Dendrite arms are greatly magnified.

must flow through the mushy zone is often quite long. Resistance to this flow is an important source of microporosity. In the early stages of solidification of equiaxed alloys, the dendrites are free-floating and some feeding of the shrinkage takes place by movement of both liquid and solid (i.e., by *mass feeding*). The coherent semisolid which forms in the cooler parts of the mushy zone is weak because of the interdendritic liquid and liable to tear apart (*hot tear*) under imposed thermally induced strains. (These aspects of solidification behavior are discussed in later chapters.)

A physical picture is also readily developed of solidification of castings or ingots when heat flow is not unidirectional. The plate casting of Fig. 5-29 is a simple example. Again, the dendrites are sketched much larger than they would actually be. When an alloy of wide freezing range and high conductivity (such as Al–4.5% Cu) is cast in a sand mold, only very small temperature differences exist across the thickness or even along the length of the plate. The small temperature differences result in only very small differences in fraction of solid across the thickness or length of the plate. The entire plate solidifies as a nearly uniform *mush*. For very high-quality aluminum castings, heavy chilling is employed both to reduce dendrite arm spacing and to promote *directional solidification*, as shown in Fig. 5-29b. When an alloy of narrow freezing range or low thermal conductivity (e.g., low-alloy steel) is poured in the same mold, temperature differences across the thickness and along the length of the casting may become significant with respect to the range of solidification. Then, solidification behavior is rather different, as shown also in Fig. 5-29. In this figure, the wide-

TEMPERATURE AND FRACTION OF SOLID 165

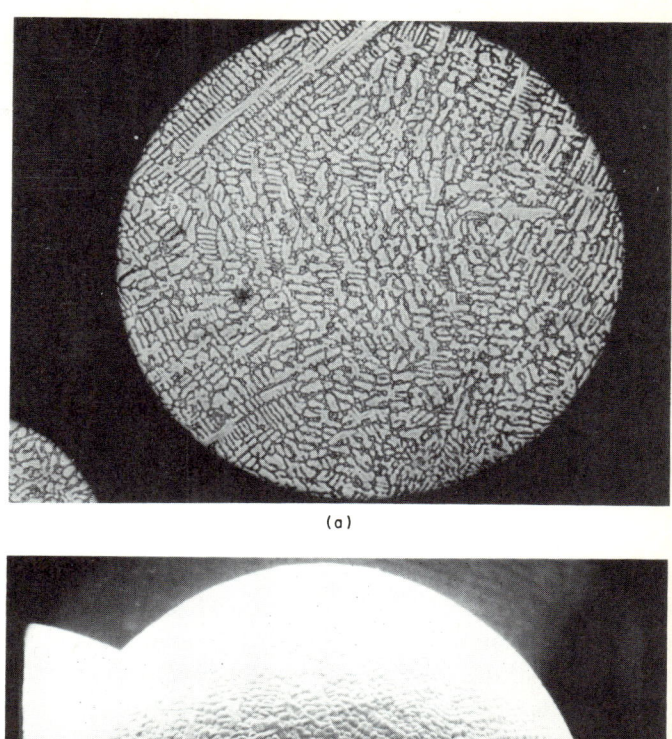

(a)

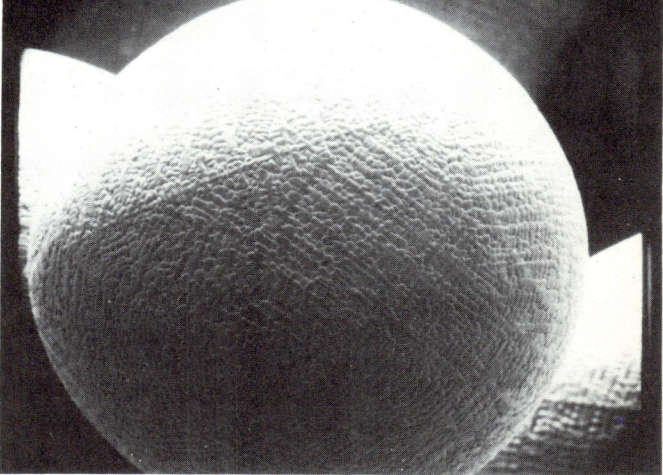

(b)

**FIGURE 5-30**
Atomized liquid droplet. (a) Photograph of cross section (magnification ×200); (b) scanning electron micrograph of droplet surface (magnification ×240).

FIGURE 5-31
Schematic illustration of solidification of a static-cast ingot, columnar grains. (Dendrite arms are greatly magnified.)

freezing-range alloy is shown with equiaxed grains and the narrow-freezing-range alloy with columnar grains. This is often the case, but either alloy can have either type of structure. The important aspect of the figure is the distribution of liquid and solid during solidification. In the foundry and solidification literature many examples are to be found of the application of these ideas to specific alloys and foundry problems.

Another example of a casting that solidifies with low thermal gradient is a small liquid droplet produced by various methods, including *atomization* of a molten stream by impingement of a gas jet. The droplet is typically the order of 50 μm diameter or

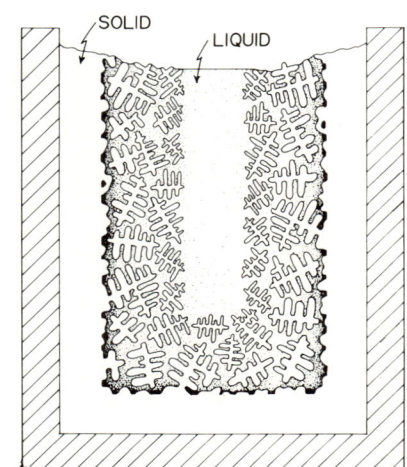

FIGURE 5-32
Schematic illustration of solidification of a static-cast ingot, equiaxed structure. (Dendrite arms are greatly magnified.)

less. The rate of heat removal during solidification is controlled by convection or radiation so that there are only very small temperature differences within the droplet. The droplet solidifies in a mushy manner as does the sand-cast plate of Fig. 5-29a. Figure 5-30 is an example of the structure of a droplet so formed. The rough surface is due to liquid existing at the droplet surface until near the end of solidification; as solidification shrinkage took place, this liquid was drawn inward.

At another extreme of size, Figs. 5-31 and 5-32 are schematic illustrations of solidification of a large static ingot. In most commercial alloys, the liquid-solid mushy zone comprises a substantial portion (but not all) of the ingot cross section during much of solidification. The dendrite arms shown in this region are very greatly magnified. Several thousands of secondary dendrite arms typically would be present across the width of the mushy zone than the few shown. Here, as in the other examples cited, the fraction of solid at any given location in the ingot can be calculated, at least approximately, knowing only the temperature at that location.

## SOLIDIFICATION OF UNDERCOOLED MELTS

In usual commercial practice, sufficient heterogeneous nuclei are present so that supercoolings observed before solidification begins are rarely more than a few degrees centigrade. When, however, clean pure materials or alloys are solidified out of contact with effective nuclei, very large supercoolings can be obtained. These approach 20 percent of the melting point for typical metals (e.g., in excess of 300°C for iron and nickel) and are even higher for a few materials, including phosphorus. Experimentally, the easiest way to obtain large undercoolings in bulk metal specimens is to melt them in an inert atmosphere, or vacuum, using as crucible a clean glassy material or a crystalline material which is coated with a viscous liquid slag.

Growth of undercooled pure materials is dendritic, with the dendrites resulting from simple thermal supercooling rather than *constitutional supercooling*, which leads to the dendrites discussed up to this point. Dendrites which are formed from thermal undercooling alone have been termed *thermal dendrites*. Many experimental measurements have been made on undercooled materials of dendrite growth velocity as a function of bath undercooling. Experimental measurements on ice, tin, nickel, cobalt, and germanium have been summarized by Chalmers.[47] Other more recent measurements include those on bismuth[48] and phosphorus.[49] The measurements on phosphorus by Glicksman and Schaeffer are of particular interest since only in this material have dimensionless undercoolings greater than unity been obtained, where the dimensionless undercooling $\Delta U$ is

$$\Delta U = \frac{c_s \Delta T}{H} \qquad (5\text{-}14)$$

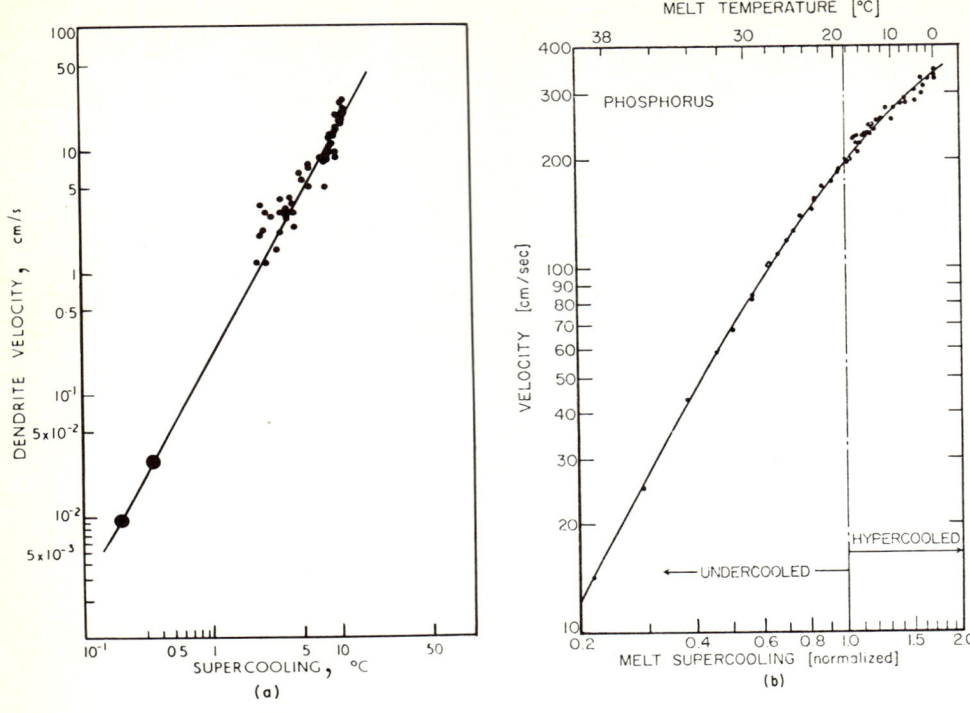

FIGURE 5-33
Dendrite velocity versus melt undercooling for (a) tin and (b) phosphorus. (*From Glicksman and Schaeffer.*[48,49])

and $c_s$ is specific heat of the solid material, $\Delta T$ is undercooling, and $H$ is heat of fusion. Thus, in this material, sufficient undercooling has been obtained prior to nucleation so that no further heat need be extracted from the bulk material to obtain complete solidification. Figure 5-33 shows some typical measurements of dendrite-tip velocity versus undercooling for a metal and for phosphorus. For dimensionless undercoolings less than unity, the curves are of approximately the form

$$R = a\,\Delta T^n \qquad (5\text{-}15)$$

where $R$ is dendrite-tip velocity, $a$ is a constant, and $n$ is not far from 2. Most experiments performed have shown the exponent $n$ to be between 1.5 and 3.

Many theoretical attempts have been made to calculate growth velocity of undercooled melts. The problem can be simply stated although not so simply solved. Figure 5-34 shows a dendrite tip of a pure material growing into a pure melt undercooled an amount $\Delta T$. The equilibrium melting point of the tip is depressed an amount

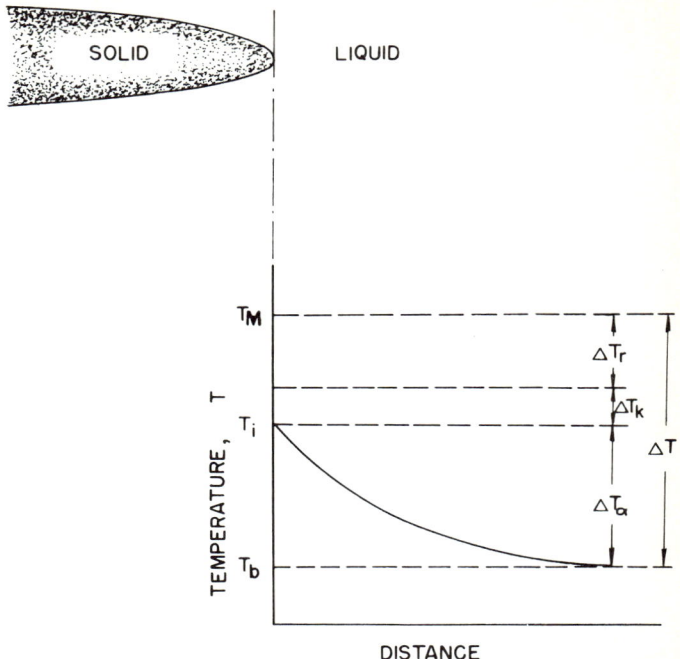

**FIGURE 5-34**
Thermal profile in front of a dendritic tip growing in a melt undercooled an amount $\Delta T = T_M - T_b$.

$\Delta T_r$ owing to the radius of curvature at the tip. Since the interface is moving, the actual interface temperature $T_i$ must be still lower by an amount $\Delta T_k$, the undercooling required to transport atoms across the liquid-solid interface. Finally, as the dendrite grows, it releases heat of fusion, which is dissipated into the undercooled melt. This heat diffuses down a temperature gradient into the melt, resulting in an undercooling $\Delta T_\alpha$; thus,

$$\Delta T = \Delta T_r + \Delta T_k + \Delta T_\alpha \quad (5\text{-}16)$$

Ideally, one would now like to find a steady-state (*shape-preserving*) solution which would account for the three effects of radius of curvature, interface kinetics, and heat diffusion and provide a unique solution of growth rate $R$ as a function of total undercooling $\Delta T$; this has not been done.

Rigorous solutions have been obtained for the much simpler case where $\Delta T_r$ and $\Delta T_k$ are assumed to be negligible, that is, for the *isothermal dendrite*. In this case, a parabola of circular cross section provides such a solution, as shown by Ivantsov.[50]

Parabolas of elliptical cross section also provide solutions, as shown by Horvay and Cahn.[51] In either case, the problem is to solve the second law of heat conduction [Eq. (1-29)] with the boundary conditions that the temperature at the liquid-solid interface over the whole dendrite is constant and equal to the temperature within the dendrite. Diffusion fields of neighboring dendrites are assumed not to overlap. Finally, since the surface of the dendrite is isothermal, the gradient of temperature at the liquid-solid interface over the whole dendrite is normal at the interface. Thus,

$$\rho_s H \frac{d\mathbf{N}}{dt} \mathbf{N} = -k_L \mathbf{N} \nabla T \quad (5\text{-}17)$$

where $k_L$ is thermal conductivity of the liquid, and $\mathbf{N}$ is the unit vector perpendicular to the liquid-solid interface. The solution for the paraboloid of revolution is

$$-Pe^P Ei(-P) = \Delta U \quad (5\text{-}18a)$$

$$P = \frac{R r_t}{2\alpha_L} \quad (5\text{-}18b)$$

where $P$ is dimensionless growth velocity, $\Delta U$ is dimensionless undercooling as defined by Eq. (5-14), $r_t$ is dendrite-tip radius, and $\alpha_L$ is thermal diffusivity of the liquid; Fig. 5-35 plots this result. Growth velocity is linear with undercooling at low undercoolings and increases to infinity as a dimensionless undercooling of unity is approached; however, the solution yields only the product $R r_t$ and does not uniquely give either. Thus, for a given bath undercooling, thin dendrites growing rapidly or thick dendrites growing slowly both satisfy the diffusion conditions. How does the dendrite choose? One would expect to find a rigorous answer to this question through a perturbation analysis such as that of Mullins and Sekerka for plane front solidification, but as yet no one has attempted this for a dendrite.

Meanwhile, numerous attempts have been made to solve the dendrite problem by including the effect of capillarity in calculations and then assuming the dendrite chooses that radius of curvature $r_t$ which maximizes growth velocity for a given undercooling.[52-55] An equivalent assumption is that of minimization of $\Delta T$ for a given $R$. Only approximate solutions have been obtained to this problem, the simplest of which is to assume that the dendrite remains parabolic and isothermal so that the undercooling for heat dissipation $\Delta T_\alpha$ is given by Eq. (5-17). Next, the temperature of the dendrite is assumed to be depressed by an amount $\Delta T_r$ calculated from Eq. (5-8), taking as the radius of curvature that at the tip $r_t$. Finally, interface kinetics are neglected. Calculations are readily carried out in this way but, unfortunately, do not agree well with experiments. One reason for the discrepancy is that the interface kinetic undercooling $\Delta T_k$ [Eq. (5-16)] is significant at the high velocities encountered

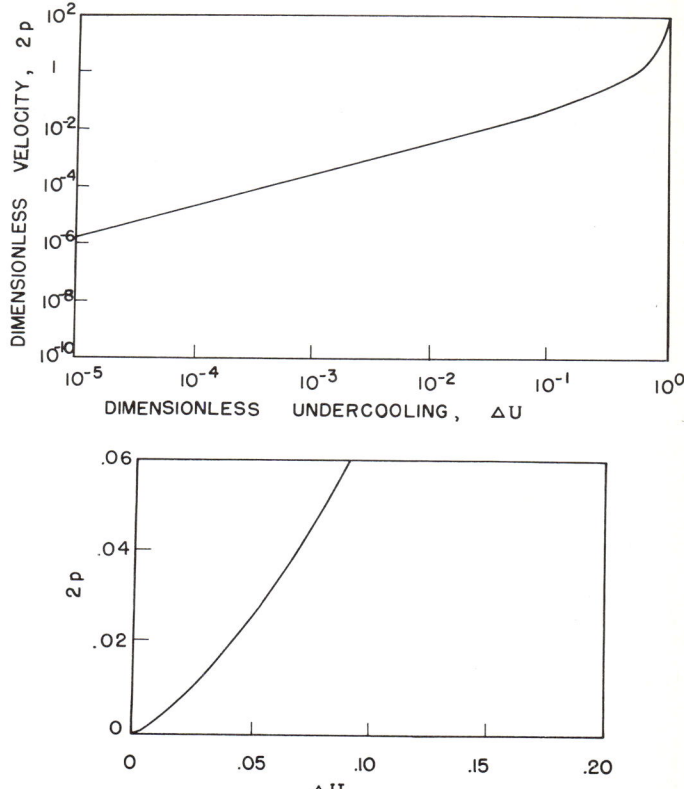

FIGURE 5-35
Dimensionless velocity versus dimensionless undercooling for paraboloid of revolution. *Top:* log scale; *bottom:* linear scale. (*From Horvay and Cahn.*[51])

in undercooled liquids. Once kinetic and curvature effects are included, the dendrite is clearly no longer isothermal and so the heat-flow equation [Eq. (5-18)] must be modified.

The various recent analyses of the dendrite problem include consideration of both these effects, but agreement of theory with experiment is still not good when reasonable values of the kinetic coefficient are included.[54–57] Possible further reasons for the discrepancy are that the assumption of paraboloidal shape is itself an approximation and there is no obvious reason why a dendrite should choose that radius corresponding to minimum undercooling. The theories cited assume no interaction of thermal fields of neighboring dendrites. Finally, as proposed by Chalmers[47] and subsequently shown experimentally by Morris and Winegard,[56] growth of a dendrite need not be at steady state at all. Instead, its shape and velocity can fluctuate markedly during growth, influenced by the formation of secondary arms behind the growing tip. In this case, any theory based on steady-state growth can be at best an approximation. The *dendrite-tip problem* is far from being rigorously solved.

## STRUCTURE OF UNDERCOOLED MELTS

Grain growth occurs so rapidly after solidification in pure metals that structural studies on undercooled melts are always made on material containing at least some alloy. A number of such studies, beginning with those of Walker, described by Chalmers, show that an abrupt reduction of grain size occurs at a critical undercooling.[47,48,57,58] The undercooling is surprisingly uniform at about 175°C for a variety of iron- and nickel-base alloys. A possible reason for the abrupt reduction is that at this undercooling cavitation occurs which promotes multiple nucleation. The cavitation is thought to result from the rapid fluid flow required to feed the solidification shrinkage. It causes nucleation at temperatures above the usual nucleation temperature because of the high pressures resulting in subsequent collapse of the bubble. Calculations of Horvay[59] show that this is a possible mechanism.

An alternate effect is also expected to affect the grain size of undercooled melts. The dendrites which form at large undercoolings must be highly unstable near the end of recalescence. Dendrite remelting must then begin to occur and may result in some *grain multiplication*, as discussed earlier in this chapter. In one experiment, Glicksman and Schaeffer[48] have apparently shown that this grain multiplication sometimes results in catastrophic disintegration of the dendrites which formed first.

It is not difficult to undercool commercial alloys of iron, nickel, or copper, and it should be possible commercially to produce ingots of substantial size in this way. The technique would indeed be attractive in manufacturing if dimensionless undercoolings $\Delta U$ greater than unity could be obtained in these materials. In this case, extremely short solidification times would be obtained and a fine structure of high homogeneity should result. Unfortunately, such large undercoolings have not been obtained, and so a substantial fraction of liquid always remains after recalescence is complete. Solidification of this last liquid then takes place as in usual nonundercooled castings and ingots. Dendrite coarsening occurs during this period, and final dendrite arm spacing at each point within the ingot depends on the time required to complete solidification at this point. The dendrite arm spacing of the final ingot appears to be not greatly different from that of nonundercooled ingots.[58,60]

# REFERENCES

*1* BOWER, T. F., and FLEMINGS, M. C.: *Trans. AIME*, **239**:1620 (1967).
*2* JACKSON, K. A., HUNT, J. D., UHLMANN, D. R., and SEWARD III, T. P.: *Trans. AIME*, **236**:149 (1966).
*3* KATTAMIS, T. Z., and FLEMINGS, M. C.: *Trans. AIME*, **233**:992 (1965).
*4* BOWER, T. F.: Sc.D. thesis, M.I.T., Cambridge, Mass., 1965.
*5* MORRIS, L. R., and WINEGARD, W. C.: *J. Crystal Growth*, **6**:61 (1969).
*6* BOWER, T. F., BRODY, H. D., and FLEMINGS, M. C.: *Trans. AIME*, **236**:624 (1966).
*7* AHEARN, P. J., and FLEMINGS, M. C.: *Trans. AIME*, **239**:1590 (1967).
*8* BRODY, H. D., and FLEMINGS, M. C.: *Trans. AIME*, **236**:615 (1966).
*9* NOVIKOV, I. I., LYUTSAU, V. G., and ZOLOTOREVSKY, V. S.: *Phys. Metals Metallog. (USSR) (English Transl.)*, **16**:2 (1962).
*10* SUBRAMANIAN, S. V., HAWORTH, C. W., and KIRKWOOD, D. H.: *J. Iron Steel Inst.*, **206**:1027 (1968).
*11* SHARP, R. M., and HELLAWELL, A.: *J. Crystal Growth*, **6**:253 (1970).
*12* DROUZY, M., and MASCRE, C.: *Mem. Sci. Rev. Met.*, **53**:241 (1961).
*13* FLEMINGS, M. C.: in "Solidification," American Society for Metals, Metals Park, Ohio, p. 311, 1971.
*14* DOHERTY, R. D., and MELFORD, D. A.: *J. Iron Steel Inst.*, **204**:1131 (1966).
*15* FLEMINGS, M. C., POIRER, D. R., BARONE, R. V., and BRODY, H. D.: *J. Iron Steel Inst.*, **208**:371 (1970).
*16* SINGH, S. N., BARDES, B. P., and FLEMINGS, M. C.: *Met. Trans.*, **1**:1383 (1970).
*17* KIRKWOOD, D. H., and EVANS, D. J.: "The Solidification of Metals," Iron and Steel Institute Publ. No. 110, p. 108, 1968.
*18* ROHATGI, P. K., and ADAMS, JR., C. M.: *Trans. Met. Soc. AIME*, **239**:1729 (1967).
*19* ROHATGI, P. K., and ADAMS, JR., C. M.: *Trans. Met. Soc. AIME*, **239**:1737 (1967).
*20* SPEAR, R. E., and GARDNER, G. R.: *Trans. AFS*, **71**:209 (1963).
*21* GRANGER, D. A., and BOWER, T. F.: "Cast Structure in Two D.C. Semicontinuous Cast Copper Alloys" (to be published).
*22* SUZUKI, A., SUZUKI, T., NAGAOKA, Y., and IAWATA, Y.: *Nippon Kingaku Gakkai Shuho*, **32** (1968); also Brutcher transl., 7804 (1969).
*23* KATTAMIS, T. Z., COUGHLIN, J., and FLEMINGS, M. C.: *Trans. Met. Soc. AIME*, **239**:1504 (1967).
*24* JACKSON, K. A.: *Solidification Cinemas*, Bell Laboratories.
*25* CHERNOV, A. A.: *Kristallografiya*, **1**:583 (1956).
*26* KLIA, M. O.: *Kristallografiya*, **1**:577 (1956).
*27* KAHLWEIT, M.: *Scripta Met.*, **2**:251 (1968).
*28* PAPAPETROU, A.: *Z. Krist.*, **A92**:89 (1935).
*29* JACKSON, K. A., HUNT, J. D., UHLMANN, D. R., and SEWARD III, T. P.: *Trans. Met. Soc. AIME*, **236**:149 (1966).
*30* PESTEL, G., LANGENBERG, F. C., and HONEYCUTT, C. R.: U.S. Patent No. 2,963,758, 1960.

31 UHLMANN, D. R., SEWARD III, T. P., and CHALMERS, B.: *Trans. Met. Soc. AIME*, **236**:527 (1966).
32 TILLER, W. A., and O'HARA, S.: "The Solidification of Metals," Iron and Steel Institute Publ. No. 110, p. 27, 1968.
33 KATTAMIS, T. Z., HOLMBERG, U. T., and FLEMINGS, M. C.: *J. Inst. Metals*, **55**:343 (1967).
34 CHURCH, N., WIESER, P., and WALLACE, J. F.: *Trans. AFS*, **74**:113 (1966).
35 CHALMERS, B.: "Principles of Solidification," John Wiley & Sons, Inc., New York, 1964.
36 SARATOVKIN, D. D.: "Dendritic Crystallization," Consultant Bureau, New York, 1959 (transl. by J. E. S. Bradley).
37 BUCKLEY, H. E.: "Crystal Growth," John Wiley & Sons, Inc., New York, 1951.
38 WEINBERG, F., and CHALMERS, B.: *Can. J. Phys.*, **30**:488 (1952).
39 MIKSCH, E. S.: *Trans. AIME*, **245**:2069 (1969).
40 EDMUNDS, G.: *Trans. AIME*, **116**:114 (1945).
41 COUGHLIN, J. M.: S.B. thesis, M.I.T., Cambridge, Mass., 1965.
42 GOULD, G. E.: *Cobalt*, **23**:82 (1964).
43 VERSNYDER, F. L., and PIERCEY, B. J.: *S.A.E. J.*, **8**:36 (1966).
44 VERSNYDER, F. L., and SHANK, M. E.: *Mater. Sci. Eng.*, **6**:213 (1970).
45 FLEMINGS, M. C., and MEHRABIAN, R.: *Trans. AFS*, **78**:388 (1970).
46 BOWER, T. F., GRANGER, D. A., and KEVERIAN, J.: in "Solidification," American Society for Metals, Metals Park, Ohio, p. 385, 1971.
47 CHALMERS, B.: "Principles of Solidification," John Wiley & Sons, Inc., New York, 1964.
48 GLICKSMAN, M. E., and SCHAEFFER, R. J.: "The Solidification of Metals," Iron and Steel Institute Publ. No. 110, p. 43, 1968.
49 GLICKSMAN, M. E., and SCHAEFFER, R. J.: *J. Crystal Growth*, **67**:297 (1967).
50 IVANTSOV, G. P.: *Dokl. Akad. Nauk. SSSR*, **58**:567 (1947).
51 HORVAY, G., and CAHN, J. W.: *Acta Met.*, **9**:695 (1961).
52 TEMKIN, D. E.: *Dokl. Akad. Nauk. SSSR*, **132**:1307 (1960).
53 BOLLING, G. F., and TILLER, W. A.: *J. Appl. Phys.*, **32**:2587 (1961).
54 KOTLER, G. R., and TARSHIS, L. A.: *J. Crystal Growth*, **5**:90 (1969).
55 TRIVEDI, R.: *Acta Met.*, **18**:287 (1970).
56 MORRIS, L. R., and WINEGARD, W. C.: *J. Crystal Growth*, **1**:245 (1967).
57 COLLIGAN, G. A., and BAYLES, B. S.: *Acta Met.*, **10**:895 (1962).
58 KATTAMIS, T. Z., and FLEMINGS, M. C.: *Trans. AIME*, **236**:1523 (1966).
59 HORVAY, G.: *Proc. 4th Natl. Congr. Appl. Mech. (ASME)*, 1315 (1962).
60 KATTAMIS, T. Z.: *Z. Metallk.*, **61**:856 (1970).
61 POLICH, R. F., and FLEMINGS, M. C.: *Trans. AFS*, **73**:28 (1965).
62 JACKSON, K.: in "Solidification," American Society for Metals, Metals Park, Ohio, p. 000, 1971.
63 MILLER, W. A., and CHADWICK, G. A.: "The Solidification of Metals," Iron and Steel Institute Publ. No. 110, p. 49, 1968.

## PROBLEMS

**5-1** An ingot of Fe–25% Ni is cast and unidirectionally solidified.
  (a) What would be the segregation ratio if there were no diffusion in the solid?
  (b) Show that diffusion in the solid would be expected to reduce the observed segregation ratio significantly.
  (c) If the isotherms move proportionally to the square root of time, how do solidification time and dendrite arm spacing vary with distance from the chill?

  Data: $\quad D_s = 10.96\, e^{-38,062/T}\quad$ (diffusion coefficient of Ni)
  $\quad\quad\quad\quad d^2 = 4 \times 10^{-8} t_f$

**5-2** A 20-cm-thick plate is cast of Al–10% Mg at its liquidus temperature in (a) a sand mold and (b) a water-cooled copper mold. Assume the temperature distribution in the solidifying metal is approximately the same as it would be in pure aluminum (except that the freezing point is lowered by addition of solute).
  (a) For each case, what is the fraction of solid at the center of the plate when the outside is just completely solid?
  (b) What is the thickness of the liquid-solid mushy zone for the chill-cast plate at the end of solidification? At the beginning of solidification?

**5-3** Al–10% Mg alloy is solidified unidirectionally at constant temperature gradient $G = 50°C/cm$ and constant rate $R = 0.01$ cm/s. Assume the Scheil equation applies within interdendritic *volume elements*.
  (a) During solidification at steady state, plot the following versus distance in the liquid-solid mushy zone. Show the numerical values of the following quantities plotted at liquidus and solidus temperatures:
    (i) Temperature
    (ii) Liquid composition
    (iii) Composition of the liquid-solid interface
    (iv) Average composition of the solid
    (v) Average composition of the liquid plus solid
    (vi) Fraction of liquid

**5-4** Can you explain qualitatively the reason why solidification with low $(df_s/dT)_{f_s}$ seems to favor cells or stubby dendrites over fully developed cells or dendrites with higher-order branches?

**5-5** Columnar dendrites of Al–4.5% Cu alloy typically appear as in Fig. 5-7. List several ways in which you could alter solidification of this alloy to achieve a cellular structure.

**5-6** An isolated dendrite tip is growing in an essentially isothermal melt of Al–4.5% Cu alloy. Sketch the solute profile in front of the dendrite tip (i.e., make a sketch for this case comparable to Fig. 5-31).

**5-7** Calculate the solute buildup in front of the dendrite tip of Prob. 5-6, assuming a tip radius of 1 $\mu$m, a growth rate $R$ of $10^{-3}$ cm/s, and $D_L$ of $5 \times 10^{-5}$ cm²/s. [*Note:* Use

Fig. 5-35, but with $\Delta U$ now being the dimensionless supersaturation at the tip $\Delta C_L / C_L^*(1-k)$, where $C_L^*$ is composition of the liquid at the dendrite tip. Assume $D_L = 5 \times 10^{-5}$ cm$^2$/s.]

5-8  Sketch the grain structure you would expect to find on a cross section of the continuous-cast steel slab of Fig. 1-19.

5-9  Would you expect the surface of an aluminum die casting to look like that of the spherical droplet in Fig. 5-30? Why, or why not?

5-10  List some ways in which you can make dendrite arms in a cubic metal grow in other than the usual crystallographic direction. Explain the mechanism.

5-11  A rapidly solidified Al–4.5% Cu dendritic structure is reheated and held in the liquid-solid region at an intermediate fraction of solid. Show schematically how you would expect the structure to appear after 1, 5, and 10 min.

# 6

# SOLIDIFICATION OF POLYPHASE ALLOYS: CASTINGS AND INGOTS

## PERITECTIC SOLIDIFICATION

In the peritectic reaction $L + \alpha \rightarrow \beta$, liquid plus one solid phase react to form a second solid phase on solidification. The reaction is isothermal in binary alloys and can be written as $L + \alpha \rightarrow \beta$. Figure 6-1 shows a portion of the lead–bismuth diagram as an example of a binary phase diagram containing a peritectic reaction. On equilibrium solidification, alloys between 23 and 36 percent bismuth undergo the peritectic reaction at 184°C. In an alloy of exactly the peritectic composition, 33 percent bismuth, the reaction is complete at 184°C, with $\beta$ being the only phase existing at lower temperatures.

In dendritic solidification of most alloys, incomplete diffusion in the solid prevents the peritectic reaction from taking place to an appreciable extent. When the second phase forms, it "surrounds" the first and the peritectic reaction can proceed only by diffusion of solute through the second phase. As example, consider solidification of a Pb–20% Bi alloy, making the same assumptions used in developing the Scheil equation [Eq. (5-1)]. These include negligible diffusion in both solid phases and equilibrium at liquid-solid interfaces. Now, dendrites of $\alpha$ phase form first, with the

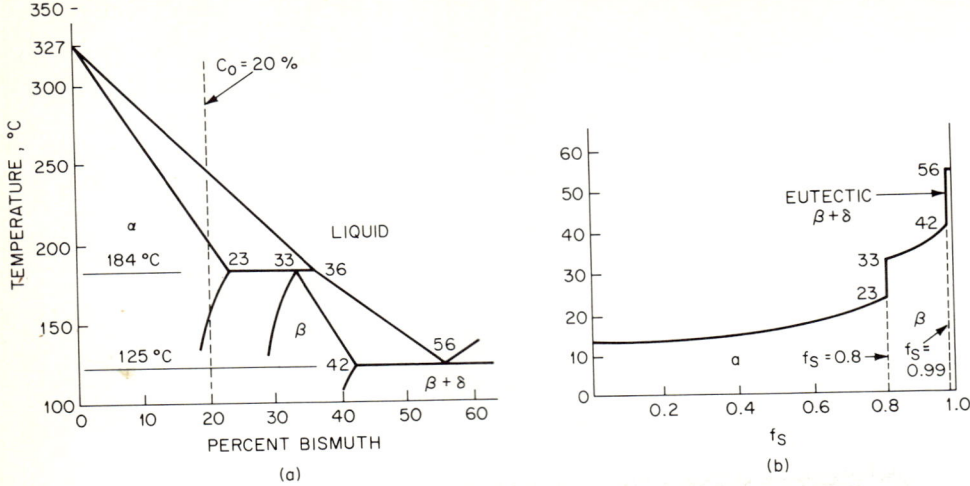

FIGURE 6-1
Pb–Bi phase diagram and solute-redistribution curve for Pb–20% Bi.

composition that solidifies increasing gradually from $kC_0$ to 23 percent bismuth at the peritectic temperature. The second phase $\beta$ then begins to form, surrounding the $\alpha$ phase and preventing the peritectic reaction from taking place. As the temperature falls, $\beta$ continues to grow until eutectic forms at the completion of solidification. The resulting structure is shown schematically in Fig. 6-2. Qualitative calculation of solute distribution is made using the differential form of Eq. (5-1) (since $k$ is not constant throughout solidification) and integrating between appropriate limits. Results of this calculation are shown for the Pb–20% Bi alloy in Fig. 6-1b. Figure 6-3 is a schematic microstructure of the solidified alloy. Examples of similar structures of other alloys, some with limited amounts of peritectic reaction occurring, are to be found in the literature.[1,2]

If one were to solidify the Pb–20% Bi alloy in a crystal-growing furnace at increasingly high values of $G/R$, the temperature of the dendrite tips would drop progressively below the liquidus temperature and the severity of microsegregation would progressively decrease. These changes can be quantitatively described, as was done for simpler alloys in Chap. 3. At sufficiently high $G/R$ and no convection, the alloy would solidify with a plane front at its equilibrium solidus, the resulting solid being of uniform composition $C_0$. It is also possible to solidify Pb–Bi alloys of greater than 23 percent Bi with a plane front. Following Chap. 4, the structures of alloys between the limit of solid solubility of $\alpha$ and the peritectic composition (23 to 33 percent Bi) would be a two-phase composite of $\alpha + \beta$. Alloys between 33 and 36

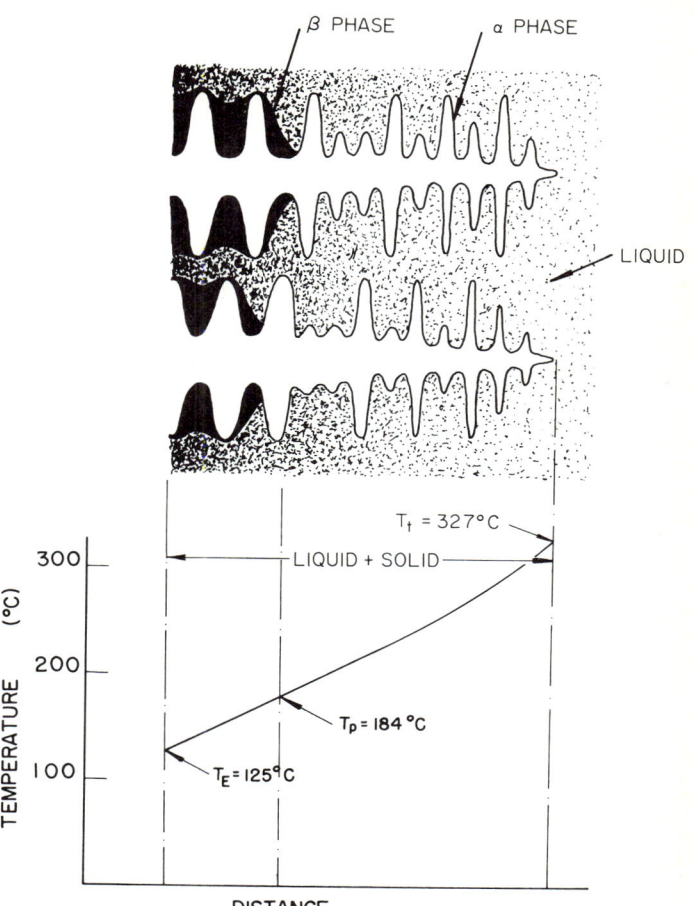

FIGURE 6-2
Schematic diagram of columnar solidification of Pb–20% Bi alloy.

percent Bi would be single-phase, and alloys above 36 percent Bi would be a eutectic-like composite of $\beta + \delta$.

The most important commercial material that solidifies with peritectic reaction is carbon steel. Thermal analysis indicates the peritectic reaction in this material proceeds to a much greater extent than in most other alloys.[3] The reason for this is the high diffusivity of carbon in the solid. In carbon steel, the first phase to form is the $\delta$ iron (see the phase diagram of Fig. 6-6). The $\delta$ phase disappears completely by peritectic reaction on cooling into the austenite region. The fact that $\delta$ iron existed at

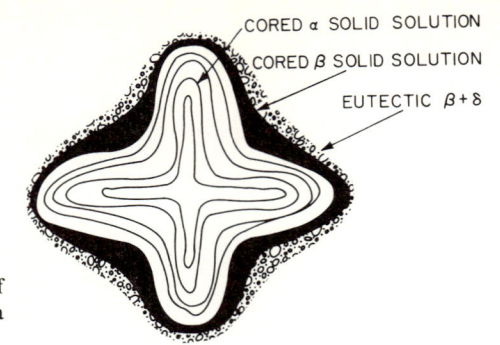

**FIGURE 6-3**
Schematic diagram of microstructure of cast Pb–20% Bi alloy (section through a primary dendrite arm).

all during solidification is not apparent from the microstructure of cast steel except by inference from the grain structure, as discussed in Chap. 5. Volume changes accompanying the reaction have been proposed as a contributing cause of hot tearing, but there is no evidence to support this theory.

## EUTECTIC SOLIDIFICATION

Many important commercial alloys solidify with a eutectic as part or all of their structure. Figure 6-4 illustrates three different ways in which the eutectic may form during columnar solidification. In each case, it is assumed a hypoeutectic alloy is solidifying with low $G/R$ so that dendrites grow with their tip temperature near the liquidus temperature. In a pure binary alloy, the eutectic front at the roots of the dendrites would be planar, as sketched in Fig. 6-4a. Usually, however, some impurities are present, and at sufficiently low $G/R$ the eutectic front breaks up into colonies or cells, as discussed in Chap. 4 and shown schematically in Fig. 6-4b. With still more third element, or lower $G/R$, the two-phase cells take on a dendritic appearance or new eutectic grains nucleate, as shown in Fig. 6-4c. Of course, if $G/R$ is greatly increased above that encountered in castings and ingots, a fully eutecticlike composite is obtained.

Some representative eutectic structures solidified at low $G/R$ are shown in Fig. 6-5, one a *regular* (in this case, rodlike) structure and the other a faceted–nonfaceted eutectic. Another commonly observed structure is the *divorced* eutectic. In divorced eutectics, the eutectic is present in so small a volume fraction that the plate- or rodlike structure does not form. Instead, the minor phase of the eutectic forms and grows in islands that often appear unconnected on a polished surface. The other phase forms by continuing growth of the dendrites.

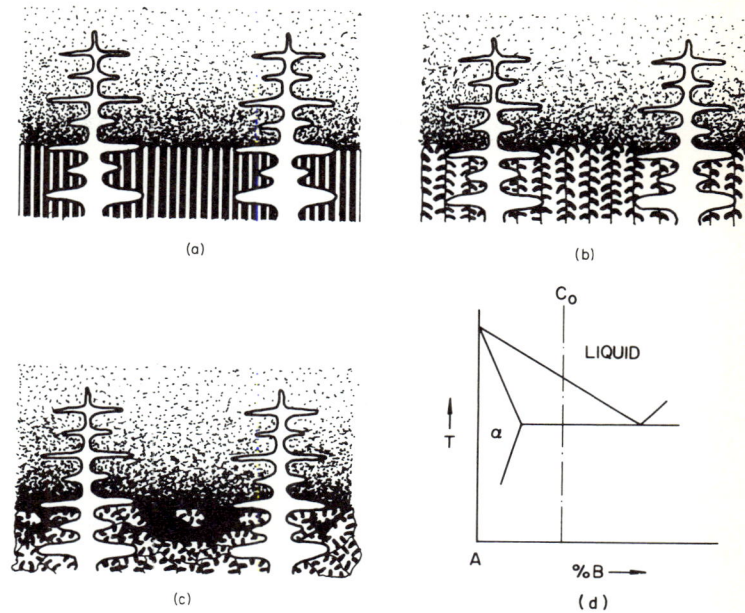

**FIGURE 6-4**
Solidification of eutectics in a directionally solidified hypoeutectic alloy. (a) Eutectic plane front; (b) eutectic cells; (c) eutectic grains. Structures (b) and (c) result at low $G/R$ when impurities are present. Phase diagram and alloy composition are shown in (d).

One eutectic of particular commercial importance is that between aluminum and silicon. Aluminum alloys containing about 7 percent Si comprise an important family of high-strength structural-casting alloys. Alloys in the neighborhood of 12 percent Si (approximately eutectic composition) are widely used in die casting. Both the eutectic and hypereutectic alloys find wide use for applications requiring wear resistance at high temperatures. At the cooling rates encountered in sand casting, the eutectic structure of Al–Si alloys is normally of the faceted–nonfaceted type. Increasing the cooling rate or adding small amounts of several elements (one of them sodium) modifies the structure to one which is rodlike (Fig. 6-5). The silicon phase in both structures is twinned on one or more {111} slip systems. The twinning permits orientation changes during growth and must also permit the silicon phase to grow by a twin-plane reentrant-edge mechanism, as discussed in Chap. 9. A possible qualitative explanation of the plate-rod transition observed in this alloy is as follows. At low growth rates (and low temperature gradients), the faceted silicon phase leads the aluminum-rich phase and assumes its platelike morphology. At higher growth rates

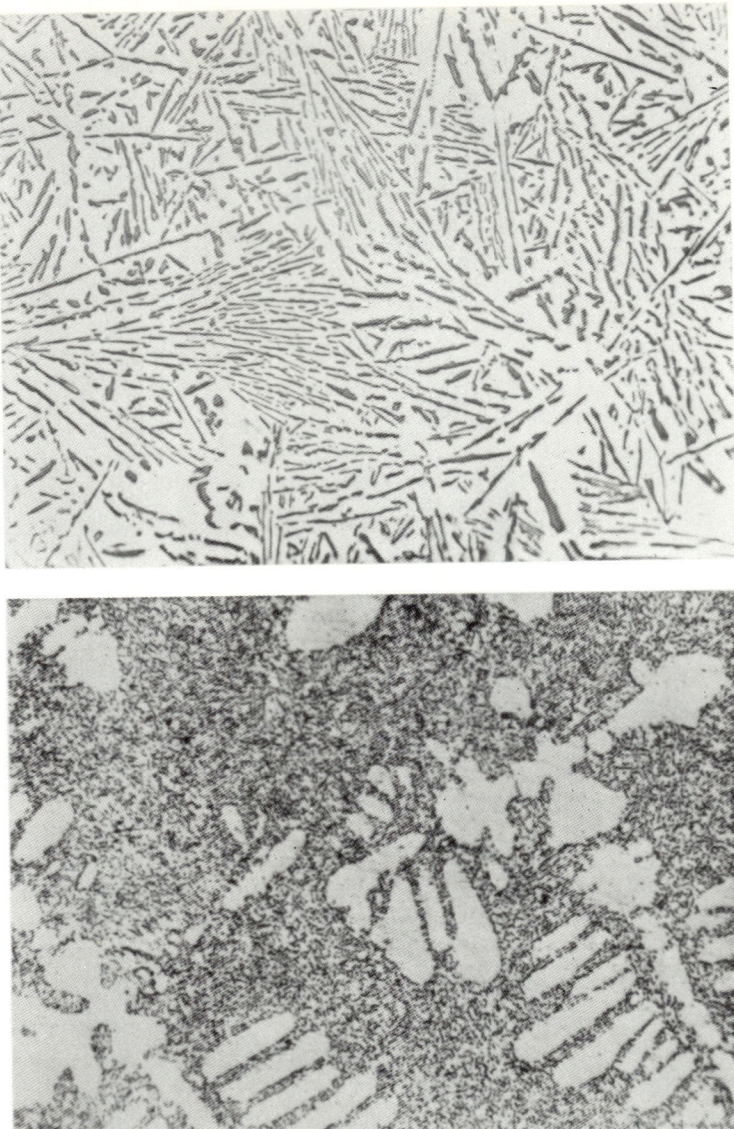

**FIGURE 6-5**
Aluminum-silicon eutectic. *Top:* Silicon present in faceted platelike form; *bottom:* silicon present as rods; primary aluminum dendrites also present. (Magnification ×200.) (*From Day and Hellawell.*[4])

and temperature gradients, the kinetic undercooling required for growth of the silicon phase increases to such an extent that the aluminum phase grows ahead of the silicon, leaving the silicon phase to grow in holes or dimples in the growth front. The resulting diffusion-controlled morphology is now rodlike. Trace elements such as sodium, which also lead to the plate-rod transition, could do so by poisoning the growth steps, thus reducing the effectiveness of the twin plane and causing the silicon phase to lag the aluminum.[4,5]

## CAST IRONS

The commercially most important metallic eutectic is that at about 4 percent carbon in the iron–carbon system. Gray, white, and ductile cast irons are all close to this composition. A typical gray cast iron is in the range of 2 to 4 percent carbon, with 1 percent or more silicon and a small amount of manganese. The solidification and structures of cast irons can be understood qualitatively in terms of the simple iron–carbon binary diagram, assuming complete diffusion of carbon in the austenite during and after solidification.

As an example,[6] an Fe–3% C alloy can solidify according to the equilibrium iron–graphite system (solid lines, Fig. 6-6) or the metastable equilibrium iron–cementite system (dashed lines). At the eutectic temperature, either the equilibrium Fe–G eutectic or the metastable equilibrium Fe–$Fe_3C$ eutectic (ledeburite) forms, depending on the temperature of formation. If eutectic solidification takes place much below the metastable eutectic, growth rate of ledeburite is so much more rapid than Fe–G eutectic that this structure prevails. On the other hand, if eutectic solidification occurs above the Fe–$Fe_3C$ eutectic line (and, of course, below the Fe–G eutectic), ledeburite cannot form and the structure solidifying is Fe–G. A number of factors influence the temperature where eutectic solidification occurs. One of these is *cooling rate*; slower cooling rates favoring solidification according to the equilibrium Fe–G system.

Figure 6-7 shows schematically the structure formation in the two types of cast irons. On the left, the cooling rate is sufficiently rapid so that solidification (and subsequent cooling) is according to the Fe–$Fe_3C$ system. Assuming equilibrium pertains otherwise, austenite dendrites form above the eutectic temperature. At the metastable eutectic, ledeburite forms. On subsequent cooling to the eutectoid temperature, some additional carbides form as the composition of the austenite follows the metastable solubility line $n$–0. Below the metastable eutectoid temperature, the austenite (now of eutectoid composition) is unstable and transforms to pearlite on cooling. The resulting structure is that of *white cast iron*. It is brittle but with excellent wear resistance and finds use in applications such as rolls for paper manufacture.

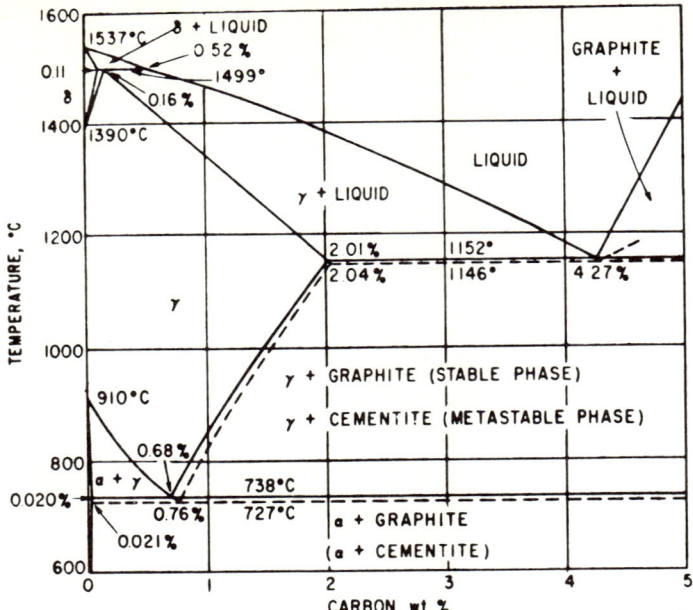

**FIGURE 6-6**
Iron–carbon phase equilibrium diagram. Dashed lines represent phase boundaries for metastable equilibrium with cementite. (*From Ref.* 17.)

This structure is also the starting material for producing *malleable* iron, discussed in Chap. 10. In practice, strong carbide-formers are added to cast iron to promote formation of white iron when this material is the desired end product. To produce the white structure for making malleable iron, carbide-formers can be used only in moderation or when the times required for annealing become intolerably long.

The same 3% C alloy, when cooled more slowly, solidifies according to the stable iron–carbon diagram, although transformation at the eutectoid temperature may still be according to the metastable diagram. The result, sketched in Fig. 6-7f, is *pearlitic-gray cast iron*, a relatively hard, strong form of cast iron widely used in many engineering applications, with graphite "flakes" in the structure which grew cooperatively with austenite at just below the iron–graphite eutectic temperature. If this same cast iron is cooled still more slowly, both solidification and subsequent cooling proceed completely according to the equilibrium phase diagram. At the eutectoid transformation, instead of forming $Fe_3C$, the carbon diffuses to the nearest graphite flake and precipitates there as additional graphite. The resulting structure is graphite

CAST IRONS    185

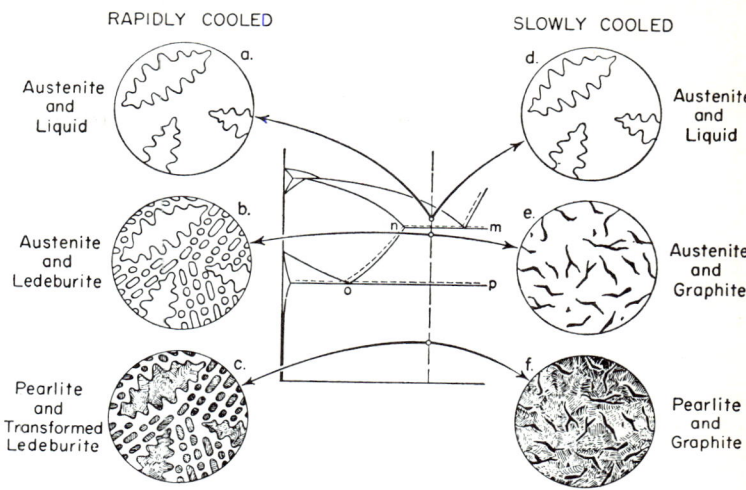

**FIGURE 6-7**
Solidification of a hypoeutectic cast iron; (*a*), (*b*), and (*c*) show formation of a *white cast iron*; (*d*), (*e*), and (*f*) show formation of a *pearlitic-gray cast iron*. (*From Taylor, Flemings, and Wulff.*[6])

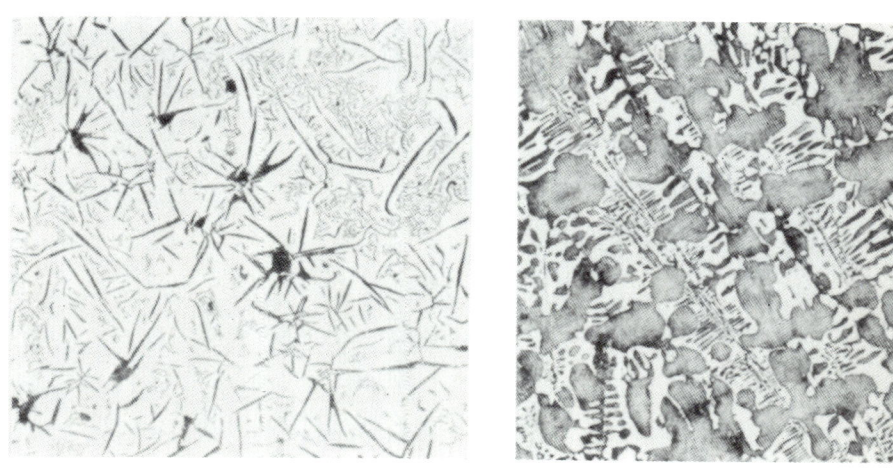

FERRITIC GRAY
CAST IRON, 50X.

WHITE CAST IRON, 100X.

**FIGURE 6-8**
Photomicrographs of two types of cast iron. (*From Ref. 40.*)

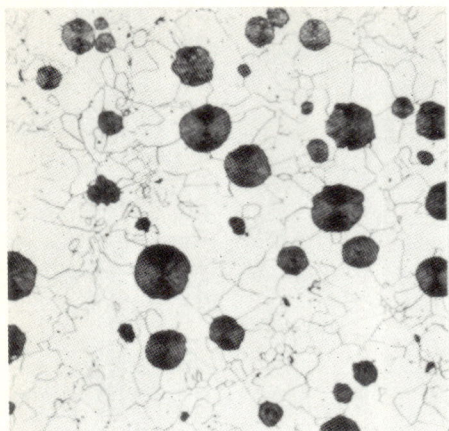
FERRITIC DUCTILE IRON, 100X.

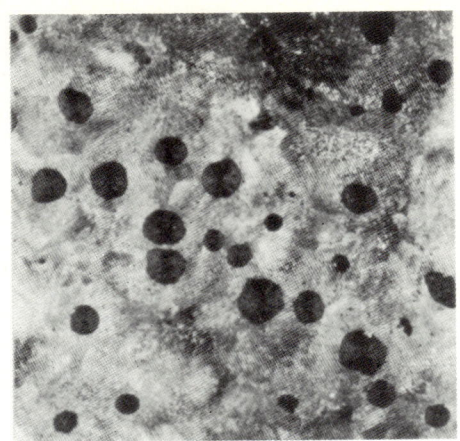
PEARLITIC DUCTILE IRON, 100X.

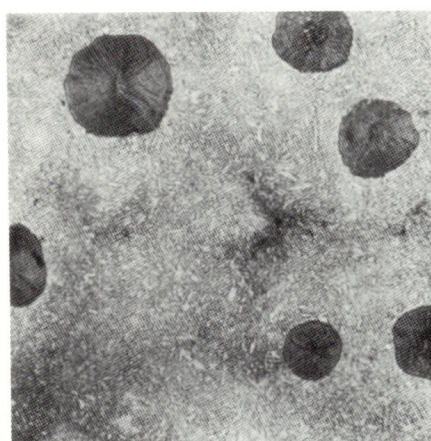
QUENCHED AND TEMPERED DUCTILE IRON, 200X. (TEMPERED MARTENSITE MATRIX).

FIGURE 6-9
Structure of three types of ductile cast iron. (*From Ref.* 40.)

flakes embedded in a very low-carbon iron (ferrite) and is termed *ferritic-gray cast iron*. It is a soft, highly machinable iron. The cast iron may be used in this condition, or, after machining, it may be heat-treated to form pearlitic-gray iron. Figure 6-8 shows microstructures of two of the types of cast iron discussed above.

Addition of a small amount of magnesium to a relatively pure Fe–3% C alloy results in a form of cast iron that can be made to solidify according to the Fe–G equilibrium diagram as does gray cast iron. However, there is one important difference in this new type of iron, termed *ductile iron*: instead of solidifying as plates from

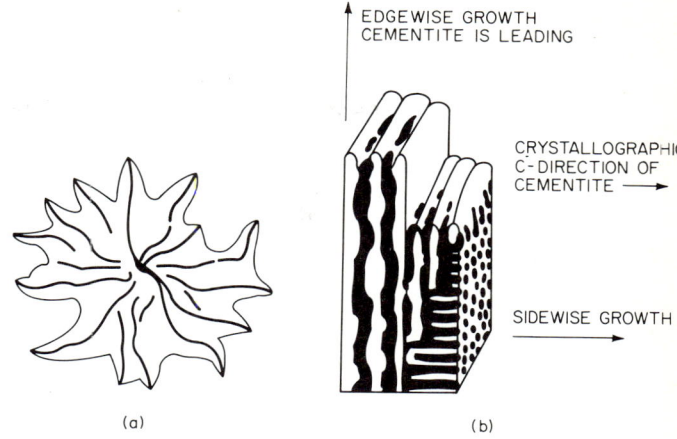

**FIGURE 6-10**
Schematic illustration of growth of eutectics in the iron–carbon system. (*a*) Grain of Fe–G eutectic; (*b*) Fe–Fe$_3$C eutectic. (*From Hillert and Subba Rao.*[10])

the melt, the iron forms as nearly perfect spheres with their basal planes oriented perpendicular to the radii of the nodules. Figure 6-9 shows photomicrographs of the material. The development of this alloy was first announced in 1948, and it has now become an engineering material of major importance. It combines many of the engineering advantages of steel with the processing economies of cast iron.

## EUTECTIC SOLIDIFICATION IN CAST IRONS

The silicon and other elements added to cast iron result in a eutectic which solidifies over a range of temperatures. Eutectic *grains*, or *colonies*, form during solidification, as sketched in Figs. 6-4c and 6-10a. Within each of the grains, graphite forms and grows; and, as it does so, it branches frequently and changes its crystallographic orientation, producing a fully interconnected platelike structure with the plane of the plate parallel to the basal plane of the hexagonal crystal. Possible mechanisms which could produce this branching with orientation change are twinning and autonucleation.

The fineness of eutectic structures in cast irons is determined by local interface velocity as in other eutectics. In eutectics of multicomponent alloys, local interface velocity is not determined by the rate of heat extraction alone. It is an inverse function of the number of grains from which solidification is proceeding in a local region. Casting practices or alloy elements which promote nucleation of new eutectic colonies result in coarser graphite spacing as well. Silicon is perhaps the most effective nucleating agent for graphite and is often added to cast iron to promote formation of a graphitic structure in preference to that of white cast iron. Actually, it accomplishes this task both by nucleating the graphite and by widening the thermal gap between the transformation temperatures of the stable and metastable diagram. As an example, there is a 7°C temperature difference between the stable and metastable eutectic temperature in the binary iron–carbon system. Addition of 2 percent silicon increases this difference to 35°C. This increased temperature difference provides a much greater thermal region over which the stable eutectic grows in preference to the metastable eutectic.[7]

Magnesium as well as other elements alter the solidification of graphite in Fe–C melts such that the graphite forms as spherulites, not plates. However, in spite of decades of study, there is no firm agreement on how this is accomplished. It is known that in the outer part of graphite spherulites, the graphite basal plane is oriented perpendicular to the sphere radius although this is perhaps not true at the sphere center. Probably, the magnesium alters the growth kinetics in some way, as by poisoning preferred growth sites. In any case, the result is graphite of essentially spherical external form which grows from a central nucleus and is highly complex internally.[8]

It is known that graphite spherulites form directly from the liquid in hypereutectic irons. In alloys of hypoeutectic composition, it is most generally believed that the spherulites form in the liquid, are thereafter quickly surrounded by austenite, and subsequently grow by diffusion of carbon through the austenite. However, it may also be true that some grow directly from the melt, even in hypoeutectic irons, which could be accomplished if the growing graphite is "pushed" ahead of the austenite as both phases grow.

When solidification is such that the Fe–Fe$_3$C eutectic forms instead of the Fe–G eutectic, the structure which grows is as shown in Fig. 6-10b. Growth proceeds preferentially by edgewise growth of cementite plates, with slight orientation changes during growth permitting the plates to spread in a fanlike fashion. Spaces between these fanlike cementite plates fill in with a rodlike structure so that the final structure possesses two different eutectic morphologies.[9,10]

## SOLUTE REDISTRIBUTION IN TERNARY ALLOYS

Solute redistribution in castings and ingots of ternary alloys is treated in the same general way as described in Chap. 5 for binary alloys.[11,12] For many alloys, it is a good approximation to assume negligible diffusion of solute in the solid. This assumption plus others used in deriving the Scheil equation [Eq. (5-1)] permit simple quantitative treatment of the problem. The result is analogous to the case of two-component alloys and shows that the interdendritic liquid is always enriched in solute during solidification until it reaches a minimum in the liquidus surface. As an example, in any alloy of the simple ternary system of Fig. 4-25, the interdendritic liquid is enriched during solidification until it reaches the ternary eutectic point. In the general case, a single phase forms first, and as this phase grows the liquid composition and temperature change, defining a *solidification path* on the liquidus surface. When this path strikes a line of twofold saturation, a second phase begins to form and the solidification path then follows the line of the ternary eutectic.

For the initial stage of solidification when only a single phase $\alpha$ is forming, the three-component analog of the Scheil equation is simply

$$\frac{df_L}{dC_{Lm}} = -\left(\frac{1}{1 - k_{\alpha m}}\right)\frac{f_L}{C_{Lm}} \qquad (6\text{-}1)$$

where $C_{Lm}$ is liquid composition of alloy element $m$, and $k_{\alpha m}$ is equilibrium-partition ratio between $\alpha$ and liquid with respect to alloy element $m$. An identical equation can be written for the other alloy element, element $n$, relating $f_L$, $C_{Ln}$, and $k_{\alpha n}$. The only practical limitation in using Eq. (6-1) is that it is necessary to know tie lines of the ternary system in order to calculate $k_{\alpha m}$ and $k_{\alpha n}$, and these are not known for many systems. Combining Eq. (6-1) with its equivalent for alloy $C_{Ln}$ yields

$$\frac{dC_{Lm}}{dC_{Ln}} = \frac{(1 - k_{\alpha m})}{(1 - k_{\alpha n})}\frac{(C_{Lm})}{(C_{Ln})} \qquad (6\text{-}2)$$

and this equation defines the solidification path along the liquidus surface to the line of twofold saturation.

When the line of twofold saturation is reached, an extra term must be added to Eq. (6-1):

$$\frac{df_L}{dC_{Lm}} = -\left(\frac{1}{1 - k_{\alpha m}}\right)\frac{f_L}{C_{Lm}} - \left(\frac{k_{\beta m} - k_{\alpha m}}{1 - k_{\alpha m}}\right)\frac{df_\beta}{dC_{Lm}} \qquad (6\text{-}3)$$

where $f_\beta$ is the weight fraction of second phase $\beta$, and $k_{\beta m}$ is equilibrium partition ratio between $\beta$ and liquid with respect to element $m$. A similar equation is written for element $n$. This equation is readily derived following the procedure used to develop Eq. (2-3) but with two phases $\alpha$ and $\beta$ forming. The derivation is left to the reader as a problem at the end of this chapter.

As an example of the application of the foregoing, we shall consider the aluminum-rich corner of the Al–Cu–Ni ternary, Fig. 6-11. An alloy of composition $C_0$, as shown in this figure, solidifies first with formation of primary $\alpha$ aluminum. The com-

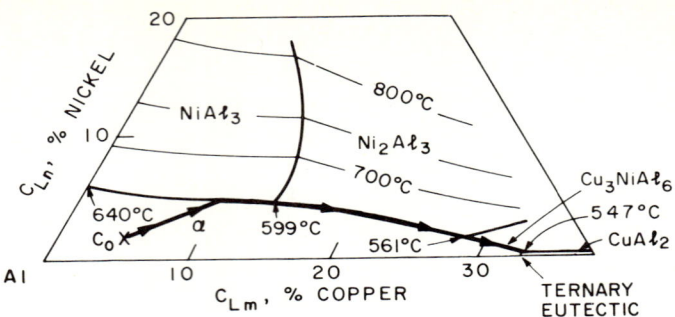

**FIGURE 6-11**
Liquidus surfaces of aluminum-rich corner of Al–Cu–Ni phase diagram. Arrows show solidification "path" of Al–4.5% Cu–2% Ni alloy. (*From Flemings and Mehrabian.*[11])

position of interdendritic liquid then moves (as solidification proceeds) approximately radially away from the aluminum-rich corner, following a solidification path defined by Eq. (6-2). The composition of the solid forming at a given liquid composition is

$$C_{\alpha m} = k_{\alpha m} C_{Lm} \quad (6\text{-}4a)$$

$$C_{\alpha n} = k_{\alpha n} C_{Ln} \quad (6\text{-}4b)$$

Since there is no diffusion in the solid, Eqs. (6-2) and (6-4) describe the final microsegregation (coring) in the solid phase. Calculated results are shown in Fig. 6-12 for copper in the $\alpha$ phase. Only a negligible amount of nickel dissolves in this phase ($k_{\alpha n} \simeq 0$).

When the solidification path strikes the line of twofold saturation, the binary eutectic $\alpha + \mathrm{NiAl}_3$ forms and grows as the liquid composition moves along the line of twofold saturation to the first peritectic temperature 599°C. At this temperature, just as in the analogous binary case, one phase ($\mathrm{NiAl}_3$) stops forming and in its place a new phase ($\mathrm{Ni}_2\mathrm{Al}_3$) forms. Thus, the binary eutectic $\alpha + \mathrm{Ni}_2\mathrm{Al}_3$ forms between 599 and 561°C, and the binary eutectic $\alpha + \mathrm{Cu}_3\mathrm{NiAl}_6$ forms between 561°C and the ternary eutectic temperature 547°C. At 547°C, the remaining liquid solidifies as ternary eutectic. The amounts of the different phases forming are calculated from Eq. (6-3) and are shown in the insert of Fig. 6-12 for the alloy used as example. The final microstructure of this alloy contains five phases: $\alpha$, $\mathrm{NiAl}_3$, $\mathrm{Ni}_2\mathrm{Al}_3$, $\mathrm{Cu}_3\mathrm{NiAl}_6$, and $\mathrm{CuAl}_2$. The $\alpha$ phase comprises approximately 0.8 of the final structure; the bulk of this (approximately 0.6 of the final structure) is primary $\alpha$. The remainder solidifies during formation of the eutectic.

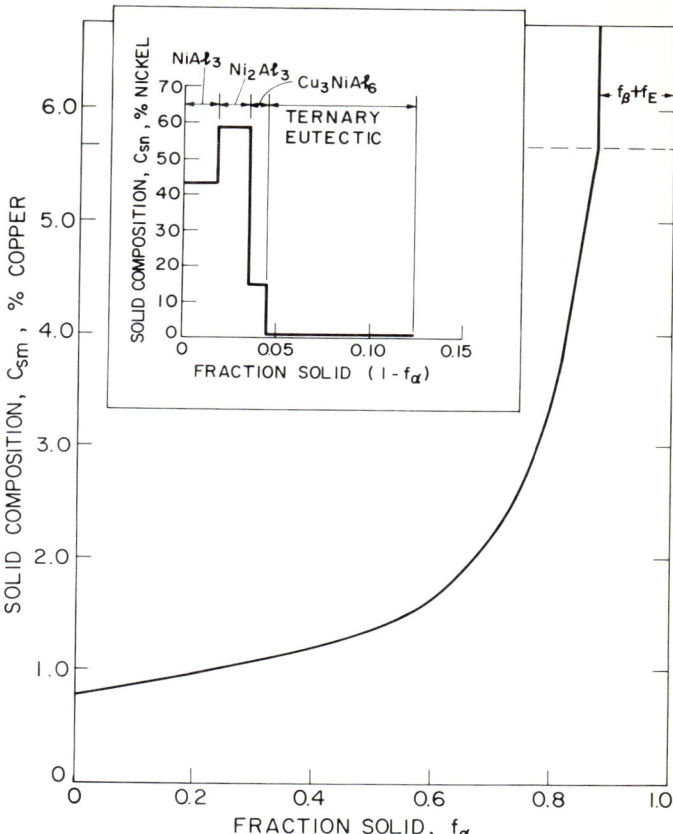

**FIGURE 6-12**
Microsegregation in Al–4.5% Cu–2% Ni alloy. $f_\alpha$ is primary $\alpha$ and that forming as part of the binary eutectics; $f_\beta$ is $NiAl_3$, $Ni_2Al_3$, and $Cu_3NiAl_6$; $f_E$ is ternary eutectic. (*From Flemings and Mehrabian.*[11])

## MONOTECTIC SOLIDIFICATION

Monotectics, like eutectics, can be solidified with aligned rodlike structures provided solidification is conducted with high enough temperature gradients and low enough growth rates, as discussed in Chap. 4. In dendritic solidification of castings and ingots, similar rodlike monotectics can also form in interdendritic spaces. Forward and Elliott[13] suggest the rodlike silicates they observed form in this way, and Frederiksson and Hillert[14] propose rodlike manganese sulfides form by this mechanism. An alternate common structure of monotectics is typified by the *divorced* spherical morphology of lead in leaded bronzes and by sulfide inclusions in Fe–O–S alloys. In these

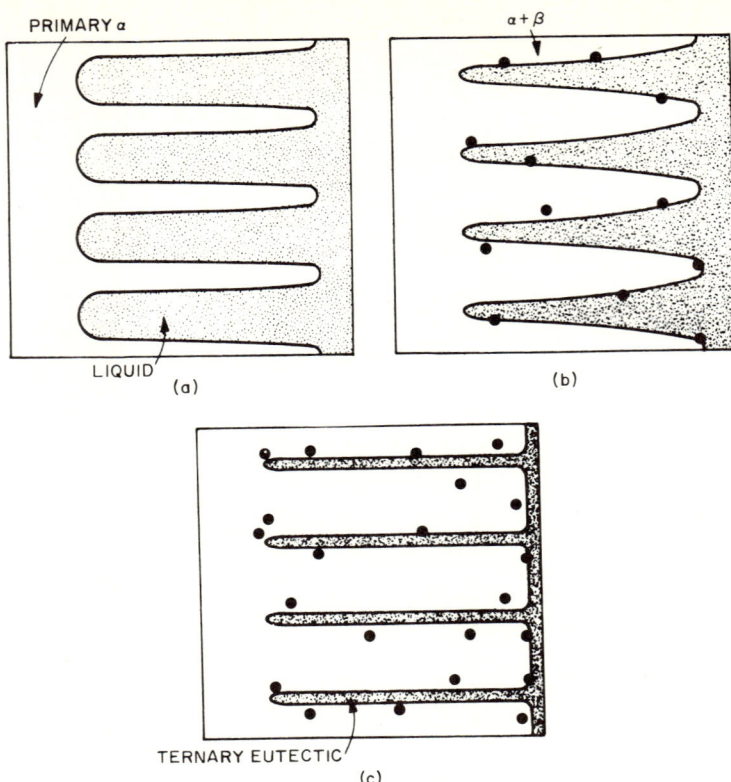

**FIGURE 6-13**
Schematic illustration of a small *volume element* of a ternary alloy during solidification, (*a*) Primary α phase forms; (*b*) divorced pseudobinary eutectic or monotectic forms; and (*c*) ternary eutectic solidifies. (*From Flemings and Mehrabian.*[12])

alloys, the primary phase forms dendritically, as sketched in Fig. 6-13, until the monotectic line of twofold saturation is reached. Thereafter, the monotectic forms as small droplets $L_2$ from the mother liquid $L_1$ while the dendrites continue to grow. At least in the Fe–O–S system, the droplets are not pushed ahead by the advancing solid but are entrapped as the dendrite grows. Each droplet is thereby "isolated" from further interaction with the mother liquid $L_1$ or with other droplets $L_2$. Yarwood et al.[15] have shown that this solidification can be treated quantitatively by equations closely analogous to Eq. (6-3). The liquid $L_2$ is treated as a phase within which diffusion does not occur since no diffusion can occur between the isolated droplets (which is discussed more fully later in this chapter).

# PRIMARY INCLUSIONS

In cast alloys often the first phase to form is not the phase that comprises the bulk of the structure. When a grain-refiner is added to molten metal, the first phase is usually a compound involving the added agent. When a deoxidizer is added, the deoxidation product is often a crystalline or glassy phase which begins to solidify before the major phase. In steel, such deoxidation products are $Al_2O_3$, resulting from aluminum deoxidation, and $SiO_2$, resulting from silica deoxidation. Also common are complex deoxidation products which are combinations of various elements present in the steel; examples are $MnSiO_3$ and $Mn(O,S)$.

Thermodynamics of deoxidation of steel have been intensively studied.[17] One simple example is deoxidation with silicon of iron containing only oxygen. In this case, the reaction is

$$\underline{Si} + 2\underline{O} = SiO_2 \qquad (6\text{-}5)$$

where $\underline{Si}$ and $\underline{O}$ are silicon and oxygen, respectively, dissolved in the melt and $SiO_2$ is the deoxidation product. Activities of the dissolved oxygen and silicon are nearly equal to weight percent at the concentration levels of interest in steel deoxidation, and so

$$K' = [\%\underline{Si}][\%\underline{O}]^2 \qquad (6\text{-}6)$$

This constant is a function of temperature. Solubilities increase with increasing temperature, as shown in Fig. 6-14. The isotherms of Fig. 6-14 define a portion of the liquidus surface of the iron-rich corner of the Fe–O–Si ternary phase diagram. The portion defined is that where $SiO_2$ is the first phase to form. The iron-rich corner of the diagram, as constructed from thermodynamic data by Elliott and coworkers,[13,16] is given in Fig. 6-15. Alloys with higher Si and O contents than the line of twofold saturation solidify with the deoxidation product as the primary phase. Except for alloys of very low silicon content, the reaction product is $SiO_2$. According to Fig. 6-15, the $SiO_2$ would be crystalline at equilibrium, but it usually forms as a liquid and remains as a glass in the final cast structure. The free-energy difference between crystalline and liquid $SiO_2$ is small, and kinetics apparently favor formation of the glassy phase.

The limited evidence available indicates that the formation of primary silica deoxidation products occurs with relatively little supersaturation by a heterogeneous nucleation mechanism. The silica inclusions form typically in the astonishingly large number of about $10^8$ per cubic centimeter in commercial steel. Their initial size is, at most, a few microns; they then grow rapidly in size with time, even when held isothermally. In one study,[18] the largest particles observed were an order of magnitude larger a minute after deoxidation than they were 10 s after deoxidation. The reason for this growth is interesting and is not what one might at first expect. Processes

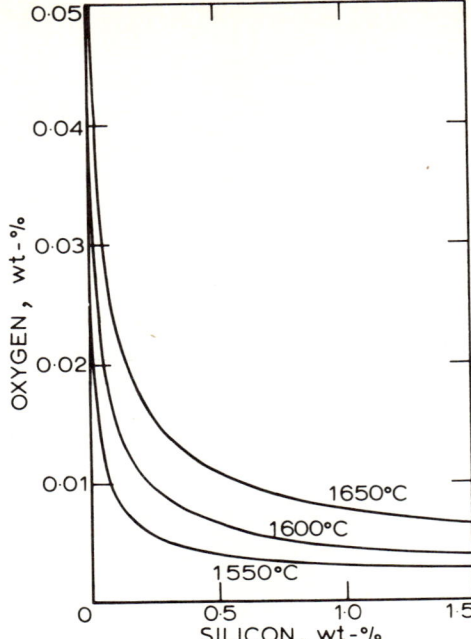

FIGURE 6-14
Silicon and oxygen contents of iron in equilibrium with solid silica. (*From Turkdogan.*[41])

involving solute diffusion through the melt do not play a very large role, but, instead, the particle coarsening comes about primarily from particles colliding and sintering. The collisions occur from two causes: First, according to Stokes' law, the larger particles rise faster than the slower ones and so may catch up to and collide with smaller ones; second, liquid shear resulting from convection also brings about particle collisions. Both these processes are analogous to the collisions that occur between raindrops, resulting in coarsening of the rain. In metallic alloys, collisions occur with high frequency because of their large number. In a melt with $10^8$ inclusions per cubic centimeter, interinclusion spacing is only a few tens of microns, and no inclusion can travel far without striking another.

The simplest kind of collision to describe quantitatively is that resulting from inclusion flotation according to Stokes' law. Following is an outline of the analysis of Gunn[19] as applied to inclusions by Lindborg and Torssell.[18] Figure 6-16 shows two inclusions rising through a quiescent bath. The inclusions are of radii $r_1$ and $r_2$, respectively, and are floating at velocities $u_1$ and $u_2$, respectively, where these velocities $u$ are given by Stokes' law:

$$u = \frac{2}{9} g_r r^2 \frac{\rho_L - \rho_i}{\eta} = k_s r^2 \qquad (6\text{-}7)$$

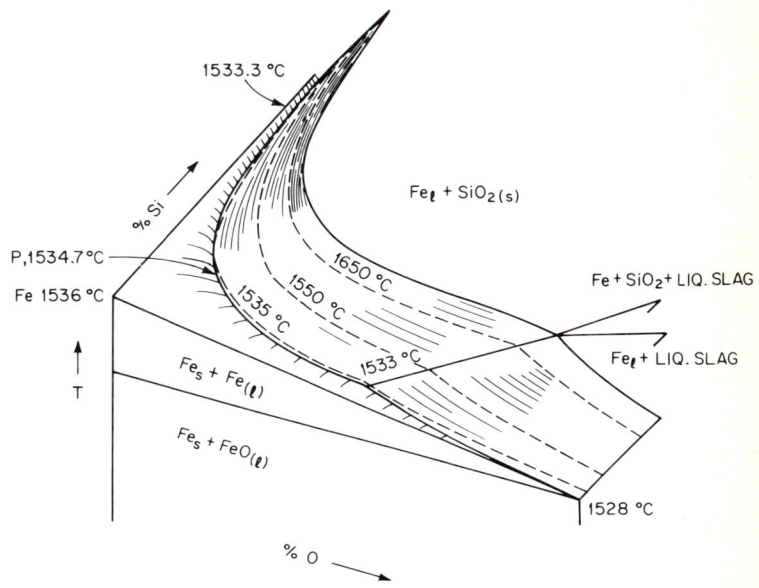

**FIGURE 6-15**
Iron-rich corner of the Fe–Si–O system. Solid lines are lines of twofold saturation; dashed lines are isotherms. Point $P$ is a maximum on the line of twofold saturation. (*From data of Elliott and coworkers.*[13,16])

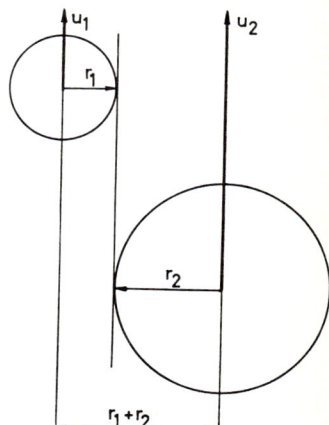

**FIGURE 6-16**
Limiting case for collision between two inclusions. (*From Lindborgh and Torssell.*[18])

where $g_r$ is acceleration due to gravity, $\rho_L$ and $\rho_i$ are densities of the liquid and inclusions, respectively, and $\eta$ is liquid viscosity. Only inclusions that are within a horizontal distance of each other, $r_1 + r_2$, can ultimately collide (Fig. 6-16). The time required for this collision to occur depends on the relative velocities of the two particles and on their vertical separation. For a collision to occur in unit time, the centers of the two inclusions must therefore lie within a certain cylindrical volume $W$, which is

$$W = k_s |r_1 - r_2|(r_1 + r_2)^3 \qquad (6\text{-}8)$$

where the *collision volume* $W$ has the units volume per unit time and $k_s$ is the rate constant for Stokes' flotation, Eq. (6-7).

If, when any two inclusions strike, they sinter rapidly together to form a single spherical inclusion, the size distribution $f(r)$ of inclusions changes according to the relation

$$\frac{df(r)}{dt} = -\int_0^\infty f(r_2) W(r_1, r_2) f(r)\, dr_2 \qquad (6\text{-}9)$$

With the foregoing three equations and knowing the size distribution of inclusions at any one instant, it is possible to calculate numerically the size distribution at any other instant, future or past. As a consequence of collisions, the size distribution is displaced toward larger particle sizes. Calculations show that simple Stokes' collisions can account for very rapid growth in particle size of inclusions in steel; this growth is further enhanced by liquid shear.

For the silica inclusions used as an example in the foregoing discussion, experiment shows that when two particles collide at steelmaking temperatures (approximately 1600°C), they usually sinter, or *coalesce*, together rapidly to form a single larger sphere. The time for such coalescence $t_s$ is given by Frenkel[20] as approximately

$$t_s \simeq \frac{r\eta_i}{\sigma} \qquad (6\text{-}10)$$

where $\eta_i$ is viscosity of the inclusions and $\sigma$ is liquid-inclusion surface energy. At 1600°C, $\eta_i = 5 \times 10^8$ P for silica.[21] Taking $\sigma$ as 1,300 erg/cm$^2$ (Turpin and Elliott[22]) and particle radius as 2 $\mu$m, this gives a sintering time of about 1 min, which is short in comparison with typical holding times of metal in the ladle. At lower temperatures, inclusion viscosity can be sufficiently high so that particles of this size do not completely spheroidize in the time available.

Primary inclusions other than silica may or may not coalesce after they collide and stick. An example is alumina inclusions. These collide and adhere as does silica, but coalescence appears to be negligible.[23] The result is that large interconnected *clusters* form, which may contain hundreds or more of individual inclusions. Similarly, primary oxide inclusions in aluminum and other lower-melting-point alloys usually do not coalesce but probably increase in size by collision and partial sintering during melting.

One aim of a good melting practice is to remove physically as many of the inclusions that form as possible. The fine deoxidation products that first form in steel and most other metals are not large enough to float completely out of the bath. The limiting Stokes' velocity of these particles is often much less than the velocities present in the bath from convection, and even if there were no convection, time for separation by floating would be impractically long. As the particles become larger, the rate of floating increases rapidly and separation of inclusions can be significant. For silica inclusions in steel, floating is rapid when the particle radius has reached about 20 $\mu$m. Significant reduction in inclusion content is then obtained by the inclusions floating upward and adhering to or dissolving in the slag at the melt surface. In aluminum and magnesium alloys, it is common practice to stir the metal in the presence of a slag (i.e., *flux*), thereby increasing collisions between the oxides themselves and between the oxides and the flux. Still another way of removing these inclusions is by filtering them from the metal. In the production of quality sand castings of aluminum and magnesium, screens (usually steel) are placed in the gating system to filter out the larger inclusions. In the production of quality ingot and continuous castings of aluminum, filtering is also employed. One method is to pass the metal through a bed of particulate refractory,[24,25] and another is to pass it through a porous refractory filter.

The morphology of primary inclusions in cast metals varies greatly, depending on interface kinetics, relative surface energies, collisions, and inclusion viscosity. Primary silica inclusions are usually spherical, as shown in Fig. 6-17a. An exception is where collisions have caused several inclusions to meet but where coalescence has been incomplete because the collisions occurred at relatively low temperatures just before or during solidification of the iron phase. In the latter case, inclusions partially sinter together to become members of *multimembered* inclusions. An example is shown in Fig. 6-17b. Primary alumina inclusions are generally more angular in appearance than the silica and form similar multimembered inclusions, with little sintering. Primary inclusions can often have also a dendritic or *cellular* morphology when their growth is limited by diffusion from the melt. Such structures are found especially at rapid cooling rates. Glassy silica inclusions form branched inclusions without crystallographic orientation.[11,26] Other common inclusions in steel, such as alumina and manganese sulfide, form crystallographically oriented dendrites.[27,28] Depending on interface kinetics, these may be either rounded or sharply faceted dendrites (Fig. 6-18). Primary inclusions in metals other than steel adopt a similar

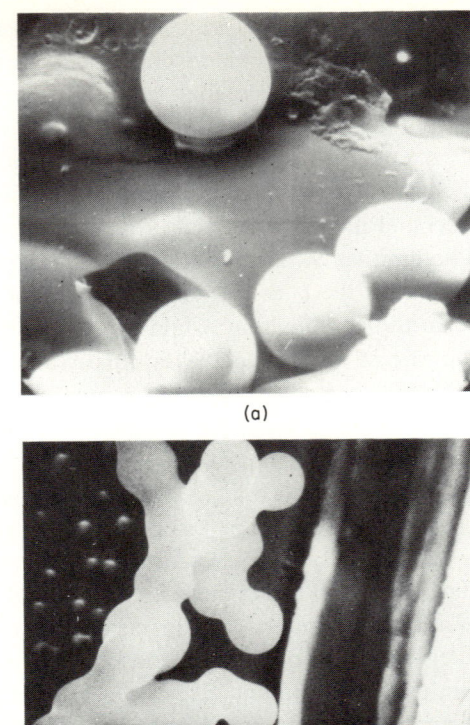

FIGURE 6-17
Single and multimembered silica inclusions in an iron base alloy. Scanning electron photomicrographs of deeply etched surface. (Magnification ×5,000.) (*From Myers and Flemings.*[26])

wide range of morphologies, depending on the same factors discussed above for steel. Some important ones are the faceted needlelike $FeAl_3$ inclusions in aluminum and the sheetlike $Al_2O_3$ "skins" in aluminum. This latter type probably results from incorporation into the melt of oxide layer from the original change in melt surface.

It was once assumed that inclusions which were observed in interdendritic spaces must have formed late in solidification while those which were observed within the dendrites formed as primary inclusions, or early during solidification of the major phase. Now, it is recognized that inclusions can be "pushed" by the thickening dendrites, and thus primary inclusions may appear predominantly in interdendritic spaces. We have seen in Chap. 4 that small included particles can be rejected by the liquid/solid interface in plane front solidification provided the rate of solidification is sufficiently low. The rate of thickening of dendrite arms is typically very much lower

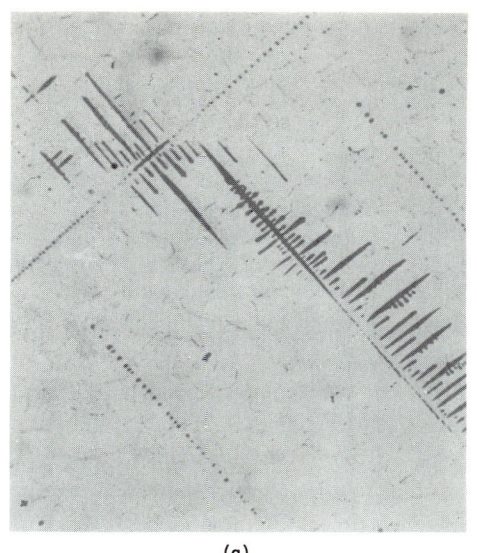
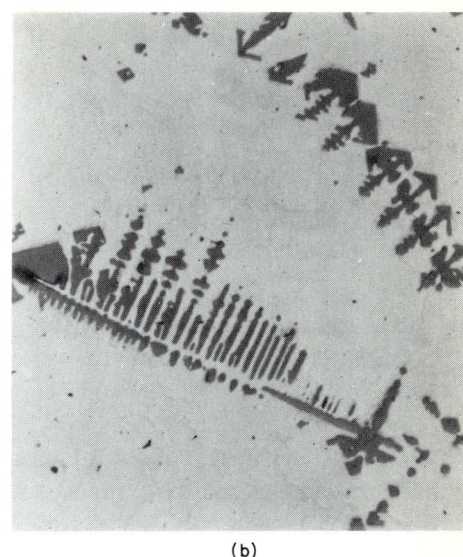

(a)  (b)

FIGURE 6-18
Dendritic primary manganese sulfide inclusions in iron (magnification ×100). (a) Nonfaceted inclusions in low-carbon iron; (b) faceted inclusions in high-carbon iron. (*From Bigelow.*[28])

than the threshold velocities observed in these experiments, and so it is not surprising that inclusions are sometimes pushed.

One type of primary inclusion that is known to be pushed by growing dendrites is silica in steel. In a study on these inclusions in an iron–copper alloy, Myers[26] has shown that, in addition to being pushed, the inclusions collide with one another and build up *multimembered* inclusions during solidification just as they do before solidification begins. The inclusions move through the interdendritic spaces owing to several causes, one being the dendrite pushing, and another, flotation. Still another cause is fluid flow within the interdendritic spaces, discussed in Chap. 7. The multimembered inclusion shown in Fig. 6-17 is an example of one that was built up primarily by collisions that occurred after formation of the primary metallic phase where coalescence was not complete.

A new process that provides an interesting example of inclusion pushing is that of injection of graphite powders in aluminum alloys. Dispersal of the powder in the metal is made possible by first nickel-coating it and then stirring the metal to prevent the powder from floating out after it is injected. The cast alloy then has finely dispersed graphite in a matrix of the aluminum alloy. It is observed that although a nearly uniform dispersion can be obtained, the graphite particles tend to be rejected by the growing aluminum dendrites.[29]

## SECONDARY INCLUSIONS

*Secondary inclusions* are those inclusions which form during or after solidification of the major phase. These result because alloy or impurity elements are usually rejected to the interdendritic spaces during solidification. As example, consider an Fe–O–Si alloy with Si/O ratio greater than that of the maximum in the eutectic valley, at 1534.7°C (Fig. 6-15). After some solidification of iron has occurred, the liquid melt reaches the line of twofold saturation. Treating this problem as that of the ternary alloys discussed earlier, solidification now occurs with simultaneous precipitation of both $SiO_2$ and Fe, and the composition of the interdendritic liquid moves along the line of twofold saturation toward the silicon-rich corner. Typically the inclusions are pushed in front of the thickening dendrite as are the primary inclusions discussed earlier. Alloys with Si/O ratio less than that of point $P$ solidify similarly, except that the composition of the interdendritic liquid moves along the line of twofold saturation toward the oxygen-rich corner, with formation of iron silicates. This discussion of solidification of Fe–O–Si alloys assumes no significant barrier to nucleation and/or growth of $SiO_2$ phase during the growth of primary dendrites. Most of the experimental results which have been obtained on Fe–O–Si alloys can be rationalized on this basis provided the inclusions are assumed to be pushed to interdendritic spaces. If the inclusions are assumed not to be pushed, then experimental results can be explained only if large supersaturation is required to nucleate or grow the $SiO_2$ phase.[11]

Another type of inclusion which forms as liquid droplets and which was mentioned earlier is oxy-sulfide inclusions in Fe–O–S alloys. These inclusions are trapped by the dendrites, as shown schematically in Fig. 6-13. The reaction is known to require little supersaturation of oxygen and silicon. Schematically, the process can be visualized with the aid of the liquidus surface (Fig. 6-19). An iron-rich alloy of O/S ratio equal to 0.5 is solidified until the line of twofold saturation is reached. Then, liquid inclusions of composition $M$ form and are entrapped by the iron dendrites. Next the interdendritic liquid moves a bit along the line of twofold saturation toward the point $P$ and inclusions form (and are entrapped) that are slightly richer in sulfur than those that first formed. Eventually, a series of liquid droplet inclusions form and are entrapped, ranging in O/S ratio from that of point $M$ to that of point $P$. Each individual inclusion solidifies as a separate *isolated* casting, following the arrows shown. As an example, liquid composition of the inclusion at $M$ moves from the monotectic line of twofold saturation to the eutectic line (as a very small amount of

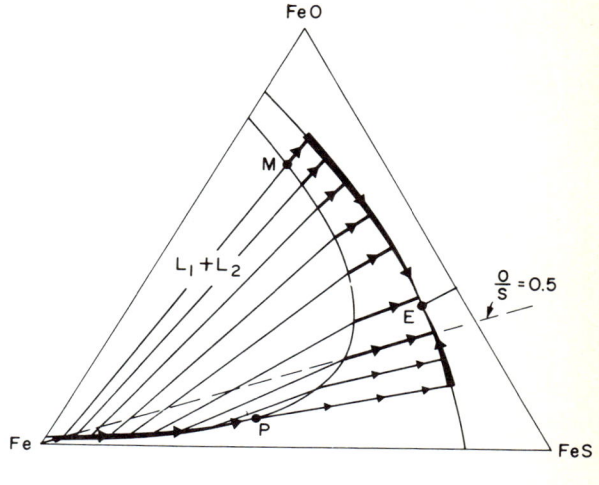

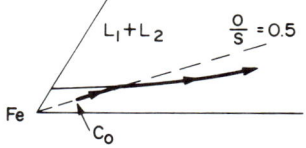

**FIGURE 6-19**
Schematic illustration of the liquidus surface of Fe–FeO–FeS system showing solidification path of an iron-rich alloy of composition $C_0$. Point $P$ is a minimum on the line of twofold saturation. Insert is enlargement of Fe-rich corner. (*From Yarwood et al.*[15])

Fe precipitates). Subsequently, it moves down this line of twofold saturation by precipitating FeO and again a very small amount of Fe. Finally, a three-phase eutectic forms that is primarily FeO and FeS.[13] It is interesting that when each small spherical casting solidifies in this way, it retains its external spherical shape (since very little Fe precipitates in the reaction), except that a shrinkage cavity forms because of the high solidification contraction of the oxides; this is shown in Fig. 6-20.

Just as in the case of primary inclusions, secondary inclusions have a variety of different morphologies, depending on interface kinetics and other solidification variables. Many solidify with a rod- or plate-type eutectic. The nonfaceted Type II sulfides in steel are generally thought to form by eutectic reaction, although Fredericksson and Hillert[14] have recently suggested it may form by monotectic reaction; Fig. 6-21a shows the structure.

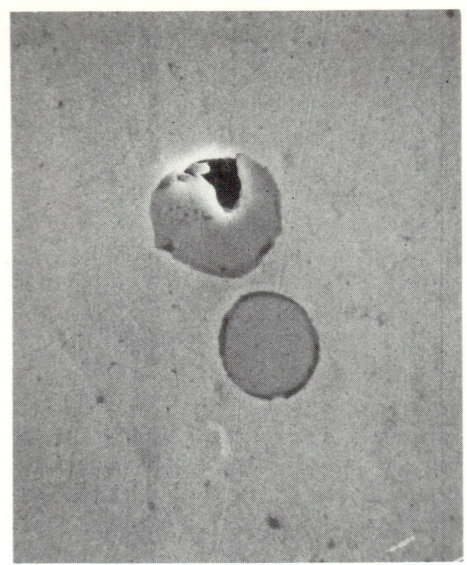

FIGURE 6-20
Shrinkage cavity in oxygen-rich inclusion in iron. Scanning electron photomicrograph. (Magnification ×2,600.) (*From Brower.*[42])

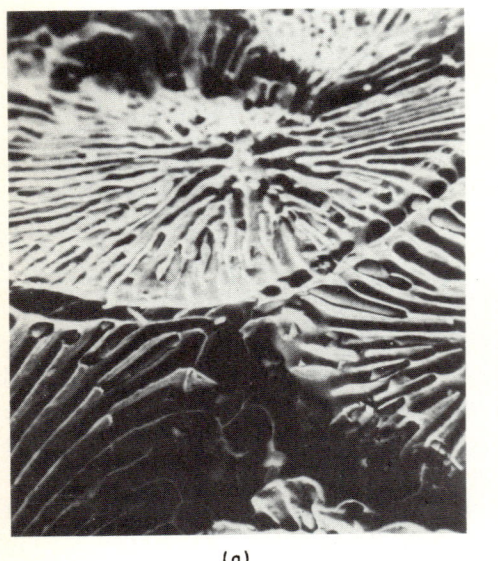

(a)

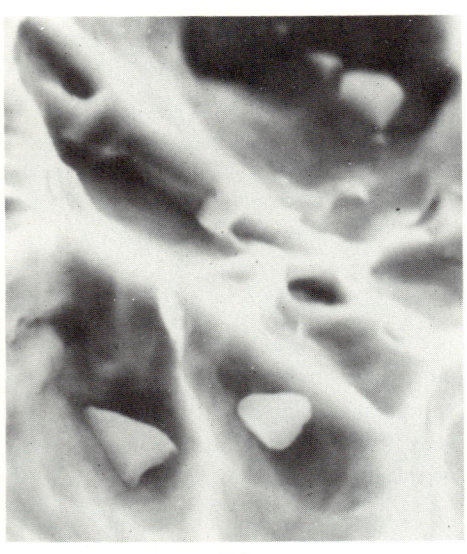

(b)

FIGURE 6-21
Two different forms of MnS secondary inclusions. (*a*) Eutectic, Type II (magnification ×700); (*b*) divorced eutectic, Type III (magnification ×500). (*From Bigelow.*[28])

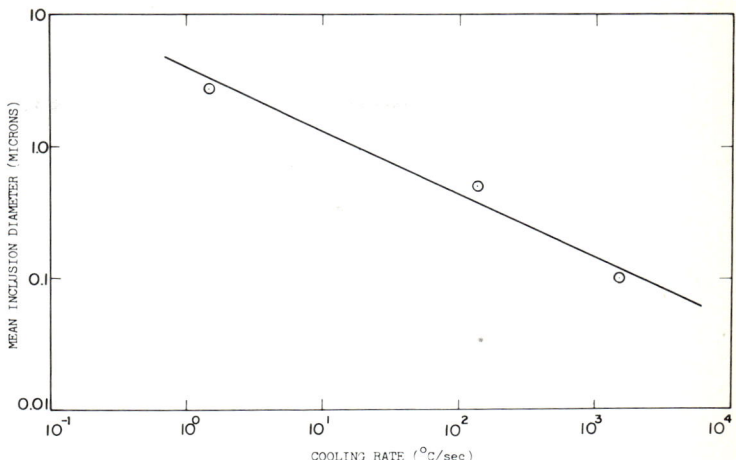

**FIGURE 6-22**
Mean measured $SiO_2$ inclusion diameter versus cooling rate for an Fe–O–Si alloy. (*From Brower.*[42])

Secondary inclusions are usually small compared with dendrite arm spacing, i.e., in the range of 0.1 to 5 $\mu$m for typical ferrous castings and ingots. Figure 6-22 shows typical data for silica inclusions in steel. As discussed in Chap. 10, rapid cooling substantially improves the properties of many alloys. One reason for this is thought to be the very fine inclusion size obtained at high cooling rates.

Faceted eutectics include the *anchor-type* sulfides in cast iron and *Chinese script* inclusions in aluminum alloys. Manganese sulfide in cast steel forms as a faceted divorced eutectic known as Type III inclusions when composition is such that interface kinetics controls sulfide growth. This type of inclusion, shown in Fig. 6-21b, is much less damaging to mechanical properties than Type II. Commercial melting practices are usually designed to avoid formation of Type II inclusions.[30] Empirically, it is known that increasing elements such as silicon, carbon, and aluminum promotes formation of the faceted, divorced eutectic rather than the unfaceted rodlike structure.

## DISSOLVED GAS

A troublesome class of impurities in cast metals is that which precipitates, alone or in combination with other elements, to form a gaseous phase. The resulting gas porosity may be a simple diatomic gas, such as $H_2$, or it may be more complex, such as CO,

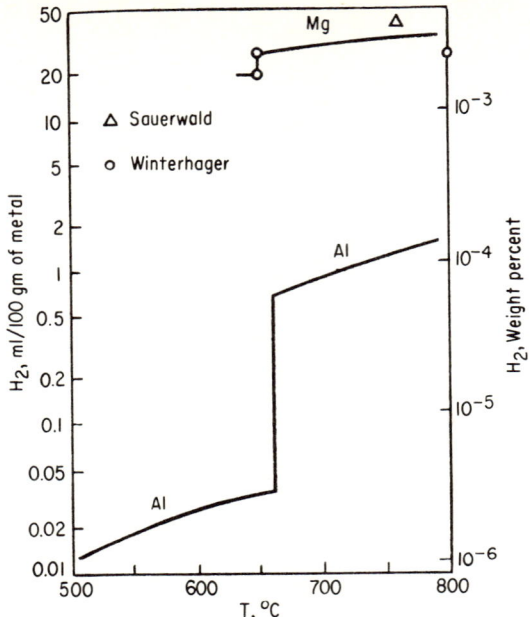

FIGURE 6-23
Solubility of hydrogen at atmospheric pressure in aluminum and magnesium. (*From Eastwood.*[43])

$H_2O$, or $SO_2$.[31,32] Hydrogen is the most common of the simple gas-forming elements. It dissolves exothermically in most metals (except titanium, zirconium, and columbium). Solubility in these metals decreases with decreasing temperature, and there is an abrupt reduction in solubility when solidification occurs, as shown in Fig. 6-23. On the vertical axis this figure shows both weight percent dissolved hydrogen and the equivalent volume of gas at standard temperature and pressure per 100 g of metal.

These latter units provide a physical description of the approximate volume of voids that can result from gas solution and subsequent precipitation. The gas dissolves atomically in both the liquid and the solid according to the reaction

$$2\underline{H} \leftrightarrows H_2 \quad (6\text{-}11)$$

where $\underline{H}$ indicates hydrogen dissolved in the liquid metal. Taking the activity of hydrogen in the metal equal to weight percent, the hydrogen dissolved in the metal at equilibrium with a partial pressure of hydrogen $P_{H_2}$ is given by

$$\%\underline{H}_L = K'_L \sqrt{P_{H_2}} \quad (6\text{-}12a)$$

$$\%\underline{H}_S = K'_S \sqrt{P_{H_2}} \quad (6\text{-}12b)$$

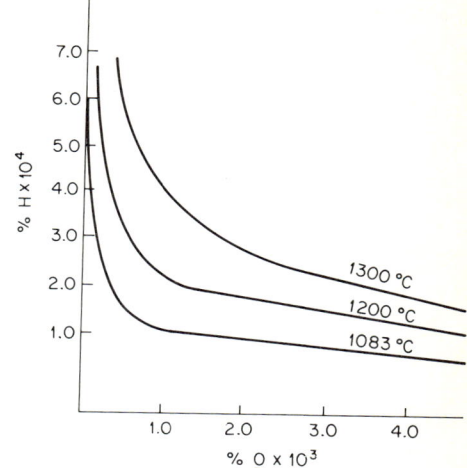

FIGURE 6-24
Equilibrium solubility of hydrogen and oxygen in copper at 1 atm pressure of water vapor.

where $K'_L$ and $K'_S$ are constants and $\%\underline{H}_L$ and $\%\underline{H}_S$ are weight percent of hydrogen dissolved in the liquid and solid, respectively, at a given temperature. The amount of dissolved hydrogen is proportional to the square root of the partial pressure of hydrogen over the melt.

In some metals, notably aluminum and magnesium, hydrogen is the only gas that dissolves to a significant extent; in others, more complex gases can form. In iron- or copper-base alloys, dissolved oxygen may be present in addition to hydrogen, and these gases can react to form water vapor according to the reaction

$$H_2O \rightleftarrows \underline{O} + 2\underline{H} \quad (6\text{-}13)$$

The appropriate equilibrium constant, written here for the liquid, is now

$$K'_L = \frac{[\%\underline{O}][\%\underline{H}]^2}{P_{H_2O}} \quad (6\text{-}14)$$

The higher the dissolved oxygen content, the lower the dissolved hydrogen content and vice versa; Fig. 6-24 shows this for pure copper.

Another complex gas of great industrial importance is CO in steel

$$CO \rightleftarrows \underline{C} + \underline{O} \quad (6\text{-}15)$$

where $\underline{C}$ and $\underline{O}$ are dissolved carbon and oxygen contents in the steel. The mass action constant here is

$$K'_L = \frac{[\%\underline{C}][\%\underline{O}]}{P_{CO}} \quad (6\text{-}16)$$

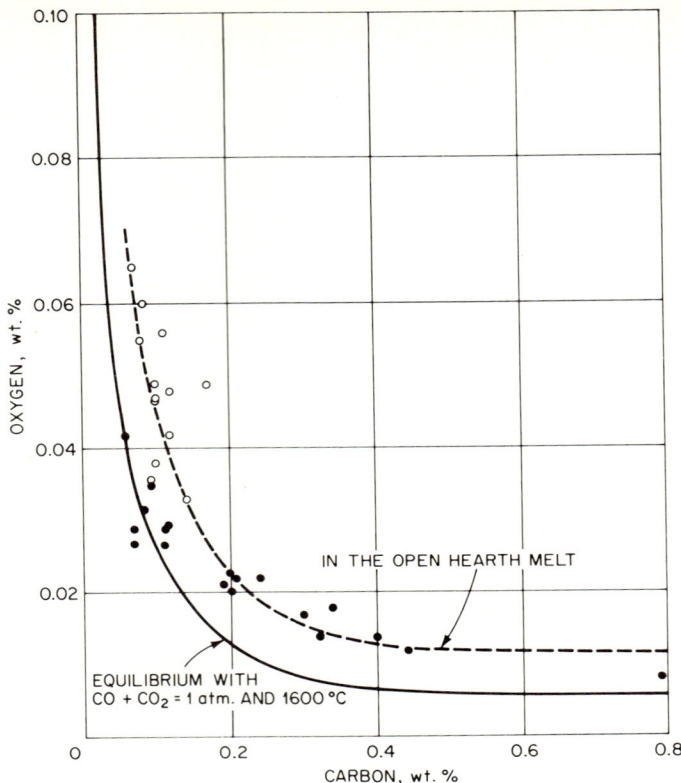

**FIGURE 6-25**
Carbon–oxygen relationship under equilibrium conditions ($P_{CO} + P_{CO_2} = 1.0$ atm and 1600°C) and in open-hearth melts. (*From Ref.* 17.)

A small amount of $CO_2$ will form at steelmaking temperatures by a similar reaction, and the resulting dissolved carbon and oxygen are shown in Fig. 6-25.

Various techniques are used commercially to reduce the dissolved gas content of molten metals. Many of these employ the principle that the solubility of the gas decreases with decreasing partial pressure of that gas. Vacuum melting, vacuum degassing, and flushing with an inert gas are degassing methods that work in this way (these are discussed more fully in the next section). A different technique is to react the gas-former with another element to precipitate it as an inclusion. Thus, silicon or aluminum is used to deoxidize steel (as discussed earlier), and phosphorus or aluminum is used to deoxidize copper-base alloys. Bronzes are often melted under oxidizing conditions to keep hydrogen content low, according to Eq. (6-13), and then are deoxidized to remove the oxygen.

## GAS REMOVAL BY BUBBLE FORMATION

Vacuum melting and vacuum degassing are rapid and effective ways of removing gases from metals. Vacuum degassing differs from vacuum melting only in that the metal is melted in air and subsequently exposed to vacuum. For best efficiency, it is desirable that bubbles form within the melt, not just that the gas diffuse out of the top exposed-melt surface. To form these bubbles, we shall see below that it is important to reduce the metallostatic *head* on the metal. This can be done by stream degassing (pouring metal through a vacuum), by processes which repeatedly draw a small amount of metal out of a large bath and expose it to vacuum, and by vigorous stirring of the melt (to bring the gas-containing liquid to the melt surface).

When a molten metal containing a dissolved gas is exposed to vacuum, a bubble will be stable provided the gas pressure inside the bubble is sufficiently high to balance external forces. These external forces are the liquid-vapor surface energy, metallostatic head, and ambient pressure. Thus, for a stable bubble,

$$P_g = P_0 + \rho_L g_r h + \frac{2\sigma_{gL}}{r} \qquad (6\text{-}17)$$

where $P_g$ is the total gas pressure in the bubble, $P_0$ is the ambient pressure, $\rho_L g_r h$ constitutes the metallostatic pressure head, and the final term is the pressure in the bubble resulting from the bubble-melt interfacial energy $\sigma_{gL}$. Assuming the gas in the bubble to be a single species, at equilibrium with the dissolved gas in the surrounding metal, $P_g$ is that gas pressure calculated according to thermodynamic relations such as Eqs. (6-12a) and (6-14). For bubbles larger than a few microns, the surface-energy term on the right is negligible. The pressure head remains an important factor in hindering bubble formation especially in the heavier metals such as steel.

For small bubbles, extremely large pressures are required. Thus, there is a *bubble-nucleation* problem, which is formally similar to that of nucleation of a solid from a liquid[33,34] (discussed in Chap. 10). Laboratory experiments on pure water and other liquids have shown that a very large barrier to bubble formation exists when effective heterogeneous nuclei are absent.[35] There is ample evidence that gas supersaturations are also sometimes obtained in molten metals. In aluminum alloys that have been filtered to remove oxide and other inclusions, large hydrogen supersaturations are obtained during solidification, but much less supersaturation is obtained when the inclusions are present.[36] In steel, large supersaturations of CO can be obtained if the crucible surface is coated with a viscous liquid refractory. Up to 15 times the concentration of carbon in equilibrium with the dissolved oxygen can be obtained in this way. However, the metal reacts violently with a vigorous carbon boil if the melt comes into contact with unglazed refractory.[37] In usual refining furnaces, the rough refractory furnace bottom provides sources of heterogeneous nuclei so that CO supersaturation is held to lower value, but even here it is significant. Figure 6-25 shows the supersaturation measured in open-hearth melts; the data for the open hearth at 1 atm external pressure are as if the metal were in equilibrium with CO at 2 to 3 atm.

One way to avoid the bubble-nucleation problem in degassing metal is to bubble an inert gas through the liquid. This method is widely employed in air-melting of nonferrous alloys and is employed to a more limited extent in melting of ferrous- and copper-base alloys. The CO boil in steel melting behaves similarly, "flushing" out other gases such as hydrogen. In inert-gas flushing, the total pressure within the bubble is that given by Eq. (6-17). The partial pressure of the gas to be removed is different. Assuming equilibrium between the bubble and the melt, this pressure is given by thermodynamic relations such as those of Eqs. (6-12a) and (6-14).

Of course, equilibrium may not always be attained between the bubble and the melt, although attempts are made to approach this equilibrium by producing as fine a bubble as possible. Results have been obtained by workers on several different metals, indicating that the rate of degassing is sometimes limited by rate of transfer of gas within the liquid or at the metal-bubble interface.[38,39] In aluminum, traces of water vapor in the inert gas slow the rate of degassing markedly, probably by forming a surface oxide film around the bubble. Addition of chlorine to the inert gas enhances rate of degassing; the chlorine probably breaks up, or *fluxes*, the oxide film.

## FORMATION OF GAS POROSITY ON SOLIDIFICATION

To illustrate the factors influencing pore formation, consider an aluminum alloy at 1 atm at its melting point. An amount of gas $\%\underline{H}_i$ is initially dissolved in the liquid; this is less than the liquid solubility but greater than the solid solubility. Assuming the solubility of hydrogen in the alloy is the same as for pure aluminum, this initial dissolved hydrogen content is between the limits of the vertical discontinuity in the curve for aluminum in Fig. 6-23. As solidification proceeds, hydrogen is rejected by the growing dendrites. The hydrogen is present in such small quantities that it does not significantly change the fraction liquid of the alloy at a given temperature. Assuming complete diffusion of hydrogen within the liquid and solid aluminum over distances the order of the dendrite arm spacing, an equilibrium lever rule is written for the dissolved gas as

$$(\%\underline{H}_s)f_s + (\%\underline{H}_L)f_L = \%\underline{H}_i \qquad (6\text{-}18)$$

where $f_s$ and $f_L$ are fractions of liquid and solid, respectively. With increasing fraction of solid, the dissolved hydrogen in the liquid and solid both increase, and, as they do so, the partial pressure of gas $P_{H_2}$ that would be in equilibrium with these dissolved gas contents also increases. When this pressure $P_{H_2}$ becomes sufficiently large, a pore

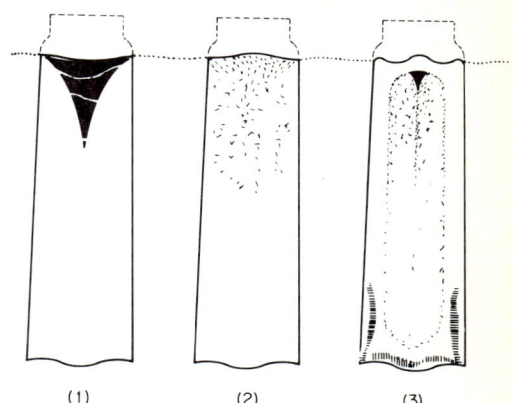

FIGURE 6-26
Three ingot structures: (1) killed; (2) semikilled; (3) rimmed. (*From Ref.* 17.)

forms. The condition is that $P_{H_2} > P_g$ as given by Eq. (6-17); or, neglecting the effect of surface tension and metallostatic head, it is necessary simply that $P_{H_2}$ be greater than ambient pressure $P_0$.

Hydrogen is a constant source of difficulty for aluminum foundrymen partly because it dissolves so readily upon reaction of molten aluminum with the water or water vapor of its surroundings. Also, the ratio of the solubility of hydrogen in the solid to that in the liquid is so small that little of the hydrogen that is present in the liquid can remain dissolved in the solid. When pores form at low fractions of solid, they are large and nearly spherical; at higher fractions of solid, they are more angular (on the size scale of the dendrite arms) and take the shape of the interdendritic spaces. When the pores form at high fractions of solid, it is difficult or impossible to distinguish between *gas-* and *shrinkage-caused* voids. (This is discussed further in Chap. 7.)

When more than one gas-forming element is present, gases form similarly to the simple case discussed above and more than one gas may form. For example, consider copper with both hydrogen and oxygen dissolved. Hydrogen will precipitate during solidification when its partial pressure [as given by Eq. (6-12a)] reaches a sufficiently high value, and water vapor will precipitate when its partial pressure [as given by Eq. (6-14)] reaches a sufficiently high value.

Most metals are cast with as low a gas content as possible in order to produce a dense casting or ingot. Exceptions are *semikilled* and *rimming* steels. In these steels, the evolution of CO during solidification is used to compensate for solidification shrinkage and thus avoid *pipe* at the ingot top. The CO blowholes thus formed are closed by subsequent working. Figure 6-26 shows ingot structures obtained at three different levels of CO evolution: ingot No. 1 is fully "killed" with a strong deoxidizer such as aluminum, and no CO evolves during solidification; ingot No. 2 is a typical semikilled ingot in which just enough CO evolves to compensate for solidification shrinkage; ingot No. 3 is a rimming ingot. In this last ingot, gas evolution is vigorous from the earliest stages of solidification. Turbulence from the rimming action is so great in the early stages of solidification in the upper portions of the ingot that the large blowholes which form are swept upward out of the ingot. In the lower part of the ingot, where rimming action is less violent, columnar grains grow and elongated CO *wormholes* form between them. Rimmed steels are made with carbon contents in the range of 0.1 to 0.15 percent and are suitable for the manufacture of steel sheet. Semikilled steels have carbon contents in the range of 0.15 to 0.30 percent and find wide application in structural shapes. Killed steels can be made with any desired carbon content and with other alloy elements that would hinder the rimming action. They are more homogeneous and sounder, and are used where these qualities are desired.

## REFERENCES

*1* SARTELL, J. A., and MACK, D. J.: *J. Inst. Metals*, **93**:19 (1964–1965).
*2* SPITTLE, J. A.: *J. Inst. Metals*, **98**:124 (1970).
*3* WEST, D. R. F.: *J. Iron Steel Inst.*, **164**:182 (1950).
*4* DAY, M. G., and HELLAWELL, A.: *Proc. Roy. Soc. (London), Ser. A*, **305**:431 (1968).
*5* STEEN, H. A. H., and HELLAWELL, A.: *Acta Met.*, **20**:363 (1972).
*6* TAYLOR, H. F., FLEMINGS, M. C., and WULFF, J.: "Foundry Engineering," John Wiley & Sons, Inc., New York, 1959.
*7* HUGHES, I. C. H.: "The Solidification of Metals," Iron and Steel Institute Publ. No. 110, p. 184, 1968.
*8* MORROGH, H.: "The Solidification of Metals," Iron and Steel Institute Publ. No. 110, p. 238, 1968.
*9* HILLERT, M., and STEINHAUSER, H.: *Jernkon. Ann.*, **144**:520 (1960).
*10* HILLERT, M., and SUBBA RAO, V.: "The Solidification of Metals," Iron and Steel Institute Publ. No. 110, p. 204, 1968.
*11* FLEMINGS, M. C., and MEHRABIAN, R.: in "Solidification," American Society for Metals, Metals Park, Ohio, 1971.
*12* MEHRABIAN, R., and FLEMINGS, M. C.: *Met. Trans.*, **1**:485 (1970).
*13* FORWARD, G., and ELLIOTT, J. F.: *Met. Trans.*, **1**:2889 (1970).
*14* FREDERIKSSON, H., and HILLERT, M.: *J. Iron Steel Inst.*, **209**:109 (1971).
*15* YARWOOD, J., FLEMINGS, M. C., and ELLIOTT, J. F. (to be published).
*16* ELLIOTT, J. F.: *Trans. Iron Steel Inst. Japan (Suppl.)*, **11**:416 (1971).
*17* "The Making, Shaping, and Treating of Steel," 8th ed., U.S. Steel, Pittsburgh, Pa., 1964.
*18* LINDBORG, U., and TORSSELL, K.: *Trans. AIME*, **242**:94 (1968).
*19* GUNN, R.: *Science*, **150**:695 (1965).
*20* FRENKEL, J.: *J. Phys. USSR*, **9**:385 (1945).
*21* FONTANA, E. H., and PLUMMER, W. A.: *Phys. Chem. Glasses*, **7**(4):139 (August 1966).

22 TURPIN, M. L., and ELLIOTT, J. F.: *J. Iron Steel Inst.*, **204**:217 (1966).
23 KNUPPEL, H., BROTZMANN, K., and FORSTER, N. W.: *Stahl Eisen*, **85**:675 (1965).
24 BRONDYKE, K. J., and HESS, P. D.: *Trans. AIME*, **230**:1553 (1964).
25 BRANT, M. V., BONE, D. C., and EMLEY, E. F.: *J. Metals*, **23**:48 (1971).
26 MYERS, M., and FLEMINGS, M. C.: *Met. Trans.*, **3**:2225 (1972).
27 REGE, R. A., SZEKERES, E. S., and FORGENG, JR., W. D.: *Met. Trans.*, **1**:2652 (1970).
28 BIGELOW, L. K.: Sc.D. thesis, Dept. of Metallurgy, M.I.T., Cambridge, Mass., 1970.
29 BADIA, F. A., and ROHATGI, P. K.: *Trans. AFS*, **76**:402–406 (1969).
30 SIMS, C. E., and BRIGGS, C. W.: *Proc. Elec. Furnace Steel, AIME*, **17**:104 (1959).
31 BEVER, M. B., and FLOE, C. F.: *Trans. AIME*, **166**:128 (1946).
32 SMITHELS, C. J.: "Gases in Metals," Chapman and Hall, Ltd., London, 1937.
33 FISHER, J. C.: *J. Appl. Phys.*, **19**:1062–1067 (1948).
34 HIRTH, J. P., and POUND, G. M.: *Progr. Mater. Sci.*, **11** (1963).
35 BRIGGS, L. J.: *J. Appl. Phys.*, **21**:721 (1953); **24**:488 (1953).
36 BRONDYKE, K. J., and HESS, P. D.: *Trans. AIME*, **230**:1542 (1964).
37 KORBER, F., and OELSEN, W.: *Mitt. K-Wilh. Inst., Eisenforsch.*, **17**:39 (1935).
38 PEHLKE, R. D., and BEMENT, JR., A. L.: *Trans. AIME*, **224**:1237 (1962).
39 DARKEN, L. S.: "Physical Chemistry of Steelmaking," p. 101, John Wiley & Sons, Inc., New York, 1958.
40 "Typical Microstructures of Cast Metals," 2d ed., Institute of British Foundrymen, London, 1966.
41 TURKDOGAN, E. T.: *J. Iron Steel Inst.*, **210**:21 (1972).
42 BROWER, JR., W. E.: Sc.D. thesis, Dept. of Metallurgy, M.I.T., Cambridge, Mass., 1969.
43 EASTWOOD, L. W.: "Gases in Non-Ferrous Metals and Alloys," American Society for Metals, Cleveland, Ohio, 1953.

# PROBLEMS

6-1 Show by calculation that the amount of eutectic given in Fig. 6-1 is the amount to be expected in a Pb–20% Bi alloy, assuming no solid diffusion.

6-2 For a unidirectionally solidified Pb–20% Bi casting, at a time when solid, solid + liquid, and liquid regions coexist, show schematically (do not calculate):
   (a) Temperature versus distance from chill
   (b) Liquid composition versus distance from chill
   (c) Fraction of solid versus distance from chill
   (d) Solid composition $C_s^*$ versus distance from chill
   (e) Minimum composition at center of the dendrite, $C_{min}$, versus distance from chill

6-3 Pb–Bi alloys containing from 0 to 56 percent Bi are grown with a plane front at steady state.
   (a) Plot solidification temperature versus percent Bi.
   (b) Indicate on your plot the phases which form on solidification and the range of compositions over which they form.

6-4  Calculate the value of $G_L/R$ required to solidify a Pb–24% Bi alloy with plane front, assuming a simple constitutional supercooling criterion for interface stability.

6-5  Explain why malleable iron castings are not produced in sections much over 1 in thick although much heavier castings are made of white iron (for applications requiring wear resistance).

6-6  How would you make a large cast-iron roll with a white-iron surface and gray-iron interior?

6-7  Explain why ledeburite eutectic shows regions that are platelike and other regions that are rodlike.

6-8  Suggest an experiment or experiments to determine whether graphite spherulites are pushed by growing dendrite arms during solidification or are entrapped by them.

6-9  What problems would you expect to encounter in trying to develop a die-casting process for cast iron?

6-10  Explain qualitatively how five phases can exist in a casting of the ternary alloy of Fig. 6-12 after solidification. Does this violate the phase rule of thermodynamics?

6-11  Find a liquidus surface projection for a real ternary phase diagram from a standard reference book. Choose three different initial compositions and show the solidification path for dendritic solidification as it occurs in usual castings and ingots. For each alloy, what phases do you expect to find in the final structure? List the assumptions you are making to answer the foregoing.

6-12  Derive Eq. (6-3) by (a) calculating the solute rejected when a differential amount of $\alpha$ and $\beta$ phase solidify, (b) calculating the solute increase in the liquid when this differential amount of solidification occurs, and (c) equating the two. Show that Eq. (6-3) is equivalent to Eq. (2-3) when $df_\beta = 0$.

6-13  Write an equation similar to Eq. (6-3) to describe monotectic solidification of Fe–O–S as sketched in Fig. 6-13.

6-14  Discuss some possible ways of removing primary $Al_2O_3$ inclusions from molten steel.

6-15  An aluminum melt contains hydrogen in the amount of 0.5 cm$^3$ (STP)/100 g of aluminum. Assume bubbles form at equilibrium with negligible effect of pressure head and surface tension, and use data from Fig. 6-23.

(a) Calculate the void volume in a 100-g casting solidified at atmospheric pressure.

(b) A widely used quality-control test for gas in aluminum alloys is the *reduced-pressure tester*. In this test a small sample of the aluminum melt is placed in a bell jar and allowed to solidify under reduced pressure. As a result of the reduced pressure the void volume is increased to an amount visible to the naked eye. Calculate the void volume in a 100-g sample of the above melt solidified in this way. The reduced pressure tester is operated at 85 mm absolute pressure.

(c) If ambient pressure is raised sufficiently high during solidification, the gas pores can be eliminated. What pressure is required?

6-16  The melt of Prob. 6-15 is degassed at its melting point by bubbling a nitrogen–chlorine mixture through it. How much nitrogen must be bubbled through per 100 g of melt to reduce the hydrogen content to 0.04 cm$^3$ (STP)/100 g metal? Assume equilibrium between bubbles and melt, with negligible effect of pressure head and surface tension. The bath is fully mixed, and hydrogen escapes from the melt only through bubbles.

(*Hint:* Calculate the amount of hydrogen removed by each small bubble of differential volume.)

6-17 A plain carbon steel is melted from a charge of pig iron and scrap steel. Iron ore is added to obtain a carbon boil and reduce the carbon content to 0.02 percent.
   (a) When the carbon content reaches 0.02 percent, what will be the oxygen content of the steel?
   (b) If the steel is cast immediately, how much of the casting will solidify before gas evolves?
   (c) If the initial oxygen concentration is below a certain level, porosity will not form. What is this level?
   (d) How much silicon must be added to accomplish the reduction of (c)?
   Assume the following: complete diffusion of dissolved $\underline{O}$ and $\underline{C}$ in liquid and solid over distances the order of the dendrite arm spacing, and negligible solubility of $\underline{O}$ in solid iron. Melting and solidification take place at 1550°C, and at this temperature the equilibrium constant for the reaction $CO_{(g)} = \underline{C} + \underline{O}$ is

$$k = \frac{(\%\underline{C})(\%\underline{O})}{P_{CO}} = 1.97 \times 10^{-3}$$

when $\%\underline{C}$ and $\%\underline{O}$ are in weight percent and $P_{CO}$ in atm. Oxygen solubility in the solid is negligible. Equilibrium constant for the reaction $\underline{Si} + 2\underline{O} = SiO_2$ at 1550°C is $K = (\%\underline{Si})(\%\underline{O})^2 = 1.14 \times 10^{-5}$. Neglect the effects of surface energy and metallostatic head.

6-18 Nearly pure copper contains $1 \times 10^{-4}$ wt% $\underline{H}$ and $8 \times 10^{-4}$ wt% $\underline{O}$. The metal solidifies dendritically with the $\underline{H}$ and $\underline{O}$ rejected to interdendritic spaces.
   (a) At what fraction of solid will pores begin to form?
   (b) What will be the final volume fraction voids present in this casting?
   (c) What gas or gases will comprise the voids?
   Assume complete diffusion of $\underline{O}$ and $\underline{H}$ in liquid and solid over distances the order of the dendrite spacing, and negligible solid solubility of $\underline{O}$. The ratio of solid solubility of $\underline{H}$ to that of the liquid is 1 : 3. Neglect the effects of surface energy and metal-lostatic head. Casting solidifies at ambient pressure. The melting point of the Cu is 1083°C. Solubility data are as follows:

| Reaction | $\log K'_L$ |
|---|---|
| $H_2 = 2\underline{H}$ | $-\dfrac{4{,}620}{T} - 3.22$ |
| $\tfrac{1}{2}O_2 = \underline{O}$ | $\dfrac{7{,}030}{T} - 2.84$ |
| $H_2O = 2\underline{H} + \underline{O}$ | $-\dfrac{10{,}640}{T} - 3.09$ |

6-19 Repeat Prob. 6-1 for $4 \times 10^4$ wt% $\underline{H}$, $1 \times 10^4$ ppm $\underline{O}$.

6-20 What gas pressure is required to form a stable bubble, 1 μm diameter, near the surface of an aluminum melt at 1 atm ambient pressure? Assume $\sigma_{gL} = 840$ dyn/cm².

# 7
# FLUID FLOW

## INTRODUCTION

An important and sometimes neglected aspect of solidification processing is *fluid flow*. In most casting processes, metal is transferred from a melting container to the mold. This must be done with great care to avoid gas pickup and oxide formation and to fill the mold completely. Once the metal is in the mold, convection continues from momentum effects and from density differences in the liquid itself. When solidification then begins, convection influences the structure and sometimes the composition of the solid which forms.

In crystal growing, forced convection is sometimes intentionally introduced to aid in solute transport from the liquid-solid interface. Most metals shrink when they solidify, and extra reservoirs of metal, termed *risers*, or *hot tops*, are used on castings and ingots to supply this feed metal. Riser design, like most other topics considered in this chapter, is a combined heat- and fluid-flow problem. Fluid flow does not necessarily stop when the supply of bulk liquid is exhausted. Fluid in interdendritic spaces is free to move, and generally does so. When it cannot flow to compensate adequately for solidification contractions, microporosity and open *hot tears* result. Most industrially important types of macrosegregation also result from interdendritic

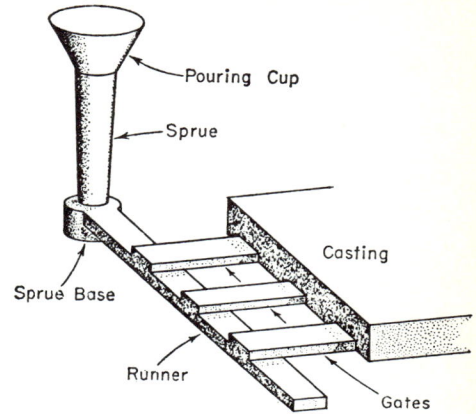

**FIGURE 7-1**
Gating system for a sand casting. (*From Taylor, Flemings, and Wulff.*[2])

fluid flow, a fact which has been understood only in the last few years. This is in spite of the fact that macrosegregation has been an important metallurgical problem since at least 1540 when Biringuccio described macrosegregation problems in connection with manufacture of bronze gun barrels.[1]

## GATING

In making shaped castings, metal is generally introduced to the mold via a *gating* system, such as the one sketched in Fig. 7-1. A function of the system is to introduce metal to the mold as smoothly as possible so that it does not react excessively with the atmosphere, mold, or mold gases. Also, good gating systems are designed to remove entrapped oxides and slag insofar as possible and to deliver metal at selected points in complex castings.[2-4]

The vertical channel through which metal enters the mold (*sprue*) is a portion of the gating system where aspiration of mold gases is particularly likely to occur, as sketched in Fig. 7-2. The aspiration is greatly reduced by tapering the sprue sufficiently so that the metal does not pull away from the mold walls as it accelerates downward. For freely falling metal, the velocity $v$ increases with the square root of falling distance

$$v = \sqrt{2g_r h} \qquad (7\text{-}1)$$

where $g_r$ is acceleration due to gravity and $h$ is distance. It is possible to design a sprue with a parabolic taper, using Eq. (7-1), such that the metal will always just fill the cross section at all points. The result is very little different from the much simpler

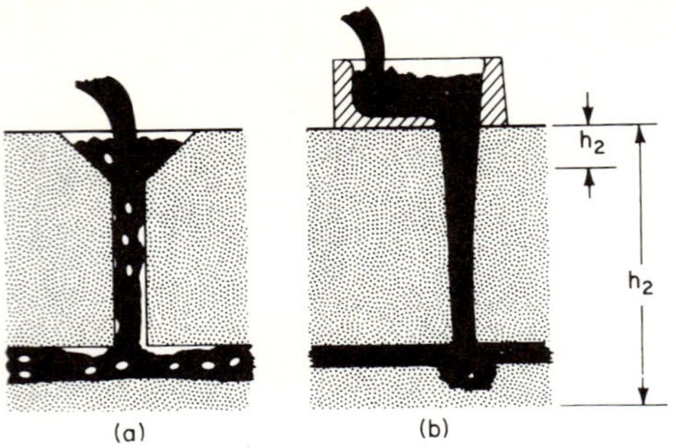

**FIGURE 7-2**
Sprue design. (a) Straight-sided sprue; (b) tapered sprue. (*From Taylor, Flemings, and Wulff.*[2])

practice of designing the sprue with a straight taper such that the cross section near the exit is kept full. From Eq. (7-1), this is seen to be

$$\frac{A_1}{A_2} \geq \sqrt{\frac{h_2}{h_1}} \qquad (7\text{-}2)$$

where $A_1$ and $A_2$ are cross sections of the sprue entrance and exit, respectively, and $h_1$ and $h_2$ are metal heads defined in Fig. 7-2. The arguments leading to Eq. (7-2) neglect frictional effects in the sprue, entrance and exit effects, and back pressure on the sprue occurring as the mold fills. The equation is a conservative estimate of the taper required.

Aspiration also occurs at abrupt changes in direction in cross section of flow channels, and one way to reduce this is by streamlining the horizontal *runner* and *gates* of the system, as shown in Fig. 7-3. A satisfactory and widely used alternate approach is, not to streamline the system, but to make the horizontal portions of the gating system large so that flow velocity in these portions is low. An example is shown schematically in Fig. 7-1; the smallest cross section is at the sprue base. Subsequent flow is through larger channels at lower velocity. A typical *gating ratio* for aluminum

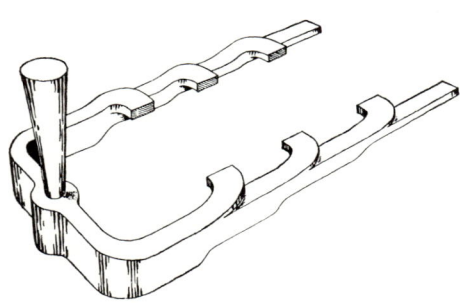

**FIGURE 7-3**
A fully streamlined gating system. (*After Grube and Eastwood.*[72])

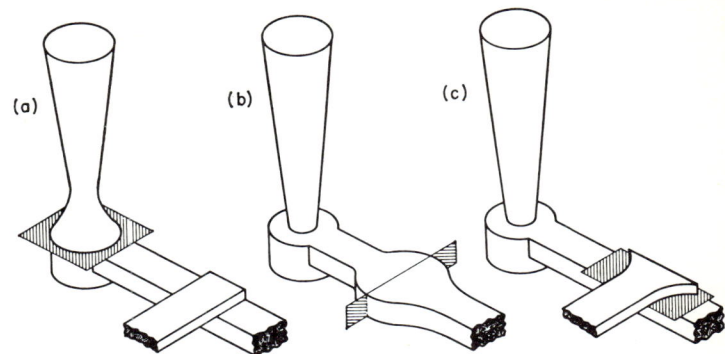

FIGURE 7-4
Some methods of placing screens in gating systems. (*From Taylor, Flemings, and Wulff.*[2])

castings is 1:3:3, meaning that the ratio of the area of the runner to that of the sprue base is 3 and the sum of the cross-sectional areas of all gates is equal to that of the runner. Gating ratios in practice for this type of gating system vary from about 1:1.5:1.5 to 1:5:5, the higher ratios being used for alloys which easily form oxides or from which it is difficult to remove entrained slag or oxides, e.g., ductile iron. A modification of the gating system discussed above (the incorporation of a filter to remove entrapped dross or slag from the molten metal) is shown in Fig. 7-4. Filters are widely used for nonferrous alloys. They are generally made of fiberglass, mica, or sheet metal.

Gating systems are used on all types of shaped castings. Many types of modern ingot-casting processes now also employ a sort of gating system. As example, in continuous casting of aluminum alloys, modern plants filter the liquid metal and then introduce it to the continuous caster through channels that provide minimum exposure of the metal to air. Many steel continuous-casting plants do the same, except that filtering is not yet employed. Most large static cast steel ingots are still filled directly from the ladle without an intermediate gating system.

In some instances, appreciable heat is lost from the metal in the runner system. The rate of temperature drop can be simply calculated for sand or other insulating molds[5,6] and even more simply for metal molds where heat flow is controlled by mold-metal interface resistance. Consider metal flowing down a long cylindrical runner of radius $a$, Fig. 7-5. The average flow velocity of the metal (volumetric flow rate divided by cross-sectional area) is $v$. Heat flow is interface-controlled, and so radial temperature gradients in the metal and mold are negligible. As an element of liquid metal enters the runner and moves downstream at velocity $v$, it loses heat

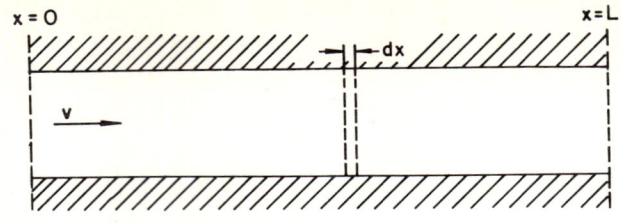
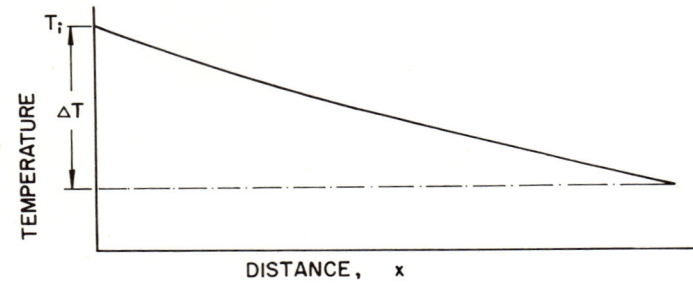

**FIGURE 7-5**
Temperature drop in metal flowing in a runner. *Top:* Sketch of runner; *bottom:* metal temperature.

radially so that the instantaneous longitudinal temperature distribution in the liquid metal is as shown in Fig. 7-5. A heat balance is readily written for the moving element, neglecting the small heat transport by conduction in the direction of flow:

$$\text{Heat leaving element} = \text{heat entering mold}$$

$$\rho_L c' \frac{dT}{dt}(\pi a^2 \, dx) = -h(T - T_0)(2\pi a \, dx) \qquad (7\text{-}3)$$

where $\rho_L$ and $c'$ are density and specific heat of the liquid metal, respectively, $T$ is metal temperature, $T_0$ mold temperature, and $h$ is mold metal interface resistance to heat transfer. Thus, the rate of temperature change of the flowing metal is

$$\frac{dT}{dt} = -\frac{2h(T - T_0)}{a\rho_L c'} \qquad (7\text{-}4)$$

The time $t_r$ for an element of metal to go from $x = 0$ to $x = L$ is

$$t_r = \frac{L}{v} \qquad (7\text{-}5)$$

and the temperature drop experienced by the element $\Delta T$ is obtained by integrating Eq. (7-4) from $t = 0$ to $t = t_r$. For $\Delta T$ small compared with $T - T_0$, the result is simply

$$\Delta T = \frac{2h(T_i - T_0)L}{a\rho_L c' v} \qquad (7-6)$$

where $T_i$ is the temperature of the metal as it enters the channel.

## FLUIDITY

In shaped-casting processes, metal enters the mold through one or a number of gates. Should it then fail to fill the cavity before it solidifies, it is said to be insufficiently *fluid*. Thus, the term *fluidity* has come to mean something quite different to the foundryman than to the physicist. To the physicist, it is the reciprocal of viscosity; to the foundryman, it is an empirical measure of a processing characteristic. The foundryman measures this property in one of several types of fluidity tests. In these tests, hot metal is caused to flow in a long channel of small cross section (channel is at room temperature). The length the metal flows before it is stopped by solidification is the measure of fluidity. Two common types are the fluidity spiral and the vacuum fluidity test sketched in Fig. 7-6.

Let us now look at the flow and solidification of a pure metal poured at its melting point in a fluidity-test channel. The instant this metal enters the channel, solidification begins at the channel entrance. As it proceeds down the channel, solidification begins in these locations also. However, because freezing began first at the channel entrance, it is here that the flow is choked off. The total length the metal flows before stopping is the fluidity. Figure 7-7 illustrates the foregoing process. Fluidity determined as above is influenced by (1) metal variables such as temperature, viscosity, and heat of fusion; (2) mold and mold-metal variables such as heat-flow resistance at the interface, mold conductivity, density, and specific heat; and (3) test variables such as applied metal head and channel diameter.

To illustrate the important parameters, a simplified analysis can be made, neglecting friction and acceleration effects and assuming no separation of the flow stream occurs. We shall carry out this analysis here for the case (sketched in Fig. 7-7) of a pure metal poured with no superheat and assume further that heat flow is $h$-controlled. Solidification at each point in the channel entrance occurs at a rate that is independent of fluid flow. Thus, solidification of each element of the channel of length $dx$ solidifies completely in a time given by Eq. (1-19), with $V/A$ equal to $a/2$ for a channel of circular cross section of radius $a$. The element at exactly the channel

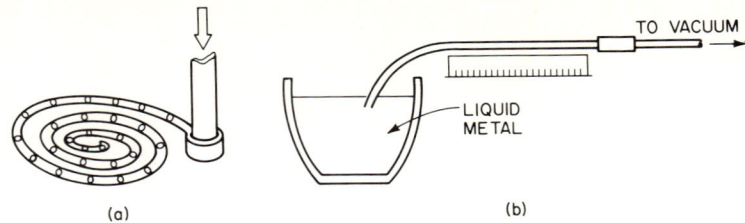

**FIGURE 7-6**
Two types of fluidity tests. (*a*) Fluidity spiral—cavity is in sand or permanent mold; (*b*) vacuum fluidity test—glass or metal tube.

entrance solidifies in a time $t_f$ given by

$$t_f = \frac{\rho_s H a}{2h(T_M - T_0)} \tag{7-7}$$

where $t_f$ is time after the metal enters the flow channel.

We now simplify the fluid-flow problem somewhat by neglecting friction and acceleration effects and by assuming no separation of the flow stream occurs. Then, for constant applied head, velocity of the stream tip is a constant $v$. The total length of flow before the channel entrance solidifies (i.e., the fluidity) is therefore

$$L_f = \frac{\rho_s H a v}{2h(T_M - T_0)} \tag{7-8}$$

where $L_f$ is fluidity. Superheat increases fluidity in a simple way when heat flow is $h$-controlled. The added length of flow due to superheat is simply that length required to dissipate the superheat according to Eq. (7-6). Combining Eqs. (7-6) and (7-8) with $\rho_s \simeq \rho_L$ gives the fluidity for superheated metal as

$$L_f \simeq \frac{\rho_s a v}{2h(T_M - T_0)} (H + c' \Delta T) \tag{7-9}$$

where $\Delta T$ is the superheat. Equation (7-9) illustrates many of the important variables influencing the foundryman's fluidity. Length of flow depends sensitively on channel size and interface heat-transfer coefficient. It increases approximately linearly with superheat. It depends on flow velocity and therefore approximately on the square root of metal head, according to Eq. (7-1).

The structure of fluidity test castings reflects the solidification mechanism discussed above. Figure 7-8 shows longitudinal cross sections from nearly pure aluminum. The shrinkage cavity is at the leading tip of the casting since this was fully

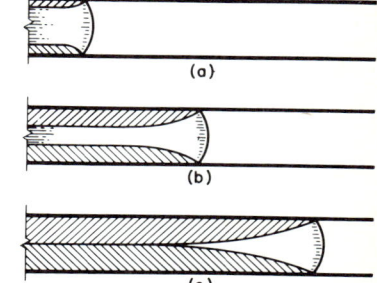

FIGURE 7-7
Flow and solidification of a pure metal in a fluidity channel, no superheat. (*a*) Beginning of flow; (*b*) during flow; (*c*) end of flow.

liquid at the time flow ceased, as sketched in Fig. 7-7. Grains throughout most of the casting point slightly *upstream*, and this is indicative of the fact that they grew while flow was occurring, as discussed later in this chapter. The regions near the channel entrance have coarse grains which do not point upstream. This is the region of the casting where superheat was dissipated; it solidified after flow ceased.

The analysis presented here has been a simplified version of the one originally developed by Ragone et al.[7] and later extended by others.[8,9] These analyses consider with more rigor the fluid flow aspects of the problem and treat cases of heat flow other than $h$-controlled. In practice, there are two important factors which influence fluidity and are not included in Eq. (7-9): surface tension, or surface films, and back pressure from mold gases. Surface tension does not have a significant effect except in very small channels (under about $\frac{1}{10}$ in diameter). However, it can prevent obtaining sharp detail in castings unless pressure head is increased to overcome it. Methods of increasing effective pressure head include vibration, centrifuging, and application of a vacuum to the mold walls of the casting (to reduce back pressure of mold gases). All of these techniques, especially the latter two, are used in making very thin-section investment castings, such as jewelry, where fine detail is important. In die casting, the pressure required is obtained in a different way by use of a mechanical piston or by gas pressure on the molten metal.

Addition of alloy elements to a pure metal always decreases fluidity (at a given superheat). Figure 7-9 shows typical results for a series of lead–tin alloys from the work of Ragone et al.[7] The reason for the decrease is that solidification no longer takes place with plane front. The dendrites that form create more resistance to fluid flow at an earlier stage of solidification. Figure 7-10 shows this schematically for columnar grains such as are generally found in fluidity test castings of dilute, nongrain-refined alloys. In grain-refined alloys, some fine grains are carried along with the tip

**222** FLUID FLOW

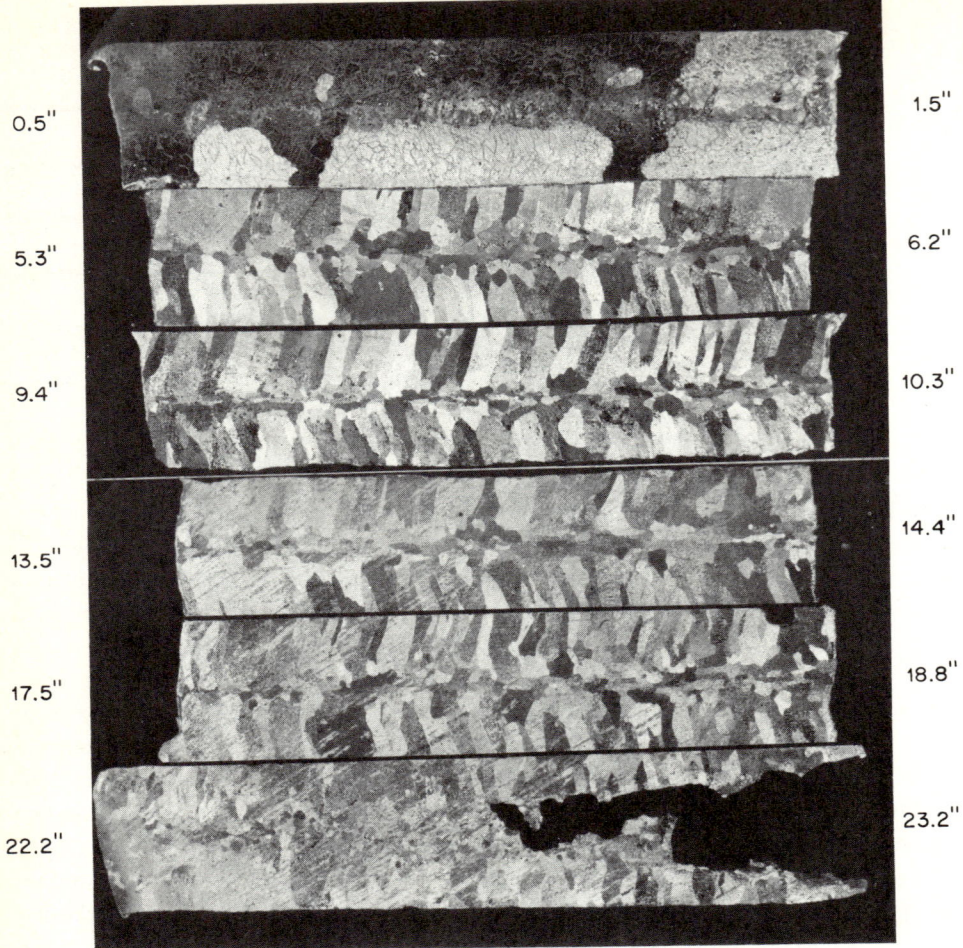

**FIGURE 7-8**
Macrostructure of nearly pure aluminum fluidity casting poured with superheat. Metal flow left to right and top to bottom. Figures at sides are distance of sections shown along channel (magnification ×4). (*From Flemings.*[9])

of the flowing stream. Thus, flow stoppage can be (as sketched in Fig. 7-11) by formation of sufficient solid at the tip to block flow.[9] As alloy composition approaches a eutectic, fluidity is generally found to increase. Early speculations were that this was because of changes in metal viscosity near the eutectic, but it is now known the increase results simply because of solidification behavior. The eutectic solidifies much like the pure metal of Fig. 7-7.

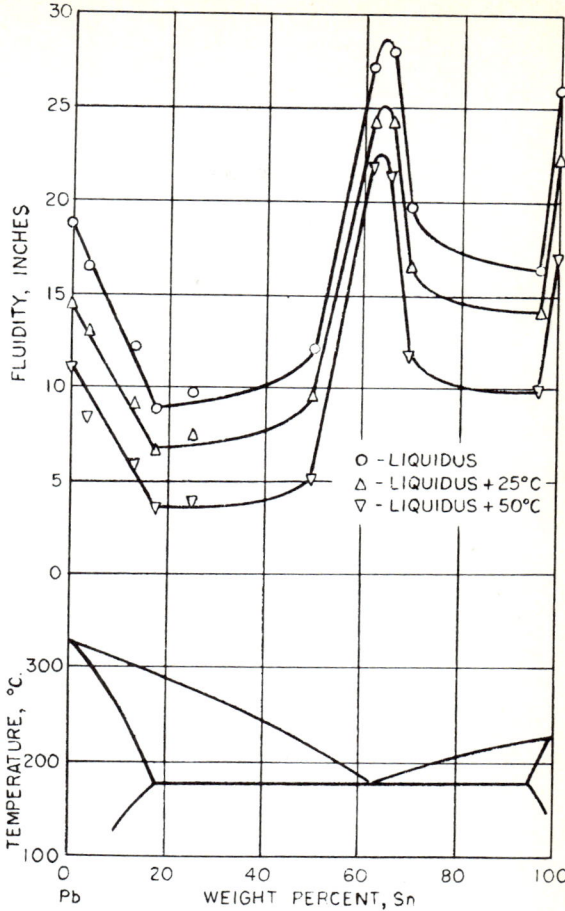

**FIGURE 7-9**
Fluidity in the lead–tin system. (*From Ragone et al.*[7])

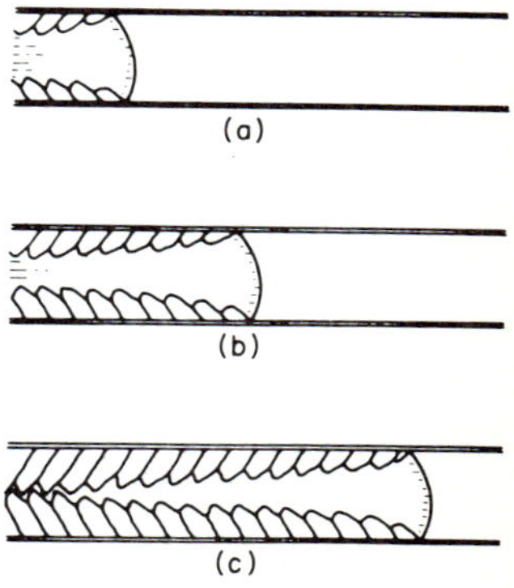

**FIGURE 7-10**
Flow and solidification of a dilute alloy in a fluidity channel. (*a*) Beginning of flow; (*b*) during flow; (*c*) end of flow.

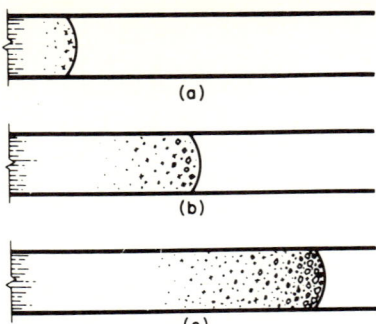

**FIGURE 7-11**
Flow and solidification of alloy of fine equiaxed grain structures. (*a*) Beginning of flow; (*b*) during flow; (*c*) end of flow.

Fluidity limits the thickness of castings that can be conveniently poured of various alloys. For sand castings, this limit is about $\frac{1}{8}$ in for nonferrous alloys and cast iron and $\frac{1}{4}$ in for steel. Much thinner castings are made in investment material through use of heated molds and application of pressure as described above. It is interesting that it is more difficult to fill thin-section-mold cavities with steel than with cast iron even though the fluidity of steel (at a given superheat) is close to that of cast iron.[3,10] This is because, with refractories and furnaces usually used for steel, it is not possible to obtain the high superheats regularly used in making thin-section iron castings.

## CONVECTION IN THE BULK LIQUID

As metal flows through gating systems for castings, it is highly turbulent. Even in well-designed systems, Reynolds numbers as high as 20,000 are common. Certainly, the flow is also turbulent as it enters the mold. Splashing and turbulence of metal entering mold cavities have been observed directly by cine radiography[11,12] and by simulation using transparent fluids.[4] In ingot casting, the most common filling method is top-pouring. The metal may drop several meters or more before striking the liquid level in the mold. One effect of all this agitation is metal damage through entrapment of gases, oxidation, and mold erosion. It also influences the structure that forms in the early stages of solidification. Generally, however, convection introduced by the momentum of pouring will dissipate before solidification has progressed very far. The dissipation of local eddies of size $l$ occurs in times the order of $t_e$ where[13]

$$t_e \simeq \frac{l^2}{v} \qquad (7\text{-}10)$$

and $v$ is kinematic viscosity. In continuous casting, convection from pouring momentum has a continuing influence on solidification behavior since solidification and pouring occur continuously. The incoming superheated liquid metal penetrates deeply into the liquid metal unless special means are provided to distribute it more evenly. The movement of this liquid alters the shape of isotherms near the liquidus temperature and to much lesser extent those near the solidus.[14] Much attention is now paid by designers of continuous-casting processes to provide favorable distribution of the incoming metal.

In nearly all industrial solidification processes, the bulk liquid (fully liquid region) undergoes thermally induced convection during solidification. Temperature differences are often small, but only very small temperature differences produce strong convection in liquid metals, semiconductors, and most oxides. The important variables influencing severity of thermal convection are contained in the Grashof number $N_{GR}$ and the Prandtl number $N_{PR}$ where

$$N_{GR} = \frac{g_r L^3 \beta_t \Delta T}{v^2} \quad (7\text{-}11a)$$

$$N_{PR} = \frac{c' \mu}{K_L} \quad (7\text{-}11b)$$

and where $g_r$ is acceleration due to gravity, $L$ is a characteristic length of the system, $\beta_t$ is volumetric coefficient of thermal expansion of the liquid, $\Delta T$ is temperature difference, $v$ is kinematic viscosity, $c'$ is specific heat, $\mu$ is viscosity, and $K_L$ is thermal conductivity of the liquid. The product of these numbers is termed the *Rayleigh number* $N_{RA}$, and this represents the ratio of buoyancy forces to viscous forces in the liquid. For an unconfined system (such as one comprising a small cooled vertical plate in an infinite liquid), laminar flow occurs for $N_{RA} \lesssim 10^8$, and turbulent flow occurs above this.

Quantitative treatment of thermal convection in liquids is to be found in standard heat- and fluid-flow texts. A growing body of literature also deals with relating these analyses to heat and fluid flow during solidification.[15-18] Unfortunately, calculation of fluid-flow behavior becomes very difficult in confined systems such as a crystal-growing boat or an ingot. Hence, most calculations are performed for idealized systems. One conclusion that can readily be made from fluid-flow theory is that very small horizontal temperature gradients result in strong convection and that, in large castings and ingots at least, the convection can be expected to be highly turbulent as the superheat is being dissipated. (This is demonstrated by Prob. 7-12 at the end of this chapter.)

One effect of thermal convection in crystal growth is to enhance mass transfer at the liquid-solid interface, i.e., to reduce the *solute boundary layer* defined in a very

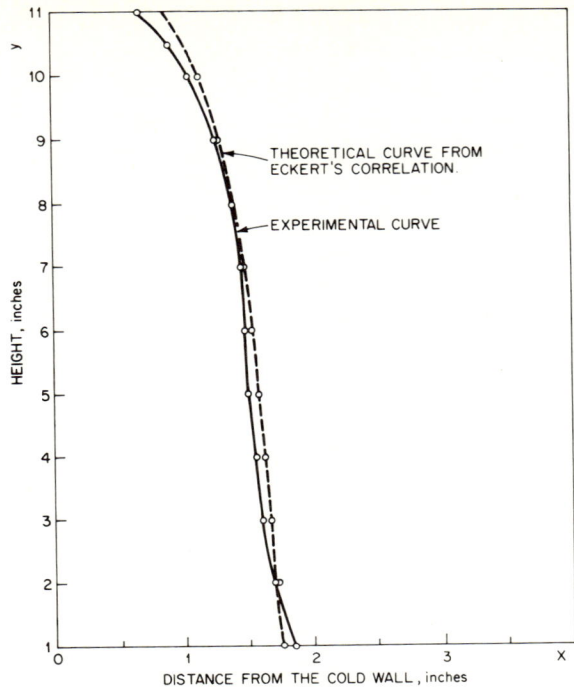

**FIGURE 7-12**
Experimental and calculated interface profile at steady state for a pure lead casting. Unidirectional horizontal heat flow. (*From Szekely and Chabra.*[20])

simple way as $\delta$ in Chap. 2. Another effect is to induce a vertical thermal gradient in horizontal crystal growth so that the interface is not exactly vertical[19,20] (an example is shown in Fig. 7-12). Still another important effect is the temperature fluctuations, first demonstrated for crystal growing by Cole and Winegard[21] and later studied by others.[22-27] The fluctuations occur when the horizontal temperature gradient exceeds a critical value. Figure 7-13 is an example of the type of temperature records one obtains from a liquid metal in a typical horizontal-boat arrangement for crystal growing. As the temperature gradient increases much above 1°C/cm, temperature fluctuations set in which are random, as shown here, or more regular, as shown by Hurle.[26] At high values of $N_{GR}N_{PR}$, the oscillations always appear irregular and presumably reflect turbulent flow.

When temperature fluctuations such as those of Fig. 7-13 reach the liquid-solid interface, they result in unsteady-state interface movement. As the amplitude of the fluctuations increases and mean growth rate decreases, the interface may periodically

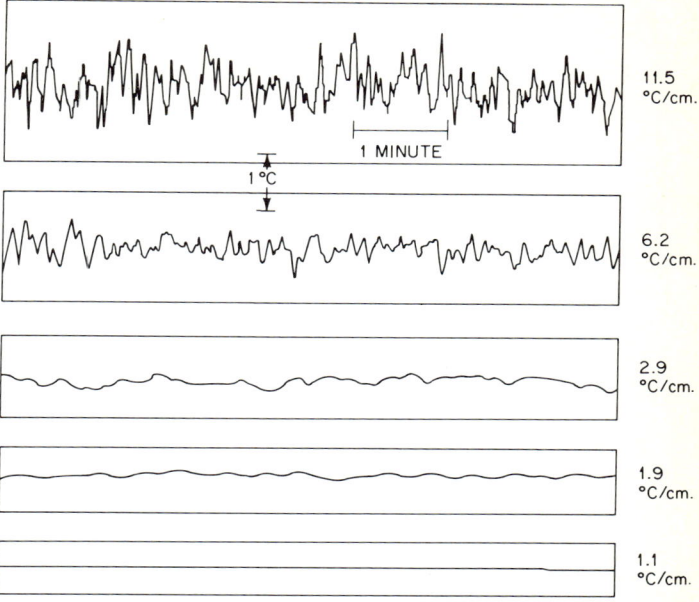

FIGURE 7-13
Typical temperature fluctuations resulting from natural convection in a horizontal boat of liquid tin. (*From Utech et al.*[25])

melt back before advancing forward again. One result of the unsteady forward movement is an increase in mean effective partition ratio along the length of the bar.[27] A more technologically important result is abrupt variation in effective partition ratio along the length of the bar as interface velocity changes. The result is abrupt changes in composition of the solid which appear as *bands* on etched structures, Fig. 2-16.

The bands in Fig. 2-16 stop forming when a magnetic field is applied to the growing crystal, because of marked reduction of the convection. Motion of a conducting fluid across lines of force decays as a result of inductive drag, and it is readily shown that this drag can be regarded as a *magnetic viscosity*. The magnetic viscosity is dominant if the Hartmann number $M$ is large compared with unity:[28]

$$M = BL\left(\frac{\sigma}{\mu}\right)^{1/2} \qquad (7\text{-}12)$$

where $B$ is the magnetic field strength, $L$ is a characteristic distance of the system, $\sigma$ is the electric conductivity, and $\mu$ is the viscosity. Quite modest field strengths (less than 1,000 G) yield Hartmann numbers greater than 100 for molten metals or semiconductors in usual crystal-growing configurations.

Thermal convection in the bulk liquid in castings and ingots leads to rapid dissipation of superheat; usually, it is essentially gone before any solidification begins. The convection also produces temperature fluctuations similar to those observed in single crystal growth. When superheat is high, fluctuations with amplitudes as great as 20°C have been measured.[29] Studies on transparent crystallizing fluids also demonstrate the convection,[30] as do studies on real ingots using radioactive isotopes.[31,32] Perhaps the most convincing evidence of the convection is the very different thermal behavior of unidirectionally solidified ingots, depending on whether they are solidified vertically upward or horizontally; the ingot solidified horizontally loses its superheat quickly, and the other does not.

Important as thermal convection is, it is often not the only type existing in casting or crystal-growing processes. Convection may be present from pouring momentum, as discussed earlier. Electromagnetic stirring occurs in crystal growing when heating is by induction without shielding. Density differences arising from solute rejected during solidification also lead to convection. In crystal growth, minimum convection is obtained by growing vertically with a positive temperature gradient upward, no transverse temperature gradient, and no electromagnetic stirring. Even here, convection can ensue if the solute-rich liquid rejected at the interface is sufficiently buoyant. A necessary condition for liquid stability is that the liquid have throughout a negative gradient of liquid density $\partial \rho_L / \partial x$ upward:

$$\frac{\partial \rho_L}{\partial x} = \rho_L \left( \beta_t G + \beta_c \frac{\partial c_L}{\partial x} \right) \quad (7\text{-}13)$$

where $\beta_t$ and $\beta_c$ are the thermal and solute coefficients of thermal expansion, respectively. Even though the condition of Eq. (7-13) is met, instability is possible if the solute gradient is destabilizing and the thermal gradient stabilizing. In this case, if a small element of fluid is displaced upward rapidly so that it maintains its solute concentration but increases temperature, it may become progressively more buoyant.

Convection in the bulk liquid of castings and ingots is little affected by the solute rejection occurring on solidification since the diffusion of this solute does not extend significantly beyond the dendrite tips. It is another matter within the liquid-solid zone back of the dendrite tips. Flow of this interdendritic fluid is strongly affected by density differences arising from segregation effects (to be discussed later in this chapter). An interesting type of enhanced convection arises near the outer extremity of the liquid-solid zone when grain structure is equiaxed. The grains here are free to move and are generally denser than the liquid. The result is downward convection of liquid and solid at the outer extremity of the liquid-solid zone.[30-33]

Effects of convection of the bulk fluid on structure of castings and ingots are, for the most part, discussed elsewhere in this book. One important effect is the grain multiplication discussed in Chap. 5. Heat pulses arriving at the dendrite tips from turbulent fluid flow accelerate melting-off of dendrite arms and carry the arms out to the bulk liquid. Many studies have now shown that convection refines the grains present, brings about the columnar-equiaxed transition, and leads to the formation of the outer chill zone in ingot solidification. Reduction or elimination of the convection leads to longer and coarser columnar grains. Enhanced convection (by electromagnetic or other stirring methods) is now used commercially by some casters to obtain very fine grain size. Bulk fluid flow also causes dendrite arms and columnar grains to point preferentially upstream[34,35] because the flow washes rejected solute preferentially from the upstream side. Still another effect of the flow is that it accelerates collision and joining of inclusions, as described in Chap. 6.

## RISERING PURE MATERIALS

Most metals and alloys contract on solidifying; the volume change results from the liquid-solid contraction, which is in the range of about 3 to 6 percent for metals and much higher for refractory oxides. To achieve sound castings and ingots in spite of this shrinkage, appendages are added which are termed *risers*, or *hot tops*. The behavior of a simple riser is seen schematically in Fig. 7-14, where a cube casting and its cylindrical riser are totally enclosed in sand. Solidification takes place simultaneously in both casting and riser, and liquid flows from the riser into the casting. A riser of optimum size is one in which the shrinkage *pipe* extends just to, but not into, the casting.

To calculate the size riser required for this casting, assume that a pure metal is poured exactly at its melting point into a sand mold and that solidification proceeds as described in Chap. 1, with negligible thermal gradients in the solidifying metal. Now visualize the casting and riser as two separate entities separated by an adiabatic surface at the casting-riser junction. Fluid can cross this surface, but no heat can flow by conduction or convection since the liquid and solid metal above and below the surface are at the melting point of the metal.

The solidification time of the cube casting $t_f$ is given by Chvorinov's rule as

$$t_f = C \left( \frac{V_c}{A_c} \right)^2 \quad (7\text{-}14)$$

where the constant $C$ is as in Eq. (1-14), $V_c$ is the volume of the cube casting, and $A_c$ is its surface area less the casting-riser contact area. A similar expression is written for the riser portion as

$$t_f = C \left( \frac{V_{rf}}{A_r} \right)^2 \quad (7\text{-}15)$$

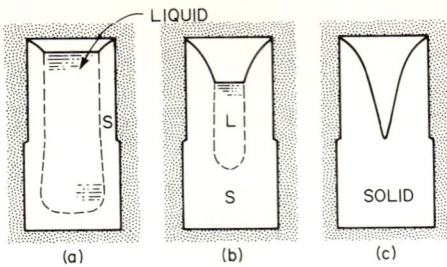

FIGURE 7-14
Solidification of a top risered cube of a pure metal. (*a*) Near beginning of solidification; (*b*) during solidification; (*c*) at end of solidification.

where $A_r$ is the surface area of the riser less that of the casting-riser contact area, i.e., the surface area of the mold cavity surrounding the riser. The top surface is included since heat is transported to this surface from the metal by convection of mold gases and by radiation. $V_{rf}$ is the finally solidified metal in the riser. This is less than the volume of the riser cavity by the amount of the solidification contraction of both the casting and the riser. It is

$$V_{rf} = V_r - \beta(V_r + V_c) \qquad (7\text{-}16)$$

where $V_r$ is volume of the riser cavity and $\beta$ is solidification contraction:

$$\beta = \frac{v_L - v_s}{v_L} = \frac{\rho_s - \rho_L}{\rho_s} \qquad (7\text{-}17)$$

$v_L$ and $v_s$ are specific volumes of liquid and solid, respectively, and $\rho_L$ and $\rho_s$ are their densities. Now, a riser of optimum size will solidify as in Fig. 7-14, with solidification times of casting and riser just equal. Thus, combining Eqs. (7-14) through (7-17) gives

$$(1 - \beta)\left(\frac{V_r}{V_c}\right) = \frac{A_r}{A_c} + \beta \qquad (7\text{-}18)$$

This equation, originally derived by Adams and Taylor,[36] is plotted in Fig. 7-15 for steel and compared with an experimentally determined curve. Steel, of course, is not a pure material, but it freezes over a relatively narrow temperature range. Agreement of the two curves is excellent and perhaps surprisingly so if one recalls the approximations of Chvorinov's rule used in deriving Eq. (7-18).

The equation provides some useful qualitative criteria for visualizing riser behavior. Relatively small risers are required for rangy thin-section castings—in the limit, these need be only a few percent of the casting volume. On the other hand, cylinders, cubes, and other "chunky" shapes require much larger risers. Depending on shape, these risers may need to be nearly as big as the casting itself. Optimum riser geometry is one which has the least surface area for a given size. For a cylindrical top riser, as in Fig. 7-14, this would be one whose height is half its diameter.

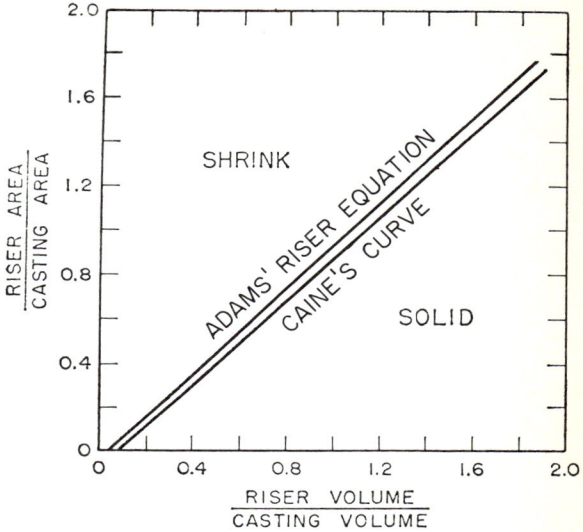

**FIGURE 7-15**
Riser requirement for steel castings. Caine's curve is from experimental data; Adams' is from Eq. (7-18). (*From Adams and Taylor.*[36])

It is possible to improve the efficiency of risers greatly by insulating them. Suppose, for example, the riser in Fig. 7-14 is completely enclosed by a thick layer of insulating material, such as plaster, while the casting remains in sand. The foregoing analysis applies except that now the constants in Eqs. (7-14) and (7-15) are not equal but differ due to the different heat diffusivities of plaster and sand. The final risering equation then becomes

$$(1-\beta)\frac{V_r}{V_c} = \left(\frac{C_c}{C_r}\right)^{1/2}\frac{A_r}{A_c} + \beta \qquad (7\text{-}19)$$

where $C_c$ and $C_r$ are the Chvorinov constants for the casting and riser, respectively. In this case, the more insulating material around the riser permits use of a much smaller riser. Figure 7-16 shows schematically the improvement of riser efficiency through use of insulating materials and also of chills.

Exothermic materials are widely used in foundry and ingot casting to achieve even greater reductions in riser size than are possible with insulating materials. The riser is surrounded with a layer of exothermic material just as if it were insulation. The exothermic material then burns at or close to the melting point of the metal so that it does not add heat to the metal. It does, however, act as an adiabatic surface (i.e., a perfect insulator) to prevent heat from leaving the riser. It is effective in this way until its temperature begins to drop by dissipation of heat to the mold or

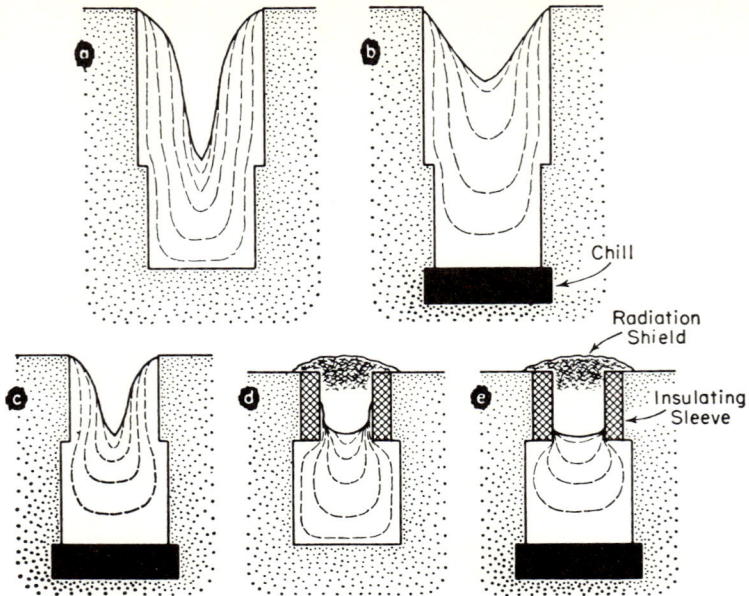

**FIGURE 7-16**
Risering a cube of a pure metal. (*a*) Open riser; (*b*) open riser plus chill; (*c*) small open riser plus chill; (*d*) riser surrounded by insulating or exothermic material; (*e*) same as (*d*) but with chill. (*From Taylor, Flemings, and Wulff.*[2])

surroundings. An alternative practice is sometimes employed for very large risers. Here, the top of the riser is left uncovered or only insulated until it is nearly solid. Then, it is covered with a granular exothermic material that burns at a temperature well above the melting point of the metal cast. The aim is to remelt most of the solid in the riser to extend the riser life.

Another aspect of riser design that is important is placement of the riser on complex castings. The problem is shown schematically in Fig. 7-17, where a *dumbbell-*shaped casting is shown to require either two risers or, if only one riser is to be employed, chilling or insulating must be employed to make solidification *directional* toward the riser. The foundryman must also consider atmospheric pressure in riser design. If a solid skin forms over the top of the riser, pressures lower than ambient develop inside the casting while the atmosphere continues to press on the outside of the skin. Atmospheric pressure eventually punctures the skin at its weakest point, resulting in casting defects in places where they are often not expected. Atmospheric pressure is used to good advantage in the blind (totally enclosed) riser that feeds metal "uphill," as shown schematically in Fig. 7-18. Much has been written on design and placement of risers, especially for steel castings.[3,37,38]

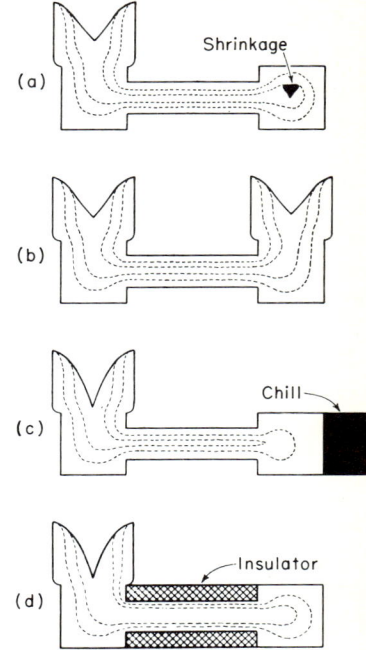

FIGURE 7-17
Risering an isolated heavy section. (*From Taylor, Flemings, and Wulff.*[2])

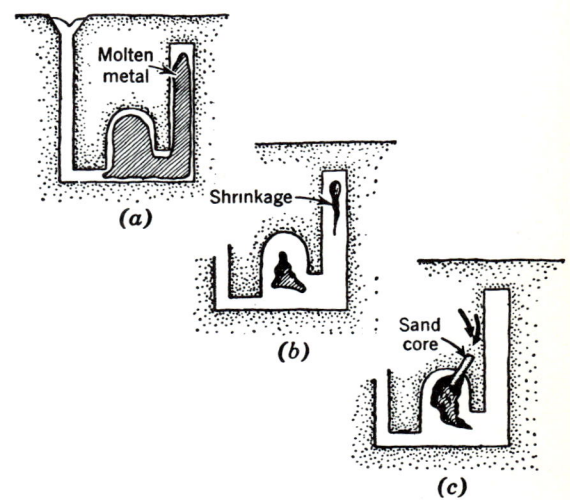

FIGURE 7-18
The blind riser. (*a*) Solid skin forms; (*b*) shrinkage cavity is in both riser and casting; (*c*) metal feeds "uphill" and shrinkage is only in the riser if atmospheric pressure acts on the liquid metal in the riser through a permeable sand core. (*From Taylor, Flemings, and Wulff.*[2])

The risering principles discussed above apply qualitatively to ingot casting as well as to sand casting, although quantitative considerations are more difficult for noninsulating molds. Design of ingot *hot tops*, including selection of exothermic and insulating materials, constitutes an important aspect of ingot solidification technology. Considerable savings can be realized by reducing hot-top sizes. Poor hot-top design leads to shrinkage and macrosegregation, as described later in this chapter.

## INTERDENDRITIC FLUID FLOW

Before turning to risering of alloys that freeze over a wide temperature range, it is useful to describe the fluid flow that occurs between dendrites in the liquid-solid region of a solidifying casting or ingot. Flow occurs here because of solidification shrinkage, contractions of the liquid and solid as they cool, and gravity. For a quantitative description of the flow, we turn again to the model of dendritic solidification developed in Chap. 5 to describe microsegregation. Figure 7-19 shows a volume element exactly like those drawn earlier, but it is drawn here in perspective. The dendrite arms shown are typically of 10 to 100 $\mu$m spacing, and so the channels available for fluid flow in and through the volume element are very small indeed. Flow through this fine mesh obeys approximately the same laws as flow through other finely porous media.[39] Thus, the mean interdendritic flow velocity **v** is linearly related to the pressure gradient as given by Darcy's law:

$$\mathbf{v} = -\frac{K}{\mu g_L}(\nabla P + \rho_L \mathbf{g}_r) \qquad (7\text{-}20)$$

where $K$ is a constant termed the *permeability* of the medium, $P$ is pressure, $\mathbf{g}_r$ is acceleration due to gravity, $\mu$ is viscosity, and $g_L$ is volume fraction of liquid. The permeability $K$ depends on pore size and geometry; it is a scalar in the special case when permeability is independent of orientation, and this is the only case we will consider. Since permeability depends on pore size, it must depend on fraction liquid. By analogy with other porous-media flow problems, it has been proposed[39,40] that this relation is approximately

$$K = \gamma g_L^2 \qquad (7\text{-}21)$$

where $\gamma$ is a constant that depends on dendrite structure and arm spacing. As solidification occurs in the volume element, and flow simultaneously occurs through it, its average density $\bar{\rho}$ changes with time as required by the condition of conservation of mass:

$$\frac{\partial \bar{\rho}}{\partial t} = -\nabla \cdot \rho_L g_L \mathbf{v} \qquad (7\text{-}22)$$

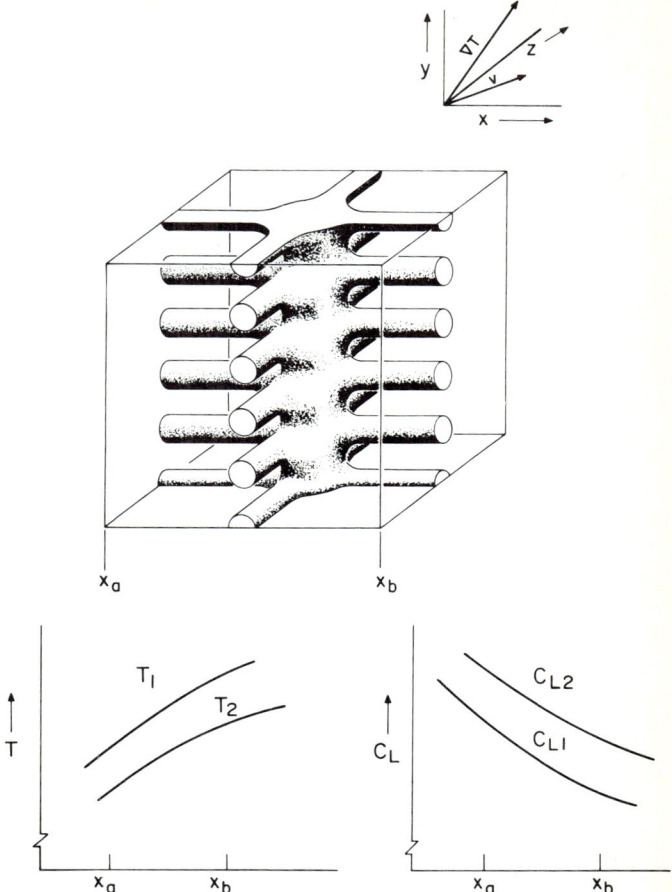

**FIGURE 7-19**
Schematic illustration of a three-dimensional volume element of a binary alloy during solidification. Isotherms are moving in the direction given by the vector $\nabla T$ (top right). Temperature and liquid composition distribution at two successive times along the $x$ axis are given at the lower left and the lower right, respectively. (*From Mehrabian et al.*[40])

where $\bar{\rho}$ is the average density of the volume element and $t$ is time. $\bar{\rho}$ is

$$\bar{\rho} = \rho_s g_s + \rho_L g_L \qquad (7\text{-}23)$$

where $\rho_s$ and $g_s$ are, respectively, the density and volume fraction of solid in the volume element.

The foregoing equations constitute the basic relations describing interdendritic flow in solidification. As an example of their application, we now look at flow and

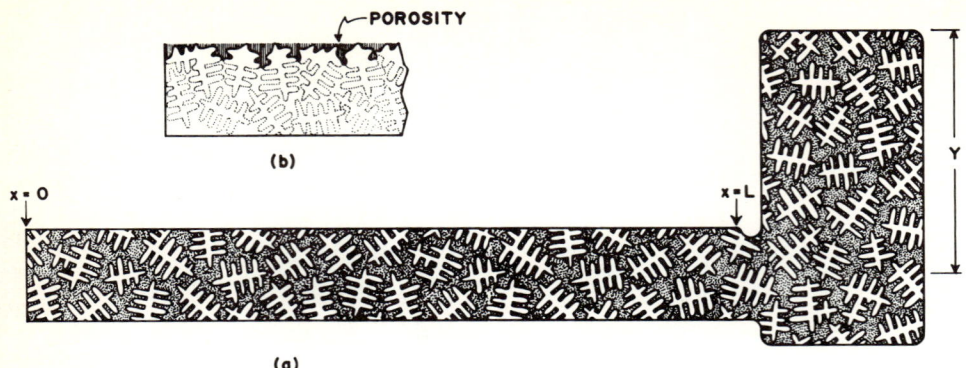

**FIGURE 7-20**
Pore formation in an end-risered plate casting of a wide-freezing-range alloy. (a) Plate during solidification; (b) surface porosity at plate extremity after completion of solidification.

pore formation in a casting of simple cross section of a mushy freezing alloy. Figure 7-20 shows a plate casting of such an alloy solidifying in a metal mold with heat transfer controlled at the mold-metal interface. Since temperature gradients within the plate are negligible, the fraction of solid at any given time is uniform throughout the plate as sketched. To compensate for solidification shrinkage, there must be a general movement of fluid through the interdendritic spaces from right to left. If resistance to this movement becomes too great, micropores form. Quantitatively, this is one of the simpler problems of pore formation to treat.

If the temperature range of solidification is not too large, the rate of solidification is given closely by a heat balance similar to that used in deriving Eq. (1-18). The result is

$$\frac{\partial g_L}{\partial t} = -\frac{A}{V}\frac{h\,\Delta T}{\rho_s H} = -a \qquad (7\text{-}24)$$

where $\Delta T$ is temperature difference between the mean solidification temperature and the mold temperature and $a$ is constant. Since heat flow is $h$-controlled, there are no significant thermal gradients in the solidifying metal and $\partial g_L/\partial t$ is independent of location in the casting. If $x$ is distance along the plate from the end furthest from the riser, Eqs. (7-22) to (7-24) reduce to

$$\frac{\partial}{\partial x}(g_L v_x) = -\frac{\beta}{1-\beta}\,a \qquad (7\text{-}25)$$

or

$$v_x = -\frac{\beta}{1-\beta}\frac{ax}{g_L} \qquad (7\text{-}26)$$

where $v_x$ is velocity of fluid flow (in the $x$ direction). Substituting Eq. (7-26) in (7-20) and integrating yields the pressure at $x$:

$$P = P_0 + \frac{\beta}{1 - \beta} \frac{a\mu}{2\gamma g_L^2} (L^2 - x^2) + \rho_L g_r Y \qquad (7\text{-}27)$$

where $P_0$ is atmospheric pressure, $g_r$ is acceleration due to gravity, and $Y$ is metallostatic head (riser height).

In Eq. (7-27) the second term on the right is the term describing pressure drop from flow; it reaches very high values at low fractions of liquid. The third term on the right is positive and is the pressure head of the molten metal in the riser. Surface shrinkage will occur in the casting when the pressure at $x$ drops below $P_0$ by an amount necessary to overcome surface tension of the liquid at the surface of the casting. Thus, pores form when

$$P \leq P_0 - \frac{2\sigma_{Lv}}{r} \qquad (7\text{-}28)$$

Surface porosity, forming as described above, is a frequent occurrence in sand, permanent mold, and die casting. To the naked eye, it often appears simply as a rough surface, but at relatively low magnifications, it is readily seen to comprise small voids which outline dendrite arms.

In castings that are more complex than the simple plate used as an example, microporosity tends to be localized at the heavier sections in the upper parts of the casting where feeding is poorest. Sometimes this localized *sponge* porosity appears to be internal, but it is nearly always found to be connected to the surface through at least a small pinhole. Formation of pores within a solidifying casting is much more difficult than at the surface because the pressure drop must be sufficient to counterbalance the atmospheric pressure acting on the riser. Thus, the absolute pressure to form an internal void by shrinkage alone is lower by $P_0$ than that required to form surface shrinkage. In addition, there may be some difficulty in nucleating internal voids, but no data are available on this for solidifying systems. When a sufficiently solid skin forms that is not punctured by atmospheric pressure, it can deform the casting surface, pushing it inward to form defects known to foundrymen as *draws*, or *sinks*.

Individual, isolated pores do form in cast metals, and some examples are shown in Fig. 7-21. They form even in unidirectional solidification at least partly because of the pressure drop resulting from flow of fluid through the liquid-solid mushy zone. In formation of internal voids in this way, the contributions of dissolved gas and pressure drop due to fluid flow are additive and it is usually difficult or impossible to state whether the porosity is primarily *gas-* or *shrinkage-caused*. Assuming no difficulty in

**238** FLUID FLOW

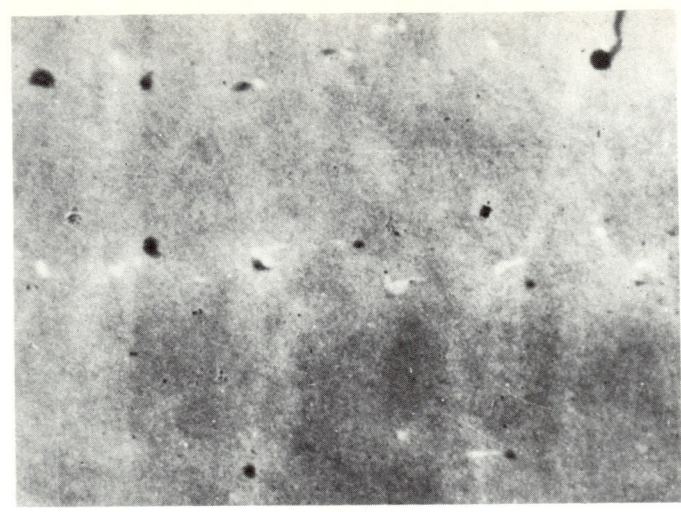

(a)

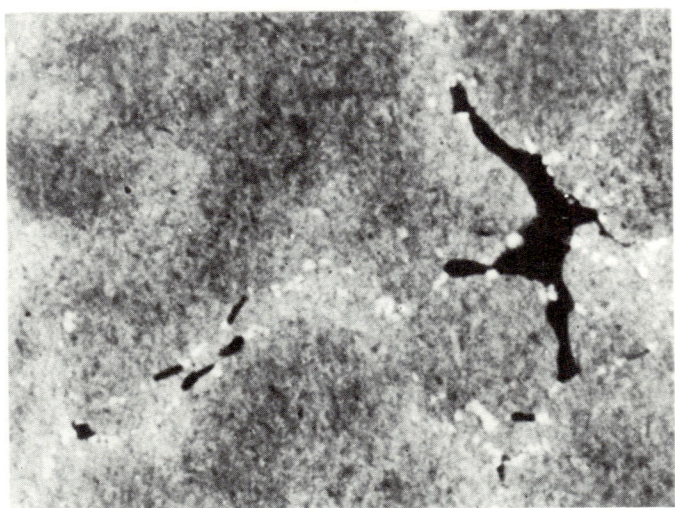

(b)

**FIGURE 7-21**
Microradiographs showing microporosity in low-alloy steel. (*a*) Sand cast, equiaxed grains; (*b*) unidirectionally solidified, columnar grains. Dark areas are microporosity. (*From Tzavaras and Flemings.*[73])

pore nucleation, the condition for formation of a pore of radius $r$ is

$$P \leq P_g - \frac{2\sigma}{r} \qquad (7\text{-}29)$$

where $P_g$ is the equilibrium partial pressure of dissolved gas at a given time during the solidification process, as described in Chap. 6. Thus, dissolved gases contribute to formation of internal microporosity, and the greater the dissolved gas content, the more likely is porosity to be internal rather than surface or surface dishing.

## RISERING ALLOYS

The technology of risering mushy freezing alloys follows the principles outlined earlier for pure metals, but there are important differences. In alloys there is only a limited *feeding distance* over which a riser can act without encountering microporosity from resistance to interdendritic fluid flow. This distance can be greatly increased by use of chills to promote *directional solidification* toward a riser. A second important difference is seen by following the progress of solidification of the top-risered-cube casting in Fig. 7-22. In the beginning of solidification, solid crystallites are free to move and some settling of the crystallites occurs. Solidification shrinkage is accounted for by movement of both liquid and solid. This is sometimes termed the *mass feeding* stage.[37,41] When the crystallites are no longer free to move, feeding is by interdendritic flow from the riser downward. Compare the different solidification behavior of this casting with that of a pure metal (Fig. 7-14). The final macrostructures of these two types of castings are shown in Fig. 7-23.

In this example of the top-risered cube, only very small temperature differences can be maintained between casting and riser and even a very small cube will take a long time to solidify if a large riser is employed. The small temperature differences mean that surface and central porosity form readily, and a casting risered as shown in Fig. 7-22 is difficult to make with a high degree of soundness. A better technique of risering is sketched in Fig. 7-24, where the riser is isolated from the casting by a narrow *neck*, thus reducing heat transport between casting and riser and permitting larger temperature differences to be sustained between casting and riser. The neck also permits the riser to be cut off at less expense. An important aspect of foundry technology is choosing the geometry of these necks so that they act as intended without solidifying before the job of the riser is completed. As with the pure metals discussed earlier, insulating and exothermic materials are widely and effectively used with mushy solidifying alloys.[42,43]

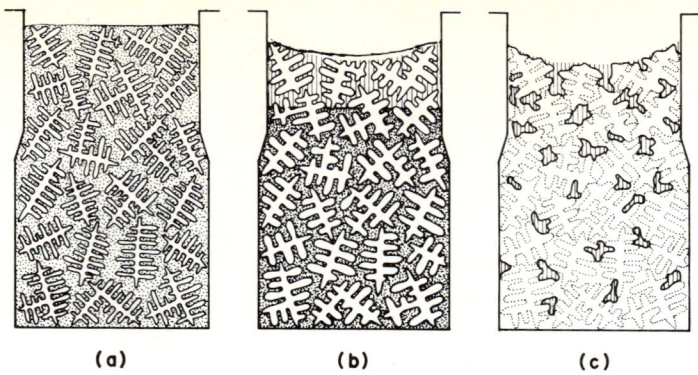

**FIGURE 7-22**
Solidification of a top-risered cube of a mushy freezing alloy (schematic). (*a*) Mass feeding stage; (*b*) interdendritic feeding stage; (*c*) solidified casting showing microporosity.

**FIGURE 7-23**
Sections of top-risered cubes of aluminum. *Left:* pure aluminum; *right:* Al–4.5% Cu alloy.

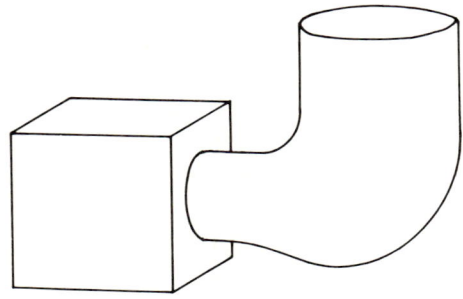

**FIGURE 7-24**
Neckdown side riser on a cube casting.

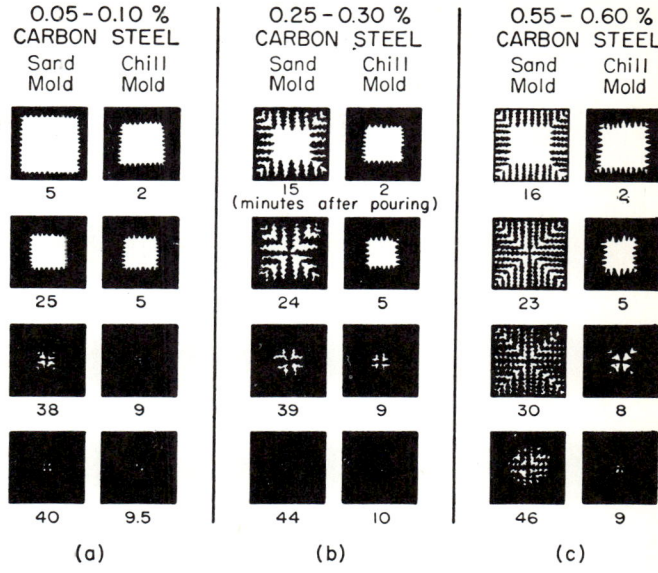

**FIGURE 7-25**
Solidification of cross sections of 7-in sq bars of cast steel. (*From Taylor, Flemings, and Wulff*[2] *after Bishop and Pellini.*[45])

Many important commercial alloys freeze with a relatively narrow mushy zone, as shown schematically in Fig. 5-29c. These include low-alloy and plain-carbon (low-carbon) steel in sand molds, and some aluminum alloys in permanent molds. The solidification behavior of plain-carbon steel is shown schematically in Fig. 7-25. Surface and internal microshrinkage can occur in these alloys, but the more troublesome porosity is that which develops at or near the centerline. This porosity is arranged interdendritically and often has a V-type distribution with the Vs always pointing away from the riser end, as sketched in Fig. 7-26. The mechanism of formation is qualitatively the same as that for internal pore formation in more mushy alloys. In the absence of dissolved gas, the pores form at local absolute pressures somewhere in the neighborhood of zero atm [Eq. (7-29)], and so there must be a pressure drop along the length of the casting of about 1 atm. It is thought that this large pressure drop leads to a sort of mass feeding late in solidification and that as the solid dendrites are forced away from the riser end, small fissures open up giving the characteristic V pattern. Much has been written, especially for steel, concerning the distance which a riser can "feed" a steel casting. One general conclusion of the experimental work conducted is that unless there is a measurable positive gradient along the casting centerline (about $1\frac{1}{2}$°C/cm), centerline shrinkage results.[44] Many

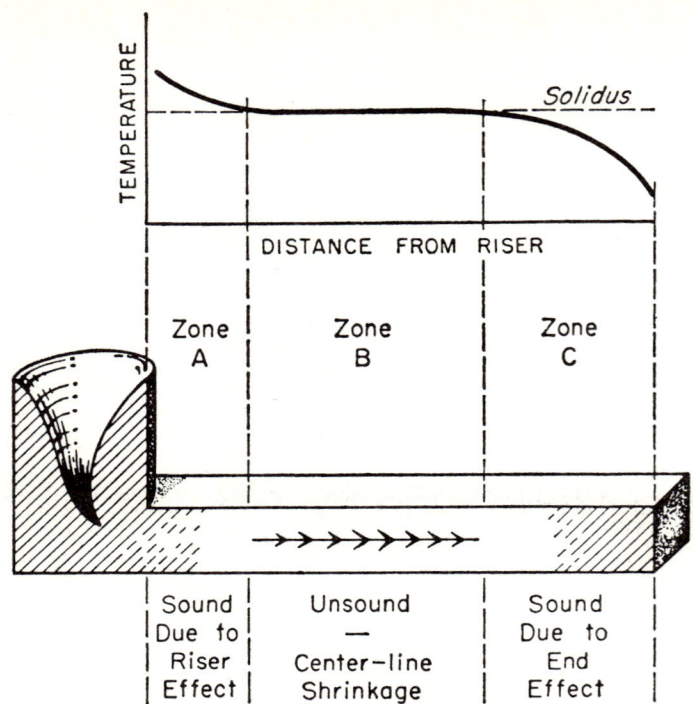

**FIGURE 7-26**
Formation of centerline shrinkage in cast steel. (*From Taylor, Flemings, and Wulff.*[2])

data are available on riser and chill placement on steel castings to obtain castings free of discernible centerline.[38]

Gray and ductile cast iron present an interesting and unusual risering problem to the foundryman. Figure 7-27 shows schematically the solidification of a hypoeutectic gray iron based on the thermal analysis data of Bishop and Pellini.[45] For the bar of 7-in-sq cross section studied, only austenite solidifies for the first 40 min. During that time, substantial metal shrinkage occurs, but this is readily fed by small risers. Thereafter, the casting is at the eutectic temperature, and eutectic of austenite plus graphite forms. During this period, little or no feed metal is required, and, in fact, depending on the metal composition, feed metal may actually be forced back up the riser as the eutectic liquid undergoes expansion on solidification.[46] This expansion, when controlled, means that only very small risers are necessary to produce sound gray-iron castings, and this is one of the reasons why gray iron is so economical to cast. When, however, the mold employed is weak, the graphite expansion often

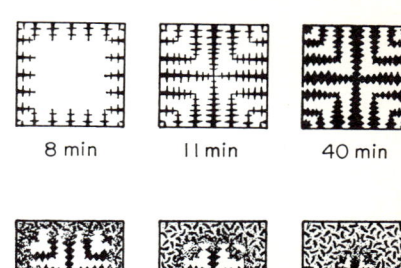

FIGURE 7-27
Solidification of sections of 7-in sq sand-cast bars of cast iron. (*From Taylor, Flemings, and Wulff*[2] *after Bishop and Pellini.*[4,5])

works against the mold walls, pressing these outward and increasing rather than decreasing the demand for feed metal. Control of mold strength is therefore an important part of cast-iron-foundry technology. Figure 7-28 is a schematic diagram illustrating how mold strength, thermal conditions, and atmospheric pressure interact to produce different types of shrinkage in cast iron.

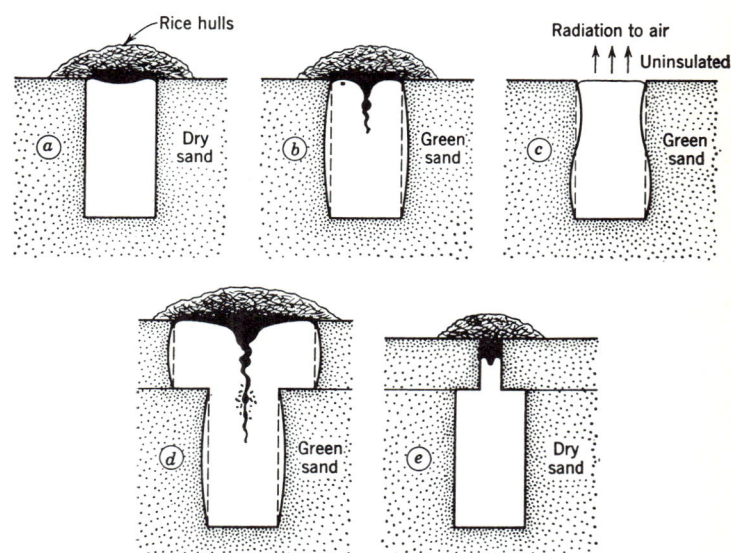

FIGURE 7-28
Schematic illustration of influence of mold-wall movement on shrinkage of gray iron. (*From Taylor, Flemings, and Wulff.*[2])

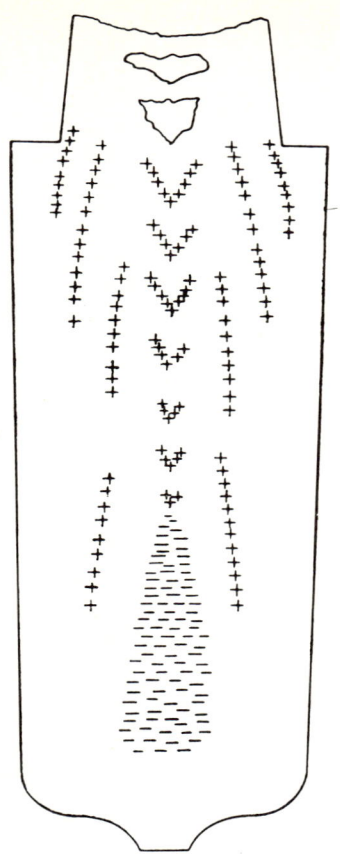

**FIGURE 7-29**
Macrosegregation in a killed steel ingot. + denotes regions of positive segregation; − denotes regions of negative segregation. (*From Ref. 47.*)

## MACROSEGREGATION

When liquid alloys of uniform composition are solidified in a casting or ingot mold, segregation arises on the scale of the dendrite arm spacing (microsegregation) and on a much larger scale (macrosegregation). Both types are inherent adjuncts to the usual casting processes. Figure 7-29 is a well-known diagram of some of the types of macrosegregation encountered in large, killed steel ingots. The centerline of the ingot is usually richer in solute than the overall average. Often macroetched sections show streaks, or *channel segregates*, arranged in V pattern and associated with porosity. Solute-rich streaks are also often found of more nearly vertical orientation in the upper and outer portions of the ingot. These are termed *A segregates*. A cone of segregation is present at the base of the ingot which is usually poor in alloy elements

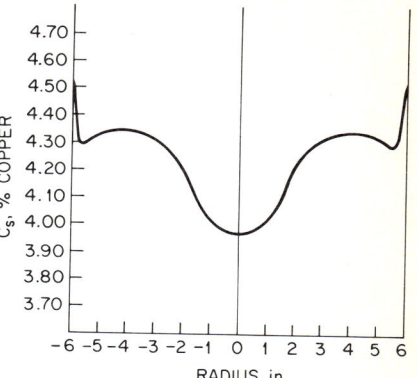

FIGURE 7-30
Segregation in a cylindrical continuously cast Al–Cu ingot produced under relatively poor casting conditions. (*Courtesy W. S. Peterson, Olin Metals Research Laboratories.*)

and rich in some types of inclusions. These and other types of macrosegregation in steel have been described in detail,[47-50] and much empirical effort has been devoted to reducing the severity of the segregation. In large ingots, it is not uncommon to find variations in carbon and other alloys elements of 30 percent or more of the amount of the alloy present (and much higher variations within local channel segregates). Good ingot design can reduce the variations substantially. Figure 7-30 shows segregation in a different type of ingot, an Al–Cu alloy continuous casting, produced under relatively poor casting conditions. The center of the ingot is low in alloy and the outer portion high. Altering casting conditions can reduce segregation substantially or cause the center, as well as the surface, to be high in alloy. In the latter case, an alloy impoverished region is found at midradius.

The cause of macrosegregation in castings and ingots is now understood to be the movement of liquid or solid within the mushy zone. Highly segregated phases are present within the mushy zone during solidification. Physical displacement of these phases leads to macrosegregation. One way in which this displacement occurs is by floating or settling of precipitated phases early in solidification. Equiaxed grains form early in solidification, and since they are not attached to other grains, they may float or settle. Inclusions may also float or settle. An especially pronounced example of this type of segregation is the *kishing* of graphite in cast iron (in which graphite flakes float to the top of castings of hypereutectic gray iron). A second way in which the displacement occurs is by movement of liquid within the liquid-solid zone as a result of thermal contractions, solidification shrinkage, and density differences in the interdendritic liquid. This liquid movement now appears to be by far the most important and general cause of segregation.[51-67] (It is discussed in quantitative detail in the next section.)

## QUANTITATIVE ASPECTS OF MACROSEGREGATION

Let us look again at interdendritic fluid flow through the volume element sketched in Fig. 7-19, assuming for the moment no solid movement. Conservation of solute mass in the volume element during solidification requires that

$$\frac{\partial}{\partial t}(\bar{\rho}\bar{C}) = -\nabla \cdot \rho_L g_L C_L \mathbf{v} \qquad (7\text{-}30)$$

where $\bar{\rho}$ and $\bar{C}$ are the average density and composition, respectively, of the volume element. Following suitable manipulations,[54] Eq. (7-30) becomes

$$\frac{\partial g_L}{\partial C_L} = -\left(\frac{1-\beta}{1-k}\right)\left(1 + \frac{\mathbf{v}\cdot\nabla T}{\varepsilon}\right)\frac{g_L}{C_L} \qquad (7\text{-}31)$$

where $\partial g_L/\partial C_L$ is the differential change in fraction liquid with differential change in liquid composition (and hence differential change in temperature) within the volume element located at $x, y, z$; $\varepsilon$ is the local rate of temperature change. Equation (7-31) is the basic *local solute redistribution equation* used to calculate macrosegregation. It is written for the general case of three-dimensional heat and fluid flow assuming constant solid (but not necessarily liquid) density during solidification, negligible net solute change from diffusion, and no pore formation. The equation describes the influence of this flow behavior on the composition of the liquid (and hence solid) at each fraction solid. It therefore also describes the overall final average composition at a given part of an ingot—and so, the *macrosegregation*. Together with Eq. (7-20), which describes the interdendritic flow, it is the basic tool used to describe macrosegregation quantitatively.

For no solidification shrinkage and no flow, Eq. (7-31) reduces to the differential form of the Scheil equation [Eq. (2-3)], and no macrosegregation results. The effect of flow is to change the fraction liquid at a given liquid composition, and hence temperature. Figure 7-19 shows the temperature and liquid composition in the volume element in the $x$ direction at two times during solidification. Flow does not alter these variables, but it does change the relative amounts of liquid and solid present. Flow from hot regions to cold increases the fraction solid and, therefore, lowers average composition $\bar{C}$ (for $k < 1$). Flow in the opposite direction increases fraction liquid and, therefore, increases $\bar{C}$. We now look at macrosegregation quantitatively in some simple laboratory experiments before returning to the industrial examples of Figs. 7-29 and 7-30. In the discussion, it is assumed that $k < 1$. For alloys of $k > 1$, segregation would be of opposite sign.

When an ingot of wide freezing range is poured against a chill (with finite $h$ resistance to heat flow), a solute-rich region is obtained in the vicinity of the chill;

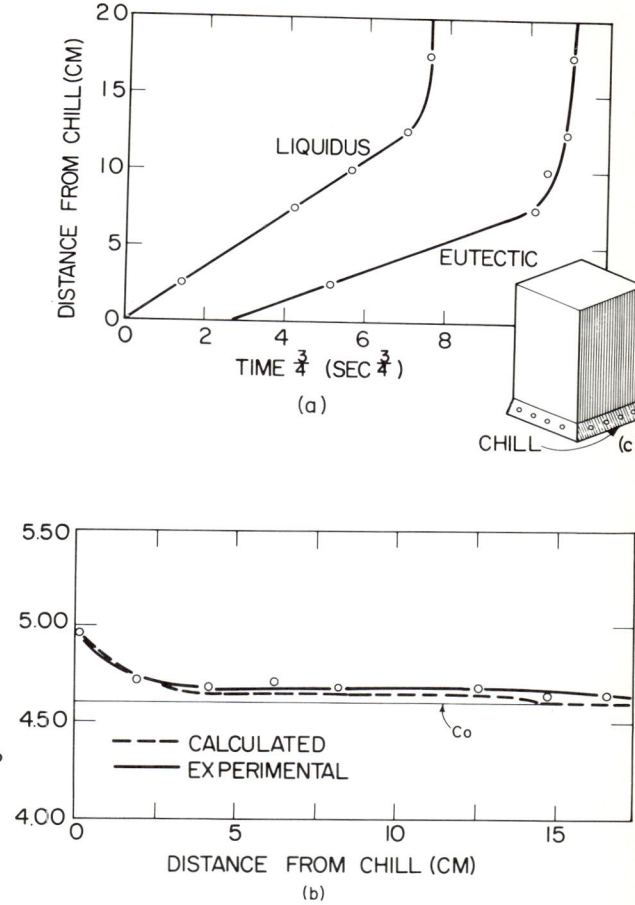

FIGURE 7-31
Macrosegregation in a unidirectionally solidified Al–4.5% Cu alloy. (a) Movement of isotherms; (b) macrosegregation; (c) ingot. (*From Flemings and Nereo.*[56])

this is called *inverse segregation*. Figure 7-31 is an example. It has long been understood that this phenomenon results from interdendritic flow of liquid metal. The maximum segregation to be expected is readily calculated with the aid of Eq. (7-31). At exactly the chill face, flow perpendicular to the chill must be zero, and so $\mathbf{v} \cdot \nabla T/\varepsilon = 0$. Assuming $\beta$ and $k$ constant, Eq. (7-31) integrates to a simple equation that is nearly, but not exactly, the *Scheil equation* [Eq. (2-4b)]:

$$C_L = C_0 g_L^{-[(1-k)/(1-\beta)]} \qquad (7\text{-}32)$$

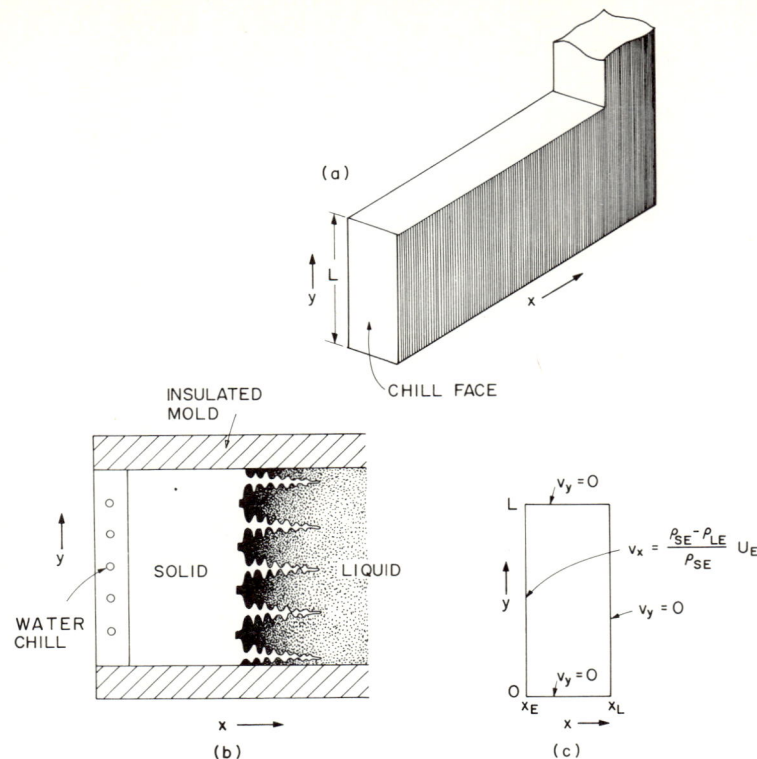

**FIGURE 7-32**
Horizontal unidirectional solidification. (*a*) Ingot; (*b*) mushy zone; (*c*) boundary conditions for interdendritic flow. (*From Mehrabian et al.*[40])

The area under a curve plotted from Eq. (7-32) as $C_s = kC_L$ and $g_s = 1 - g_L$ is the local average composition $\bar{C}_s$. It can readily be seen that for $k < 1$ this average composition is higher than $C_0$; that is, the region is rich in solute. At distances away from the chill, calculations of macrosegregation are more difficult but have been carried out by several authors.[52-56] Figure 7-31 is an example. Any abrupt variation in rate of isotherm movement causes local variations in $\mathbf{v} \cdot \nabla T/\varepsilon$ and leads to the solute *banding* often seen in castings and ingots. This type of segregation is also quantitatively described by Eq. (7-31). It therefore forms by a quite different mechanism than the solute banding in single-crystal growth discussed in Chap. 2.

An interesting difference in fluid-flow behavior occurs when an ingot is solidified unidirectionally but horizontally, as in Fig. 7-32. Now, the liquid along the horizontal axis is of varying composition and temperature and therefore varying density.

Thus, gravity provides an additional driving force for convection. If, as in the aluminum–copper system, the cooler, more solute-rich alloy is denser, then a general flow pattern exists that is downward and to the left. The flow pattern, sketched in Fig. 7-33b, is calculated using Eq. (7-20), with boundary conditions given in Fig. 7-34. Once **v** is determined in this way, then macrosegregation is readily calculated from Eq. (7-31) and is shown in Fig. 7-33. Strictly speaking, Eqs. (7-20) and (7-31) should be solved simultaneously, but little error is introduced for moderate levels of macrosegregation by *uncoupling* the equation describing fluid flow and that describing fraction of liquid at a given liquid composition.

An interesting phenomenon results when the component of fluid flow in the growth direction exceeds the isotherm velocity.[60–64] When this happens, each small element of liquid metal finds itself in progressively hotter regions as it flows. As these small elements of liquid change both in temperature and composition to conform to those of the hotter region, local melting occurs. In local regions where, by chance, a bit more melting occurs than in other regions, resistance to flow is decreased. Flow here then increases, and the greater the flow, the more the remelting. Thus, there is flow instability, and localized highly segregated *channels* result. The mathematical statement for the requirement of remelting and hence channel segregate formation is

$$\frac{\mathbf{v} \cdot \nabla T}{\varepsilon} < -1 \qquad (7\text{-}33)$$

and the foregoing analysis can be used to predict when the channels will form. Once they form, the channels make the fluid flow difficult to describe quantitatively, and attempts have not yet been made to do this. Figure 7-34 shows a channel which has formed in an aluminum–copper ingot designed to solidify with high negative value of $\mathbf{v} \cdot \nabla T / \varepsilon$. Flow is downward and to the right, and this is the orientation the channels take. McDonald and Hunt[61] show a similar example for a transparent material, where the flow occurs upward because the more segregated (although cooler) liquid is less dense than that liquid closer to the dendrite tips.

We are beginning to be able to apply our developing understanding of interdendritic flow to a description of real macrosegregation problems. As an example, the A segregates of Fig. 7-29 are channel segregates forming almost certainly just as do the channel segregates in the ingot of Fig. 7-34. Similar segregates, sometimes termed *freckles*, are seen in highly alloyed materials cast by other processes and have the same origin.[63] The centerline V-type segregates seem to be partly due to solid movement under the influence of atmospheric pressure and gravity. However, once such fissures form, they provide ideal channels for preferential flow to occur from cooler to hotter regions. The resulting segregates often are pencillike in cross section and show marked signs of local melting. Thus, both solid movement and fluid flow

**250** FLUID FLOW

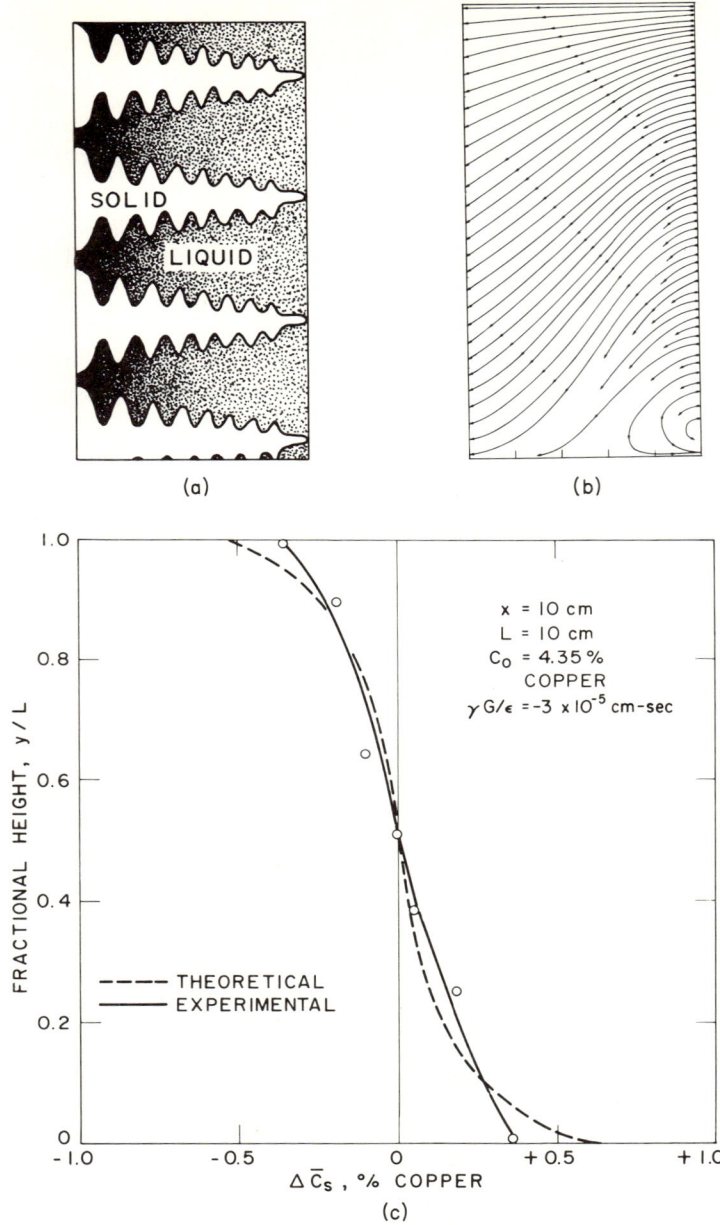

**FIGURE 7-33**
Macrosegregation in a horizontal unidirectionally solidified ingot. (*a*) Sketch of mushy zone; (*b*) calculated flow lines in mushy zone; (*c*) macrosegregation. (*From Flemings and Mehrabian.*[74])

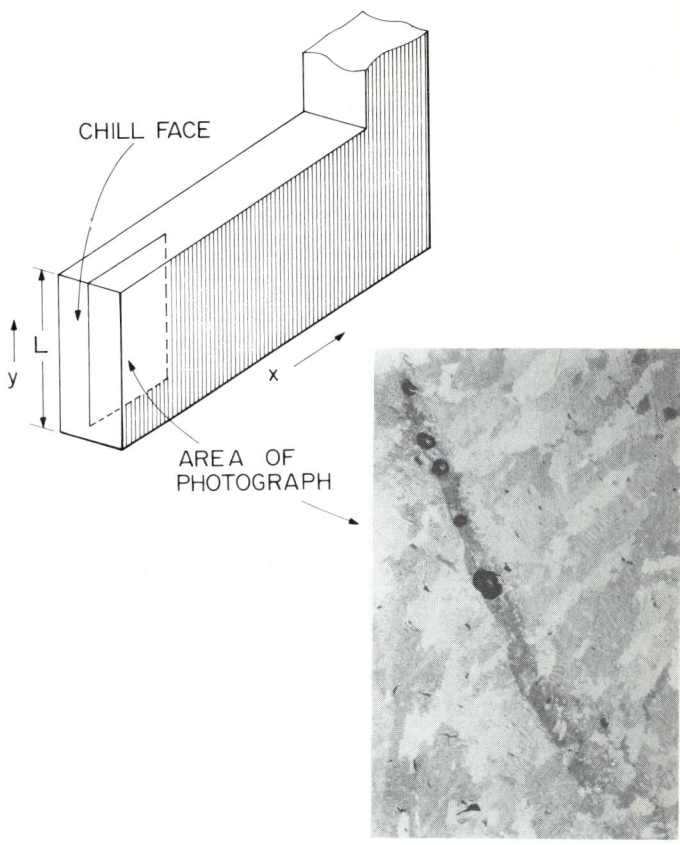

FIGURE 7-34
Channel-type segregate ("freckle") in horizontally solidified Al–20% Cu ingot. (*From Mehrabian et al.*[60])

according to Eq. (7-33) apparently contribute to these segregates. The *cone* of segregation at the base of ingots has been widely attributed to settling of crystallites, which may ultimately prove to be the correct answer. However, the final segregation can also be explained by an interdendritic fluid-flow mechanism, and the matter stands unresolved.[58,65–67]

To explain segregation such as that of Fig. 7-30 for continuous or semicontinuous casting, we turn to a simple steady-state visualization of what flow behavior might be like in the interdendritic mushy zone. If, and only if, all such flow is vertically downward, as in Fig. 7-35a, will the final composition across the ingot be uniform.

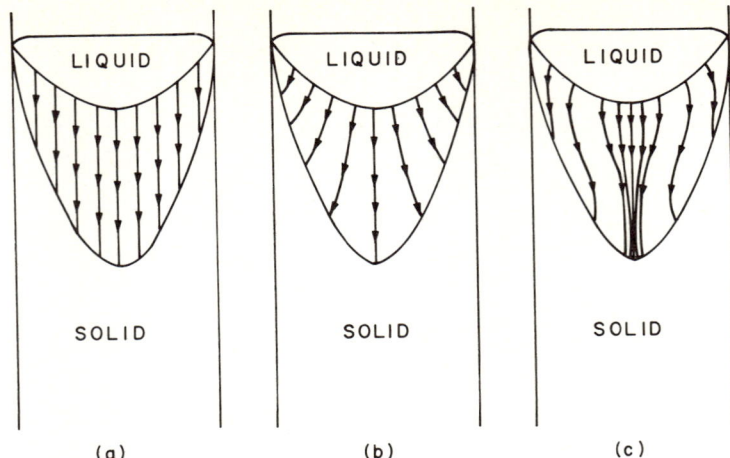

**FIGURE 7-35**
Interdendritic fluid flow in consumable electrode or continuous casting. (a) Limiting case, all flow vertical—no segregation results; (b) flow resulting in negative segregation at ingot center; (c) flow resulting in positive segregation at ingot center. (*From Mehrabian et al.*[40])

Hydrodynamic considerations do not permit this simple a flow pattern. Resistance to flow is less near the center of the ingot, and so if liquid density is nearly uniform, flow lines tend to fan outward, as in Fig. 7-35b. The result is negative segregation near the ingot center, as shown also by the real ingot of Fig. 7-30. When they fan inward, as they can do either as a result of fissuring or because liquid density increases during solidification, then positive segregation near the ingot centerline results. An important additional type of segregation results when the horizontal component of flow does not stop at the casting surface, as in Fig. 7-35b. When the semisolid external surface of the ingot begins to contract from the mold wall, it leaves a space into which the interdendritic liquid can flow. The result is a thin layer of highly segregated material referred to as *exudation*. Even if the flow does stop at the casting surface, a region of positive segregation results [see Eq. (7-32)]. This *inverse segregation* is different than, and usually a less severe problem than, exudation.

## MOVEMENT OF LIQUID PLUS SOLID

Some examples of solidification problems involving simultaneous movement of liquid and solid have already been mentioned in this chapter. One of these is mass feeding of mushy alloys, where both the liquid and solid move in the early stages of

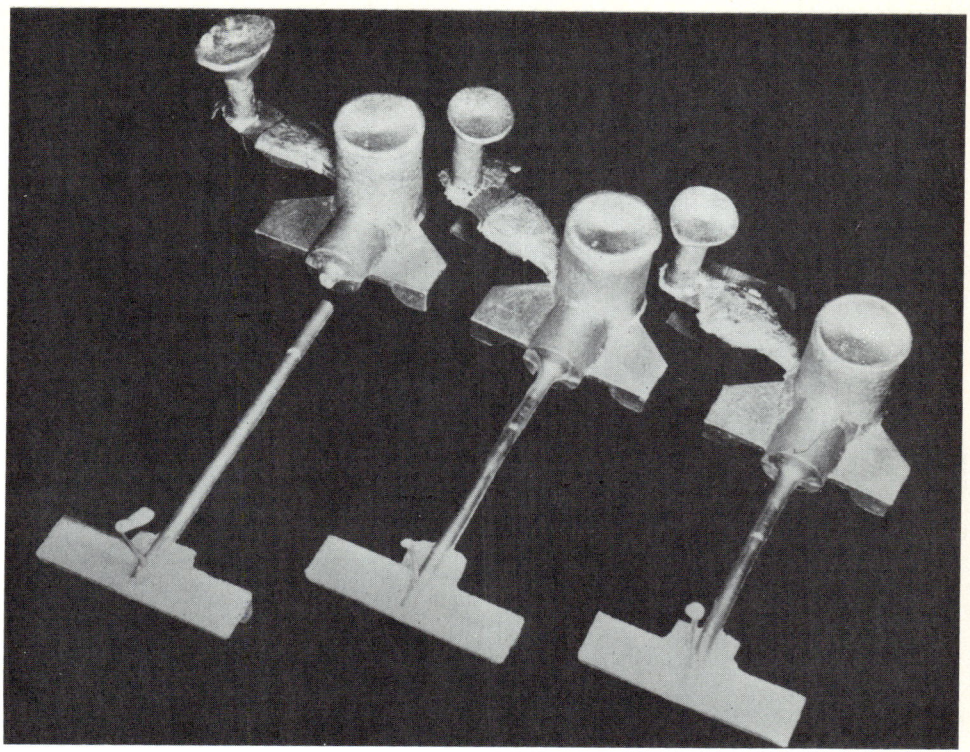

FIGURE 7-36
Restrained bar test castings. Shorter castings did not tear. (*From Rosenberg et al.*[68])

solidification to compensate for solidification shrinkage. During the mass-feeding stage, the top surface of the metal in the riser remains flat, dropping downward as solidification proceeds; this is shown schematically in Fig. 7-22.

As solidification proceeds, a point is reached where the solid is no longer able to move readily, after which the solid skeleton that has formed begins to develop strength. From this point on, solidification shrinkage must be fed by interdendritic flow. The developing strength of the solid network can now cause localized strains with resultant formation of highly segregated regions, or open *hot tears*. Consider as an example a sand-cast bar of a mushy freezing alloy. The bar is restrained at both ends as sketched in Fig. 7-36. As the thinner section cools below the temperature where it develops strength, it begins to contract. The result is strain at the weakest point, which is the *hot spot* where the thin section joins the thick section. When metal can no longer mass feed to the hot spot, the contraction strains pull the solid dendrites

**FIGURE 7-37**
Microradiograph of a thin section from the hot spot of the casting of Fig. 7-36 (Al–10% Cu). (*From Rosenberg et al.*[68])

apart at this location. If the casting is well fed, there is now a stage where liquid can flow between the separating dendrites to *heal* the incipient tears. Regions of segregation result. Mild segregation appears as solute-rich grain boundaries, as shown in Fig. 7-37. More severe segregation is readily discernible on macroetched surfaces.

As solidification proceeds, a time is reached when liquid can no longer flow to compensate for the strain. At this stage, if the strain continues, open fractures result which are termed *hot tears*. Alternately, the casting might develop enough strength at this point to resist tearing. One way to obtain an empirical measure of hot-tearing resistance of alloys is to use a restrained bar casting such as that described above. The longer the bar that can be made without tearing, the greater is the resistance of the metal to hot tearing. Figure 7-38 shows results of the work of Rosenberg et al.[68] on alloys from the Al–Cu and Mg–Al binary systems. Minimum resistance to hot tearing is found in these (and many other) alloys at compositions intermediate between the pure metal and a eutectic composition. Alloys in this composition range typically solidify such that a moderate amount of liquid crystallizes over a large temperature range at the end of solidification. Thus, there is a large temperature interval over which thermal contraction can operate to produce hot tears. Alloys with lesser

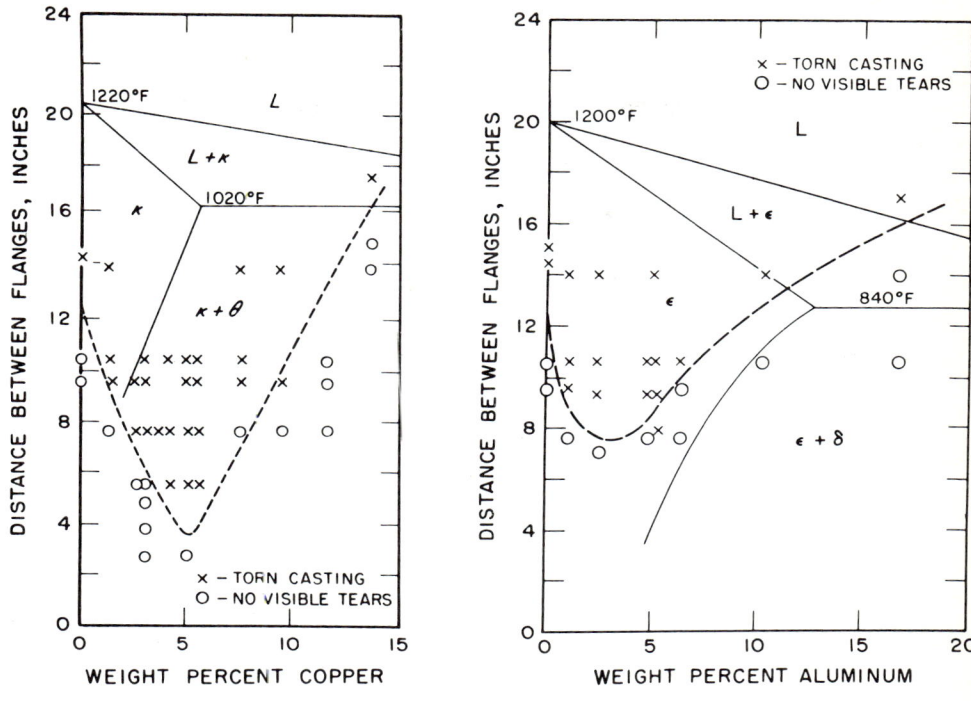

FIGURE 7-38
Hot tearing characteristics of aluminum–copper alloys (left) and magnesium–aluminum alloys (right). (*From Rosenberg et al.*[68])

amounts of solute are largely solid near their liquidus and so develop sufficient strength to resist hot tearing near the liquidus. Alloys with large amounts of solute do not develop high strength until at or near the eutectic temperature, and so hot tearing can occur only over a relatively narrow temperature region near the eutectic. Effect of alloy composition on curves of fraction of liquid versus temperature has been discussed in Chap. 5 (Fig. 5-27a).

One expects that, not only the amount of residual liquid, but also its shape, as determined by relative liquid-solid surface energies, must also affect hot-tearing resistance of a given alloy. Residual liquids which spread as a film over dendrite arms and grain boundaries prevent development of strength in the mushy zone until very near the end of solidification. Another factor affecting hot-tearing resistance is grain size: the finer the grain size, the greater the resistance. This is now understood to be because the fine-grained semisolid material is weaker than the coarse-grained material, not stronger. It develops measurable strength at a later stage of solidification, and

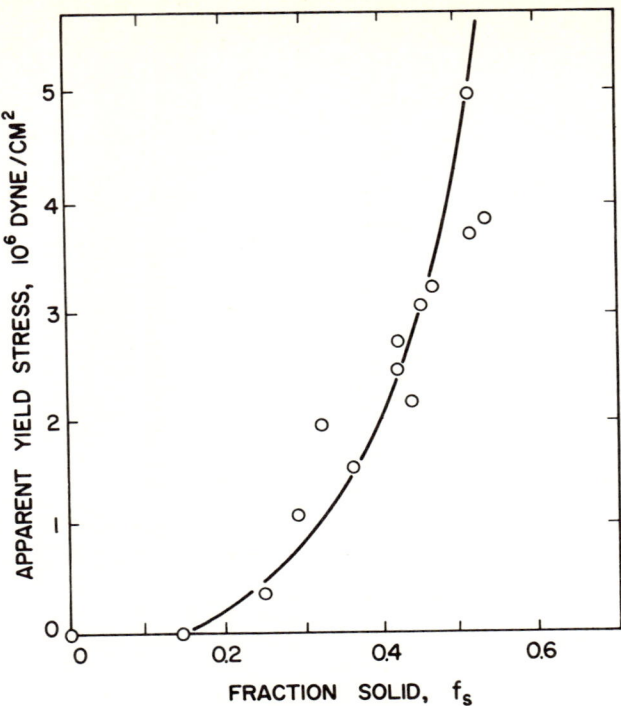

**FIGURE 7-39**
Apparent yield stress versus fraction of solid for cast Pb–15% Sn alloy. (*From Spencer et al.*[71])

until that time is free to compensate for strains by movement of both solid and liquid.[69] In practice, whether or not a casting hot-tears depends also on the casting design, thermal gradients as influenced by risering and chilling, and the strength of mold material.

A number of examples are given in the literature of measurements of strengths of partially solidified alloys.[68-71] All of these tests show the strength is very low until some given fraction of solid is reached. Usually, this is in the range of fractions solid of 0.2 to 0.4. An example is given in Fig. 7-39 from work of Spencer et al.[71] The stress plotted in this figure is the maximum (termed *apparent yield stress*), which occurs after rather large strain. At still larger strains, shear stress drops to low values. The strain occurs primarily by grain-boundary displacement, although some dendrite arm bending and breaking appear also to be involved. The reduction in shear stress at large strains occurs when the original cast structure is sufficiently disturbed so that smooth localized paths for the strain develop.

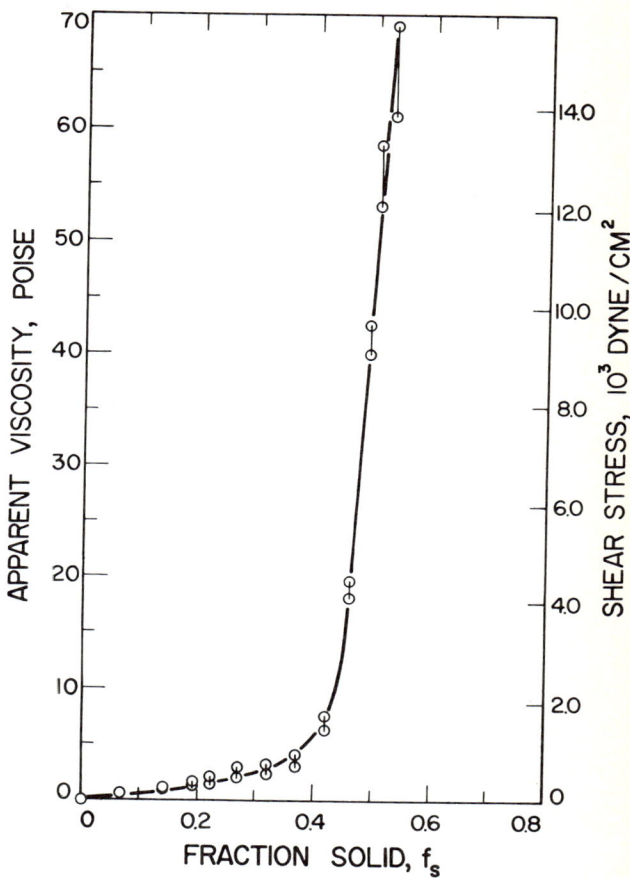

**FIGURE 7-40**
Apparent viscosity of vigorously agitated Pb–15% Sn alloy. (*From Spencer et al.*[71])

Recently, Spencer et al.[71] have shown that for several alloys, including Pb–15% Sn, vigorous agitation during solidification can prevent significant strength from developing until the fraction of solid is well above 0.4. The vigorous agitation also results in fine, nearly spherical grains. At fractions of solid below about 0.4, this liquid-solid mixture behaves as do many types of slurries at equivalent volume fractions of solid. Its effective viscosity at this fraction of solid (as long as agitation is maintained) is less than that of medium-grade motor oil. Figure 7-40 shows the viscosity of the agitated alloy as a function of fraction of solid. Also shown in this figure is the equivalent shear stress to cause steady flow, which is several orders of magnitude lower than that required when the liquid is not agitated, as can be seen by comparing Figs. 7-39 and 7-40. Perhaps this interesting property of agitated partially solidified alloys will find engineering application. One potential application is die casting of copper-base or ferrous alloys where improved die life might be expected by casting a liquid-solid slurry instead of superheated liquid metal.

# REFERENCES

1 "Vannocio Biringuccio, Pirotechnia," p. 254 (1540) (transl. by C. S. Smith and M. T. Gnudi), AIME, New York (1942).
2 TAYLOR, H. F., FLEMINGS, M. C., and WULFF, J.: "Foundry Engineering," John Wiley & Sons, Inc., New York, 1959.
3 FLINN, R. A.: "Fundamentals of Metal Casting," Addison-Wesley Publishing Company, Inc., Reading, Mass., 1963.
4 RUDDLE, R. W.: "The Running and Gating of Castings," *Inst. Metals Monogr. Rept. Ser.* No. 19, 1956.
5 HLINKA, J. W., PASCHKIS, V., and PULIR, F. S.: *Trans. AFS*, **69**:527 (1961).
6 FLEMINGS, M. C., MOLLARD, F. R., NIIYAMA, E. F., and TAYLOR, H. F.: *Trans. AFS*, **70**:1029 (1962).
7 RAGONE, D. V., ADAMS, JR., C. M., and TAYLOR, H. F.: *Trans. AFS*, **64**:640 (1956) and **64**:653 (1956).
8 NIESSE, J. E., FLEMINGS, M. C., and TAYLOR, H. F.: *Trans. AFS*, **67**:685 (1959).
9 FLEMINGS, M. C.: *Brit. Foundryman*, **57**:312 (1964).
10 KRYNITSKY, A. L.: *Trans. AFS*, **61**:399 (1953).
11 HOLMSHAW, R., and LAVENDER, J. D.: *Nondestructive Testing*, **99** (May 1969).
12 KOKUSHKIN, D. P., et al.: *Stahl (in English)*, **3**:192 (1966).
13 BIRD, R. B., STEWART, W. G., and LIGHTFOOT, E. N.: "Transport Phenomena," John Wiley & Sons, Inc., New York, 1960.
14 SZEKELY, J., and STANEK, V.: *Met. Trans.*, **1**:119 (1970).
15 SZEKELY, J., and THEMELIS, N. J.: "Rate Phenomena in Process Metallurgy," John Wiley & Sons, Inc., New York, 1971.
16 COLE, G. S.: in "Solidification," p. 201, American Society for Metals, Metals Park, Ohio, 1971.
17 SZEKELY, J., and STANEK, V.: *Met. Trans.*, **1**:2243 (1970).
18 WILCOX, W. R.: in M. Zeif and W. R. Wilcox (eds.), "Fractional Solidification," p. 157, Marcel Dekker, Inc., New York, 1967.
19 COLE, G. S.: *Trans. AIME*, **239**:1287 (1967).
20 SZEKELY, J., and CHABRA, P. S.: *Met. Trans.*, **1**:1195 (1970).
21 COLE, G. S., and WINEGARD, W. C.: *Can. Met. Quart.*, **1**:29 (1962); *J. Inst. Metals*, **93**:153 (1965).

22 MUELLER, A., and WILHELM, M.: *Z. Naturforsch.*, **19a**:254 (1964).
23 WILCOX, W. R., and FULLMER, C. D.: *J. Appl. Phys.*, **36**:2201 (1965).
24 UTECH, H. P., and FLEMINGS, M. C.: *J. Appl. Phys.*, **37**:2021 (1966).
25 UTECH, H. P., and FLEMINGS, M. C.: in H. S. Pieser (ed.), "Crystal Growth," p. 651, Pergamon Press, New York, 1967.
26 HURLE, D. T. J.: in H. S. Peiser (ed.), "Crystal Growth," p. 659, Pergamon Press, New York, 1967.
27 HURLE, D. T. J., JAKEMAN, E., and PIKE, E. R.: *J. Crystal Growth*, **3,4**:633 (1968).
28 STREETER, V. L. (ed.): "Handbook of Fluid Dynamics," McGraw-Hill Book Company, New York, 1961.
29 BOWER, T. F.: private communication.
30 REYNOLDS, J. A., and PREECE, A.: *Proc. Inst. Brit. Foundrymen*, **48**:101 (1955).
31 KOHN, A.: "The Solidification of Metals," Iron and Steel Institute Publ. No. 110, p. 357 1968.
32 MIYAGAWA, T., et al.: *Tetsu To Hagane*, **55**:338 (1969).
33 GRAY, B.: *J. Iron Steel Inst.*, **182**:366 (1956).
34 ROTH, W., and SCHIPPEN, M.: *Z. Metallk.*, **47**:78 (1956).
35 FLEMINGS, M. C., HUCKE, E. E., ADAMS, C. M., and TAYLOR, H. F.: *Trans. AFS*, **64**:636 (1956).
36 ADAMS, JR., C. M., and TAYLOR, H. F.: *Trans. AFS*, **61**:686 (1953).
37 RUDDLE, R. W.: "Solidification of Castings," 2d ed., *Inst. Metals Monogr. Rept. Ser. No.*, 7, London, 1957.
38 WALLACE, D. F. (ed.): "Fundamentals of Risering Steel Castings," Steel Founders' Society, Cleveland, Ohio, 1960.
39 PIWONKA, T. S., and FLEMINGS, M. C.: *Trans. AIME*, **236**:1157 (1966).
40 MEHRABIAN, R., KEANE, M., and FLEMINGS, M. C.: *Met. Trans.*, **1**:1209 (1970).
41 MURPHY, A. J.: "Non-ferrous Foundry Metallurgy," Pergamon Press, London, 1954.
42 CIBULA, A.: *Foundry Trade J.*, 337 (March 1967).
43 FLEMINGS, M. C., and TAYLOR, H. F.: *Foundry*, **88**:216 (1960).
44 PELLINI, W. S.: *Trans. AFS*, **61**:61 (1953).
45 BISHOP, H. F., and PELLINI, W. S.: *Foundry*, **80**:86 (1952).
46 ADAMS, JR., C. M., FLEMINGS, M. C., and TAYLOR, H. F.: *Trans. AFS*, **66**:369 (1958).
47 DERGE, G. (ed.): "Basic Open Hearth Steel Making," 3d ed., AIME, New York, 1964.
48 WEST, D. R. F.: *J. Iron Steel Inst.*, **164**:182 (1950).
49 SARATOVKIN, D. D.: "Dendritic Crystallization," Consultants Bureau Transl., New York, 1959.
50 Report on the Heterogeneity of Steel Ingots, 7th Rept., *Iron and Steel Inst.*, London, 1937.
51 ADAMS, D. E.: *J. Inst. Metals*, **75**:809 (1948).
52 SCHEIL, E.: *Metallforsch.*, **2**:69 (1947).
53 KIRKALDY, J. S., and YOUDELIS, W. V.: *Trans. AIME*, **212**:833 (1958).
54 FLEMINGS, M. C., and NEREO, G. E.: *Trans. AIME*, **239**:1449 (1967).
55 FLEMINGS, M. C., MEHRABIAN, R., and NEREO, G. E.: *Trans. AIME*, **242**:41 (1968).
56 FLEMINGS, M. C., and NEREO, G. E.: *Trans. AIME*, **242**:50 (1968).

57 HAGIWARA, I., and TAKAHASHI, T.: *Tetsu To Hagane*, **53**:27 (1967).
58 MEHRABIAN, R., and FLEMINGS, M. C.: *Trans. AIME*, **245**:2347 (1969).
59 MEHRABIAN, R., and FLEMINGS, M. C.: *Met. Trans.*, **1**:455 (1970).
60 MEHRABIAN, R., KEANE, M., and FLEMINGS, M. C.: *Met. Trans.*, **1**:3238 (1970).
61 MCDONALD, R. J., and HUNT, J. D.: *Trans. AIME*, **245**:1993 (1969).
62 MCDONALD, R. J., and HUNT, J. D.: *Met. Trans.*, **1**:1787 (1970).
63 GIAMEI, A. F., and KEAR, B. H.: *Met. Trans.*, **1**:1787 (1970).
64 COPLEY, S. M., GIAMEI, A. F., JOHNSON, S. M., and HORNBECKER, M. F.: *Met. Trans.*, **1**:2193 (1970).
65 STANDISH, N.: *Iron and Steel*, **42**:354 (1969).
66 MOMOSE, A.: *Tetsu To Hagane*, **53**:1477 (1967).
67 MARBURG, E.: *J. Metals*, **5**:152 (1953).
68 ROSENBERG, R. A., FLEMINGS, M. C., and TAYLOR, H. F.: *Trans. AFS*, **68**:518 (1960).
69 METZ, S. A., and FLEMINGS, M. C.: *Trans. AFS*, **78**:453 (1970).
70 SINGER, A. R. E., and COTTRELL, S. A.: *J. Inst. Metals*, **74**:73 (1946).
71 SPENCER, D., MEHRABIAN, R., and FLEMINGS, M. C.: *Met. Trans.*, **3**:1925 (1972).
72 GRUBE, K., and EASTWOOD, L. W.: *Trans. AFS*, **58**:76 (1950).
73 TZAVARAS, A., and FLEMINGS, M. C.: *Trans. AIME*, **233**:355 (1965).
74 FLEMINGS, M. C., and MEHRABIAN, R.: in "Solidification," American Society for Metals, Metals Park, Ohio, 1971.

## PROBLEMS

7-1   A cylinder, 10 cm diameter × 20 cm high, is cast of aluminum. Metal is top-gated through a tapered sprue, as sketched in Fig. 7-2. The cross-sectional area of the sprue at the base is 3 cm². The pouring basin height $h$ is 8 cm; the sprue plus pouring basin height $h_2$ is 30 cm. Neglecting frictional losses, what must be the area at the top of the sprue to ensure no aspiration will take place? How long will it take to fill the casting?

7-2   How long will it take to fill the above casting with the same sprue and pouring basin but with the casting moved up in the mold so that it is bottom gated?

7-3   A 50-kg aluminum permanent-mold casting is poured at 760°C at a constant rate of 2 kg/s through a runner system 2.5 cm diameter and 125 cm long. What is the temperature of the metal entering the mold after the initial 10 lb have been poured? At the end of pour? Heat flow is $h$-controlled, with $h = 0.04$ cal/(cm²)(°C)(s).

7-4   Estimate the length pure aluminum will flow along a channel, 0.2 cm diameter, in a permanent (metal) mold. The aluminum is poured with 20°C superheat. Effective metal head is maintained at 16 cm so that flow velocity is 120 cm/s. Heat flow is $h$-controlled, with $h = 0.04$ cal/(cm²)(°C)(s).

7-5   Repeat Prob. 7-3 for an equivalent sand casting, except that the metal is poured with 0°C superheat.

7-6   A 10-cm cube of pure aluminum is cast in a sand mold with a 10-cm-diameter top riser. Riser and casting are both totally enclosed in sand. How high must the riser be to obtain a sound casting?

**7-7** What would be the most efficient height-to-diameter ratio of the above riser?

**7-8** How high would the riser of the above 10-cm cube need to be if it were (*a*) perfectly insulated on top and (*b*) completely surrounded by gypsum insulation?

**7-9** Two 10-cm cubes of pure aluminum are joined by a 1-cm-diameter cylinder, as shown schematically in Fig. 7-17. If the cylinder is surrounded (*padded*) with gypsum, would you expect the risering arrangement of Fig. 7-17*d* to feed the entire casting? Support your answer by calculation.

**7-10** How high would the riser of the casting of Prob. 7-5 need to be if it were left open to the atmosphere? Assume the combined coefficient of radiation and convection to the atmosphere $h_e$ is 0.004 cal/(cm)(°C)(s) [1.9 Btu/(ft²)(h)(°F)].

**7-11** Pure iron is poured into a green sand mold to produce a cylindrical casting 0.10 cm diameter by 30 cm high. The cylinder is entirely surrounded by sand except for its cope surface, which is perfectly insulated with exothermic material. How deep will the shrinkage pipe extend into the casting?

**7-12** An ingot mold 2 m high has large, flat side walls. Steel is poured in the mold with 100°C superheat. Mold-metal interface resistance is low so that solidification begins immediately. Fluid-flow behavior near the wall is initially as it would be for a vertical plate 2 m high in an infinite metal bath. Using Eq. (7-11), show that convection will be turbulent at this early stage of solidification. Assume $\beta_t = 10^{-3}$°C$^{-1}$, $\mu = 6.7 \times 10^{-2}$ P, $\rho_L = 7.1$ g/cm³, $c' = 0.17$ cal/(g)(°C), $K_L = 5 \times 10^{-2}$ cal/(cm)(°C)(s).

**7-13** A plate is permanent-mold cast of Al–4.5% Cu alloy. The plate is ½ cm thick and 10 cm long, cast horizontally and risered on one end with a riser 5 cm high, as sketched in Fig. 7-20. Mold-metal interface resistance $h$ is 0.04 cal/(cm²)(s).

(*a*) Calculate the approximate rate of solidification using Eq. (7-24) (i.e., assuming the rate of solidification is that which would pertain if it took place at some constant temperature within the liquid-solid region).

(*b*) Calculate interdendritic flow velocity at the end of the plate near the riser at just above the eutectic temperature. Assume $\beta = 0.055$.

(*c*) Do you expect surface shrinkage to form in this plate at any temperature above the eutectic temperature?

Assume $\gamma = 6 \times 10^{-7}$ cm², $\mu = 3 \times 10^{-2}$ P, and neglect the effects of surface tension.

**7-14** What initial content of dissolved hydrogen would cause pores to form within the above casting at the end opposite the riser, at 50 percent solid? Neglect the effects of surface tension and assume the data of Fig. 6-23 apply for solubility of hydrogen in the liquid and solid phases of the alloy.

**7-15** An Al–2 wt% Ni alloy is solidified unidirectionally against a flat chill wall. Calculate the inverse segregation to be expected exactly at the chill face. Nickel is insoluble in solid aluminum and forms a eutectic at 5.7 wt% Ni, 640°C. Assume $\beta = 0.05$.

**7-16** A die casting of Al–2 wt% Ni is solidified with negligible thermal gradient. If it is perfectly fed (no pores form), the final composition will be uniform throughout. What will it be?

7-17  If an exuded layer is present on the unidirectionally solidified casting of Prob. 7-15, what will be the maximum composition this layer can have?

7-18  Show schematically how you would expect segregation to appear in a casting like that of Fig. 7-31 of an alloy which expands on solidification.

7-19  Show schematically how you would expect the segregation to appear in a casting like that of Fig. 7-33 of an alloy in which the interdendritic liquid becomes progressively less dense during solidification.

7-20  Explain under what conditions interdendritic flow can cause dendrite remelting.

# 8
# THERMODYNAMICS OF SOLIDIFICATION

## INTRODUCTION

This chapter deals with aspects of thermodynamics that have special relevance to solidification processing. Thermodynamics is at its best when nothing can happen, that is, at equilibrium. Thus, we can use it to tell us much about the nature of equilibrium phase diagrams and about the character of the liquid-solid interface at equilibrium. However, to aid in understanding the complicated nonequilibrium process of solidification, it is necessary to carry thermodynamics further than this. We do so by assuming *constrained equilibrium*, in which thermodynamics is applied locally to individual processes while other processes are assumed to occur at negligible rate or not at all. One example is application of thermodynamics to metastable phase formation. Of course, when thermodynamics is applied in this way, particular care must be taken to establish by experiment the validity of the assumptions made. Baker and Cahn[1] have recently summarized a number of aspects of thermodynamics of solidification, and portions of this chapter closely follow their treatment.

## PURE MATERIALS

For solidification of a single component at constant pressure, the phase rule tells us that only one phase can exist at equilibrium over a range of temperatures. Two phases coexist at equilibrium only at a single temperature. Figure 8-1 shows the curves of free energy versus temperature of the liquid and a stable solid $\alpha$; these intersect at the equilibrium melting point $T_M^\alpha$. Also shown in Fig. 8-1 are the free-energy curves for three metastable phases $\beta$, $\gamma$, and $\delta$. The $\beta$ and $\gamma$ phases are in metastable equilibrium with the liquid at the temperature where their free-energy curves cross that of the liquid. The $\delta$ phase has no melting point and can never form from the liquid. This phase can be formed only by some other process such as vapor deposition or electrodeposition.

When a pure liquid is cooled to its equilibrium melting point $T_M^\alpha$, the $\alpha$ phase may form; alternatively, if nucleation of $\alpha$ is suppressed, the liquid will continue to cool (*undercool* or *supercool*). At or below the melting point of $\beta$, solidification of this phase could occur. If solidification occurred at the melting point of $\beta$, it would be fully thermodynamically reversible, as in the case of a stable phase. The metastable $\beta$ phase once formed could subsequently transform to the stable phase $\alpha$, depending on nucleation and growth kinetics of $\alpha$. If the $\beta$ phase did not form, cooling would continue until $\gamma$ might form. Of course, no crystalline phase at all need form, in which case the liquid (or *glass* if its viscosity reaches a sufficiently high value) remains indefinitely as the metastable phase.

The free-energy change on forming a solid from a liquid at the equilibrium transformation temperature $T_M$ is

$$\Delta G = \Delta H - T_M \Delta S = 0 \qquad (8\text{-}1)$$

where $\Delta G$, $\Delta H$, and $\Delta S$ are molar changes in free energy, enthalpy, and entropy, respectively. At temperatures $T$ different from $T_M$

$$\Delta G = \Delta H - T \Delta S \neq 0 \qquad (8\text{-}2)$$

Neglecting the small-temperature dependence of $\Delta H$ and $\Delta S$ and combining these two equations yields

$$\Delta G = \frac{\Delta H \, \Delta T}{T_M} \qquad (8\text{-}3)$$

where $\Delta T$ is the undercooling $(T_M - T)$. Following the usual thermodynamic convention, $\Delta H$ is positive when heat is absorbed by the system (i.e., for melting); for solidification, it is negative. In Chap. 1, the reverse convention was employed in accordance with usual practice in heat-flow theory. In that chapter, the symbol for heat of fusion was $H$, with units of calories per gram. Thus, expressing $\Delta H$ in calories per mole, $\Delta H = -MH$ where $M$ is molecular weight.

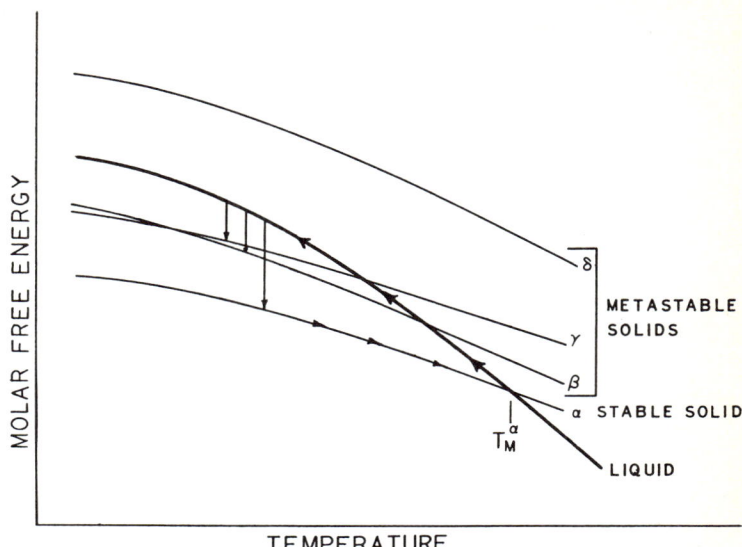

FIGURE 8-1
Free energies versus temperature for a pure material. (*From Baker and Cahn.*[1])

For materials of low viscosity, such as metals, nucleation theory seems to limit $\Delta T/T_M$ to about 0.3, and so the maximum driving force for solidification $\Delta G$ is about 0.03 eV. Much higher undercoolings are obtainable in vapor deposition, and the heat of transformation is also much higher. Thus, the driving force $\Delta G$ and the range of possible structures which can be obtained are much greater for the case of vapor deposition. Achievable rates are much lower, however. Some examples of single-component liquids which have been solidified to form metastable phases include sulfur, tin, and silica.

The free-energy curves of Fig. 8-1 and their intersections are changed by changes in ambient pressure. The change in free energy of the liquid and solid with small changes in pressure $P$ and temperature is

$$dG_L = V_L \, dP - S_L \, dT \quad (8\text{-}4a)$$

$$dG_s = V_s \, dP - S_s \, dT \quad (8\text{-}4b)$$

where $G_L$, $V_L$, and $S_L$ are molar free energy, volume, and entropy, respectively, for the liquid, and $G_s$, $V_s$, and $S_s$ are similar quantities for the solid phase. At equilibrium, $dG_L = dG_s$ and $T = T_M$. Equating the two expressions for these quantities

$$\frac{dT_P}{dP} = \left(\frac{V_s - V_L}{S_s - S_L}\right) = \frac{\Delta V}{\Delta S} \quad (8\text{-}5)$$

where $dT_P$ is the change in the equilibrium melting point from the change in applied pressure $P$. Taking the finite change in melting point $\Delta T_P$ as positive for temperature decrease, we can write approximately

$$\frac{\Delta T_P}{\Delta P} = -\frac{T_M \Delta V}{\Delta H} \qquad (8\text{-}6)$$

This equation, commonly called the *Clapeyron equation*, gives the pressure dependence of the equilibrium melting point. For liquid metals, it is quite small, i.e., on the order of $10^{-2}$°C/atm. Hence, pressure changes encountered in usual processes have little influence on melting point. A possible exception is when cavitation of the liquid metal occurs. Bubbles which form in this way subsequently collapse, and when they do, high pressures are generated. The pressures should be high enough to raise the metal melting point many tens of degrees. It has been suggested that vibration promotes nucleation in undercooled pure metals in this way.[2,3] Cavitation resulting from the vibration is thought to raise the nucleation temperature above the actual temperature of the system.

Another important influence on the melting point of a pure material is *surface curvature*. The surface curvature can be viewed as introducing an excess pressure in the solid phase (only), and we can write two expressions comparable to Eqs. (8-4), which give the change in free energy of the liquid and solid as a result of the curvature. Since the temperature differences involved are small, we can write the free-energy changes in terms of $\Delta G_L$ and $\Delta G_s$, assuming constant $S_L$, $S_s$, and $V_s$

$$\Delta G_L = S_L \Delta T_r \qquad (8\text{-}7a)$$

$$\Delta G_s = S_s \Delta T_r + 2V_s \sigma \kappa \qquad (8\text{-}7b)$$

where $\Delta T_r$ is the decrease in equilibrium melting point as a result of curvature, and $\kappa$ is the mean surface curvature

$$\kappa = \frac{1}{2}\left(\frac{1}{r_1} + \frac{1}{r_2}\right) \qquad (8\text{-}8)$$

where $r_1$ and $r_2$ are the principal radii of curvature of the interface. Equation (8-7b) assumes that $\sigma$ is isotropic and does not change as surface area changes. At equilibrium $\Delta G_L = \Delta G_s$, and so equating Eqs. (8-7)

$$\Delta T_r = -\frac{2\sigma V_s \kappa}{\Delta S} \qquad (8\text{-}9)$$

or

$$\Delta T_r = -\frac{2\sigma T_M V_s \kappa}{\Delta H} \qquad (8\text{-}10)$$

When curvature is positive, as at a dendrite tip, the equilibrium melting point is reduced, and $\Delta T_r$ in the above equations is positive. This convention is followed in accordance with the usual practice in solidification problems. $\Delta T_r$ is *undercooling due to radius of curvature*. The same convention is employed with all $\Delta T$'s in this chapter; i.e., they are positive for temperature decrease.

## BINARY ALLOYS; STABLE PHASE EQUILIBRIUM

In binary alloys, at constant pressure, we have the possibility of two-phase equilibrium over a range of temperatures and three-phase equilibrium at a single temperature. At constant temperature, the free energies of phases present in an alloy of $A$ and $B$ are a function of composition, as shown schematically in Fig. 8-2. The important condition for equilibrium is that the chemical potential must everywhere have the same value. For a binary alloy, the chemical potentials of elements $A$ and $B$ are written

$$\mu^A = \left(\frac{\partial G'}{\partial n_A}\right)_{T,P,n_B} \tag{8-11a}$$

$$\mu^B = \left(\frac{\partial G'}{\partial n_B}\right)_{T,P,n_A} \tag{8-11b}$$

Thus, the chemical potential is the change in free energy of the entire system $G'$ when an infinitesimal amount of one of the components is added reversibly (per mole added, $n_A$ or $n_B$). It is also called the *partial molar free energy*. The molar free energy $(G = G'/n_A + n_B)$ is related by the Gibbs-Duhem equation to the chemical potential giving the expressions

$$\mu^A = G - C\left(\frac{\partial G}{\partial C}\right) \tag{8-12a}$$

$$\mu^B = G + (1-C)\left(\frac{\partial G}{\partial C}\right) \tag{8-12b}$$

where $C$ is composition, mole fraction $B$. Equations (8-12) form the basis of the tangent method of determining phase compositions. They state that a tangent drawn to a molar-free-energy curve of a composition of interest intercepts the $C = 0$ vertical axis at $\mu^A$ and the $C = 1$ axis at $\mu^B$. Hence, a common tangent to two free-energy curves, as to the solid and liquid curves in Fig. 8-2, means that the chemical potentials of $A$ in solid and liquid are equal ($\mu_S^A = \mu_L^A$), as are the chemical potentials of $B$ ($\mu_S^B = \mu_L^B$). It therefore follows also that the compositions at the points of tangency

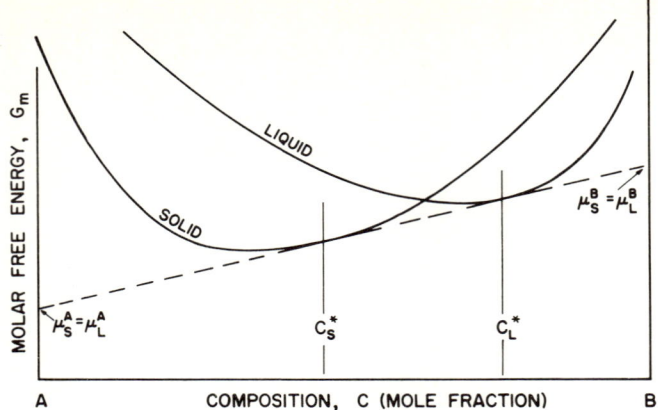

**FIGURE 8-2**
Molar free energy versus composition for a binary alloy at constant temperature.

are the compositions of the two phases at equilibrium. These are shown as $C_s^*$ and $C_L^*$ in Fig. 8-2.

The molar free energy of liquid and solid is also related to the chemical potential by the expression

$$G = (1 - C)\mu^A + C\mu^B \qquad (8\text{-}13)$$

where $\mu^A$ and $\mu^B$ are functions of the compositions of the elements. For an ideal solution

$$\mu^A = \mu_0^A + S^A \, \Delta T^A + RT \ln (1 - C) \qquad (8\text{-}14a)$$
$$\mu^B = \mu_0^B + S^B \, \Delta T^B + RT \ln C \qquad (8\text{-}14b)$$

where $\mu_0^A$ and $\mu_0^B$ are constants, the chemical potential of the pure components at unit composition. $\Delta T^A$ and $\Delta T^B$ are the reference temperature of $\mu_0^A$ and $\mu_0^B$ minus the actual temperature $T$. Equations (8-14a) and (8-14b) can be written for both the solid and liquid phases. In the notation of this text, the result is

$$\mu_L^A = \mu_0^A + S_L^A \, \Delta T^A + RT \ln (1 - C_L) \qquad (8\text{-}15a)$$
$$\mu_s^A = \mu_0^A + S_s^A \, \Delta T^A + RT \ln (1 - C_s) \qquad (8\text{-}15b)$$
$$\mu_L^B = \mu_0^B + S_L^B \, \Delta T^B + RT \ln C_L \qquad (8\text{-}15c)$$
$$\mu_s^B = \mu_0^B + S_s^B \, \Delta T^B + RT \ln C_s \qquad (8\text{-}15d)$$

Now, an equation can be readily derived giving the free energy of the liquid and solid phases as a function of composition. This is done by combining Eqs. (8-15)

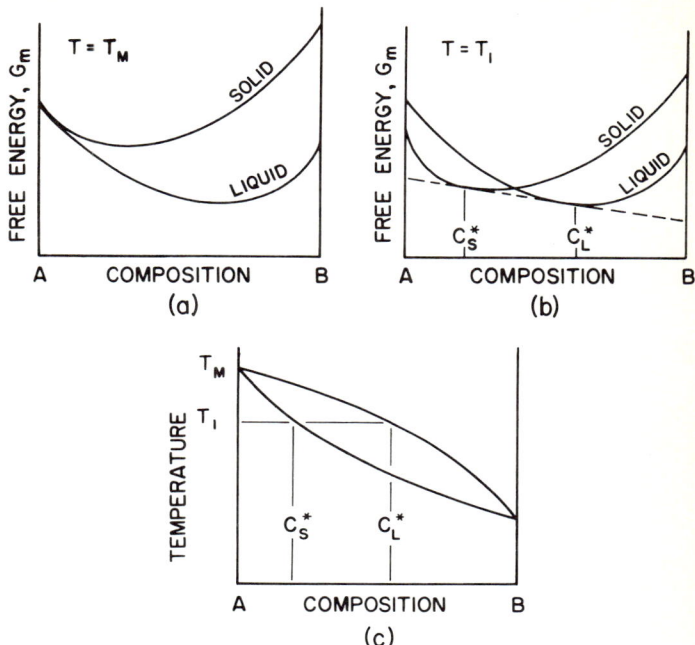

**FIGURE 8-3**
Free-energy curves at two different temperatures for an ideal solid solution with resulting phase diagram.

with Eqs. (8-13) and (8-3). The resulting curves are plotted schematically in Fig. 8-2 for the liquid and solid phases. The equilibrium interface compositions $C_s^*$ and $C_L^*$ in Fig. 8-2 constitute the liquidus and solidus of the phase diagram at temperature $T$. The loci of these points for curves drawn for different temperatures constitute the liquidus and solidus curves of the equilibrium phase diagram. Figure 8-3 shows free-energy curves for this alloy at two higher temperatures than that of Fig. 8-2. The resulting equilibrium phase diagram is also drawn. One real system which is very nearly ideal, and for which calculations have been made as described above, is the Ge–Si system shown in Fig. 8-4.

As solutions deviate from ideal, the result is the wide range of phase diagrams with which we are familiar. Figures 8-5 and 8-6 show examples of phase diagrams which result when we relax the restriction of zero heat of mixing that characterizes ideal solutions. Negative heats of mixing result in a liquidus maximum and compound formation. Positive heats of mixing result in a liquidus minimum and eutectic formation. When three phases are present at a single temperature, the free-energy

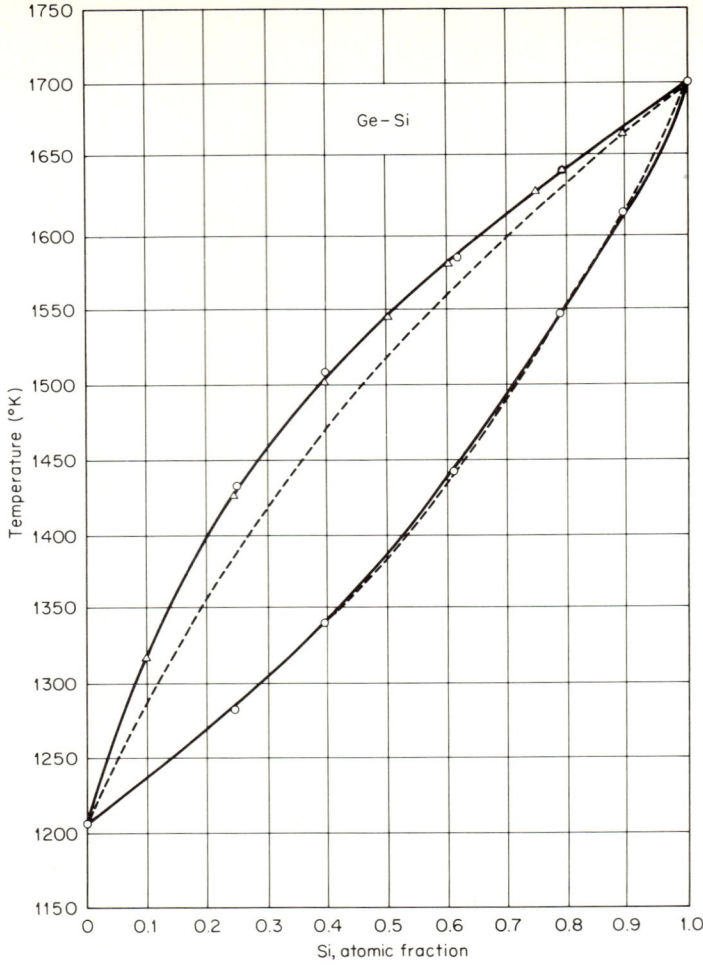

**FIGURE 8-4**
Phase diagram for the nearly ideal Ge–Si system. Solid line, experimental; dashed line, calculated. (*From Thurmond.*[17])

curves must have three points of common tangency. Figure 8-7 is a schematic illustration of this for the eutectic system of Fig. 8-5, at the eutectic temperature. In this case, a single free-energy curve can be drawn for the phases $\alpha$ and $\alpha'$. When these two phases are of different crystal structures, separate free-energy curves for the two phases are drawn as in Fig. 8-8.

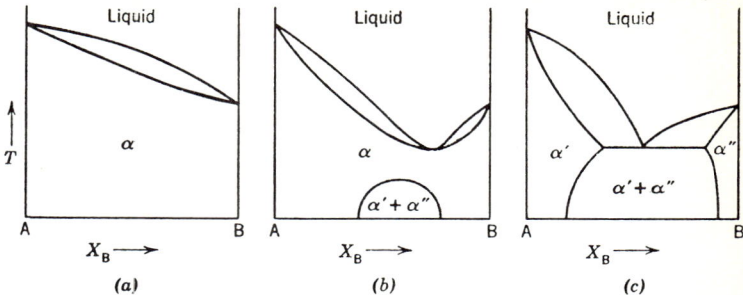

**FIGURE 8-5**
(*a*) Phase diagram for an ideal solution; (*b*) phase diagram for case where heat of mixing in the solid phase is greater than that in the liquid and both are positive; (*c*) same as (*b*) except heats of mixing are more positive. (*From R. H. Swalin.*[18])

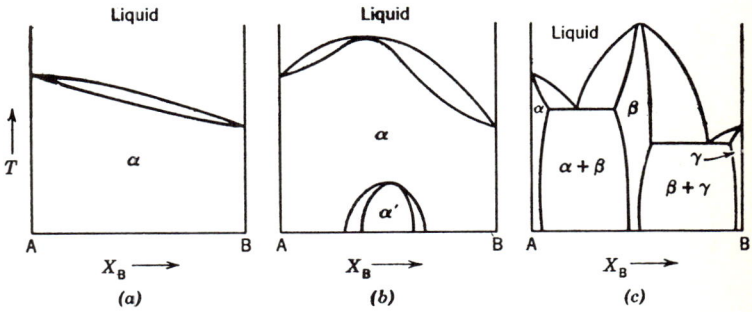

**FIGURE 8-6**
(*a*) Phase diagram for an ideal solution; (*b*) phase diagram for case where heat of mixing in the solid phase is less than that in the liquid and both are negative; (*c*) same as (*b*) except heats of mixing are more negative.

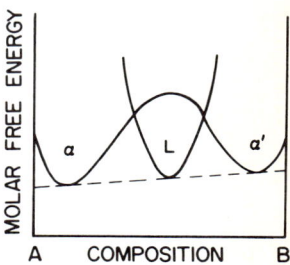

**FIGURE 8-7**
Free-energy curves at eutectic temperature for components with the same crystal structure.

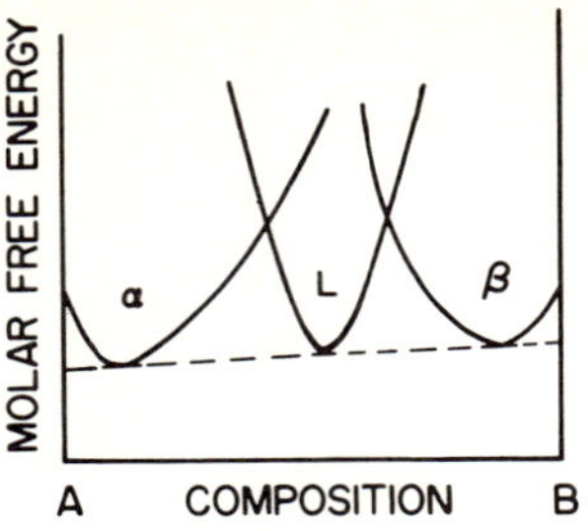

**FIGURE 8-8**
Free-energy curves at the eutectic temperature for components with different crystal structures.

## THE PARTITION RATIO

At two-phase equilibrium, $\mu_L^A = \mu_s^A$, and we use the designation $C_L^*$ for the equilibrium liquid composition and $C_s^*$ for the equilibrium solid composition. Now we can equate Eqs. (8-15a) and (b) to obtain an expression for the equilibrium partition ratio, $k = C_s^*/C_L^*$. Taking the melting point of pure $A$ as the temperature of the standard state, assuming a dilute solution and substituting Eq. (8-1), yields

$$k = 1 - \frac{m_L \Delta H^A}{RT_M^{A2}} \qquad (8\text{-}16)$$

where $m_L$ is the liquidus slope, $\Delta H^A$ is the heat of fusion of the pure solvent $A$, and $T_M^A$ is the melting point of pure $A$. Equation (8-16) is valid for both dilute ideal and dilute real solutions. It provides a convenient way (when $k$ is not too close to zero) to calculate the partition ratio of dilute solutions from quantities that are relatively easy to determine experimentally.

One sees occasional references in the literature to the equilibrium partition ratio being a function of crystal orientation. Indeed, experiments such as those cited in Chap. 2 show that the measured partition ratio does sometimes depend on orientation. Such effects must, however, result because equilibrium is not attained. The equilibrium partition ratio $k$ depends only on thermodynamic quantities that are not a function of orientation.

Equation (8-16) involves both the partition ratio and the liquidus slope. For ideal solutions, the partition ratio can be calculated independently. This is done by equating Eqs. (8-15c) and (d) at equilibrium, this time taking the melting point of the pure solute $B$ as the temperature of the standard state. The result, for a solution dilute in $B$, is

$$k = \exp\left[\frac{\Delta H^B(T_M^A - T_M^B)}{RT_M^A T_M^B}\right] \qquad (8\text{-}17)$$

where $\Delta H^B$ is the heat of fusion of pure solute $B$ and $T_M^B$ is the melting point of pure $B$.

# EFFECT OF CURVATURE

To account for effect of curvature on liquid-solid equilibrium of binary alloys, we first write four equations for the chemical potentials of the liquid and solid phases comparable to Eqs. (8-15a) to (d). The two equations for the liquid phase are unchanged; the two for the solid phase are modified by a term accounting for the excess pressure introduced by the curvature. These two equations are then written

$$\mu_s^A = \mu_0^A + S_s^A \Delta T^A + RT \ln(1 - C_s) + 2\bar{V}_s^A \sigma \kappa \qquad (8\text{-}18a)$$

$$\mu_s^B = \mu_0^B + S_s^B \Delta T^B + RT \ln C_s + 2\bar{V}_s^B \sigma \kappa \qquad (8\text{-}18b)$$

where $\bar{V}_s^A$ and $\bar{V}_s^B$ are partial molar volumes of $A$ and $B$ in the solid, respectively. Setting Eq. (8-18b) equal to (8-15c) at equilibrium

$$\frac{\Delta H^B(T_M^A - T_M^B)}{T_M^B} = RT_M^A \ln k' + 2\bar{V}_s^B \sigma \kappa \qquad (8\text{-}19)$$

where $k'$ is the partition ratio $C_s^*/C_L^*$ of the solid of curvature $\kappa$. For zero curvature, $k' = k$, and Eq. (8-20) reduces to Eq. (8-17). Substituting Eq. (8-17) in Eq. (8-19) and approximating for $k' \simeq k$, we obtain

$$\frac{k'}{k} = 1 - \frac{2\bar{V}_s^B \sigma \kappa}{RT_M^A} \qquad (8\text{-}20)$$

It can be shown that this equation is valid for a dilute real solution, as well as a dilute ideal solution. Substitution of typical numerical values in Eq. (8-20) shows that $k'$ deviates significantly from $k$ only for radii of curvature below about $10^{-6}$ cm. Thus for usual solidification problems (which generally involve radii of curvature of about $10^{-4}$ cm or larger), the partition ratio is properly assumed to be independent of curvature.

Now we can also set Eq. (8-18a) equal to Eq. (8-15a) and obtain, for dilute solutions at equilibrium,

$$\Delta S^A \Delta T^A + RT(C_L^* - C_s^*) + 2\bar{V}_s^A \sigma \kappa = 0 \qquad (8\text{-}21)$$

At the limit $C_L^* \to 0$, $\Delta T^A = \Delta T_r$, the undercooling of the pure component $A$ due to curvature. The equation obtained is exactly that given earlier as Eq. (8-9) or (8-10). Now, define $\Delta T'$ as

$$\Delta T'^A = \Delta T^A - \Delta T_r^A \qquad (8\text{-}22)$$

where $\Delta T^A$ is that of Eq. (8-21). It is the depression of the liquidus temperature from that of pure $A$ of zero curvature to that of the alloy of liquid composition $C_L$ and curvature $\kappa$. Substituting Eq. (8-22) in Eq. (8-21)

$$m_L' = -\frac{\Delta T'}{C_L} = \frac{RT_M^{A^2}(1 - k')}{\Delta H^A} \qquad (8\text{-}23)$$

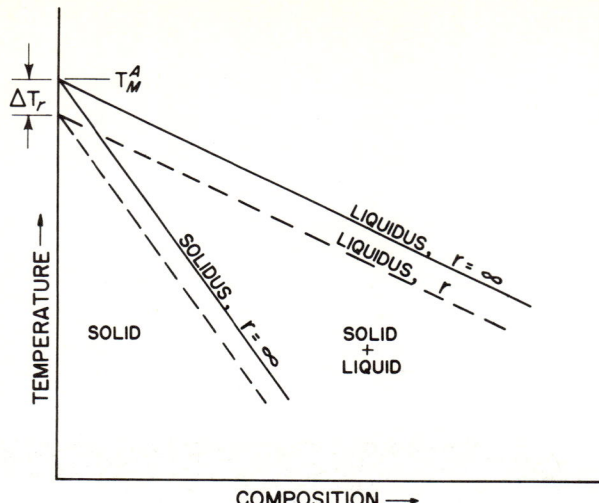

**FIGURE 8-9**
Depression of equilibrium liquidus and solidus for solid of radius $r$.

where $m_L' = -\Delta T'/C_L$ is the slope of the liquidus of a binary alloy whose solid phase has the curvature $\kappa$. According to Eq. (8-20), $k' \simeq k$, and so $m_L' \simeq m_L$, as can be seen by comparison of Eqs. (8-23) and (8-16). Thus, the effect of curvature is to shift the liquidus and solidus of the phase diagram uniformly downwards, as shown in Fig. 8-9. We have used this concept repeatedly in other chapters in discussing the effects of curvature on the solidification behavior of alloys.

## EFFECT OF PRESSURE

The effect of pressure on the liquid-solid equilibrium of binary alloys is described by writing four equations for the chemical potentials of the liquid and solid phases comparable to Eqs. (8-15a) to (d), but each with the additional term $\bar{V} \Delta P$. The two equations for the chemical potential of element $A$ are

$$\mu_L^A = \mu^{A_o} + S_L^A \Delta T^A + RT \ln(1 - C_L) + \bar{V}_L^A \Delta P \qquad (8\text{-}24a)$$

$$\mu_s^A = \mu^{A_o} + S_s^A \Delta T^A + RT \ln(1 - C_s) + \bar{V}_s^A \Delta P \qquad (8\text{-}24b)$$

Following a procedure identical to that employed for the effect of curvature, we find that the partition ratio $k''$ deviates from the equilibrium partition ratio $k$ according to

$$\frac{k''}{k} = 1 - \frac{\Delta \bar{V}^B \Delta P}{RT_M^A} \qquad (8\text{-}25)$$

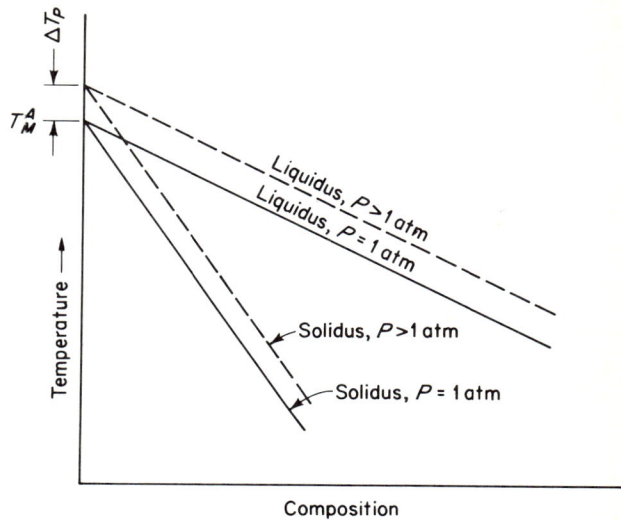

**FIGURE 8-10**
Elevation of equilibrium liquidus and solidus due to applied pressure. For alloys which contract on solidification.

where $\Delta \bar{V}^B$ is the change in partial molar volume on solidification of $B$ in dilute solutions. This equation, like Eq. (8-20), is valid for dilute real as well as dilute ideal solutions. Substitution of typical numerical values shows that $k''$ deviates significantly from $k$ only for pressures exceeding about $10^2$ atm. The liquidus slope $m_L''$ is given by

$$m_L'' = -\frac{\Delta T''}{C_L} = \frac{RT_M^A(1 - k'')}{\Delta H^A} \qquad (8\text{-}26)$$

which is the slope of the equilibrium liquidus $m_L$ when pressure is low enough that $k'' \simeq k$. Thus, for these moderate pressures and materials which contract on solidification, increasing pressure simply shifts the liquidus and solidus upwards, as sketched in Fig. 8-10. The amount of the shift is equal to that for the pure metal, Eq. (8-6).

## BINARY ALLOYS; METASTABLE PHASE EQUILIBRIUM

Metastable phase equilibrium obeys the same thermodynamic rules that apply to stable equilibrium. At this equilibrium, the temperature and chemical potentials of all phases present are equal. Metastable phase diagrams show no unusual features (except near limits of metastability, called *spinodals*, where the nucleation barrier

**276** THERMODYNAMICS OF SOLIDIFICATION

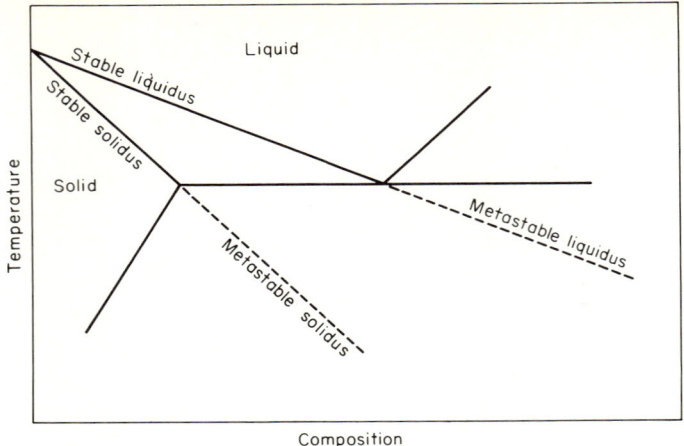

**FIGURE 8-11**
Portion of eutectic phase diagram illustrating metastability of solid and liquid.

ceases to exist). If a phase is stable in one portion of a diagram and becomes metastable in another, no discontinuity is expected in its behavior as it becomes metastable. For example, the metastable extension of the liquidus and solidus curve below the eutectic should show no break, Fig. 8-11. The free-energy versus composition curves

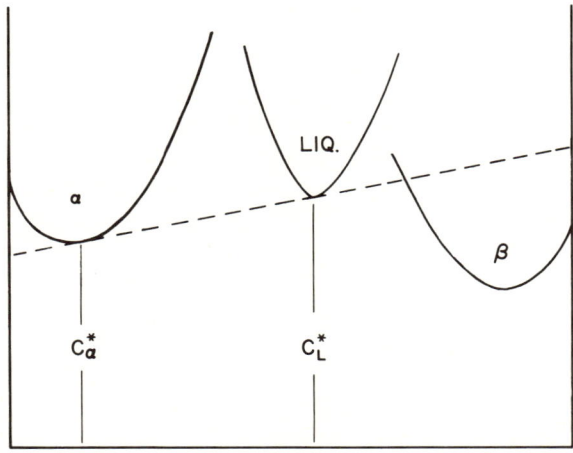

**FIGURE 8-12**
Metastable equilibrium between $\alpha$ and liquid at a temperature below the eutectic.

for this case at a temperature below the eutectic appear as in Fig. 8-12. The stable phase $\beta$ is not present, and the equilibrium compositions of those phases present ($\alpha$ and $L$) are not discontinuously different from those at just above the eutectic temperature.

There are many examples in solidification where equilibrium pertains at the liquid-solid interface even though one or more of the phases at the interface are metastable. We have used extrapolations of the liquidus and solidus (i.e., assumed interfacial equilibrium) in describing eutectic solidification in Chap. 4. In aluminum–silicon alloys, moderately fast solidification (e.g., permanent-mold casting) or minor additions of elements, including sodium, retard the growth of the silicon phase. The result is that the eutectic temperature is shifted downward significantly. Equilibrium is presumably maintained between the aluminum phase and the liquid so that the eutectic point can be visualized as being moved downward and to the right. Moderately rapid solidification suppresses altogether the equilibrium second phase in a number of alloys, including cast-iron and cadmium–antimony alloys. In the former case, the stable graphite is replaced by metastable $Fe_3C$; in the latter, stable CdSb is replaced by metastable $Cd_3Sb_2$.[4] As these alloys cool from the stable eutectic temperature to the metastable, equilibrium is expected to be maintained at the liquid-solid interface at all except extremely rapid rates of solidification.

At the very rapid solidification rates encountered in splat cooling, large increases have been obtained in the solid solubilities of the phases comprising eutectic systems.[5] The best-known example is the Ag–Cu system, Fig. 8-13, which is a simple eutectic system but which can be splat-cooled to give a continuous series of solid solutions from pure silver to pure copper. It is tempting to explain increases in solid solubility obtained in splat cooling using the same argument as above: that local equilibrium is maintained at liquid-solid interfaces, but that formation of the second phase is suppressed. A solid solution would then be obtained provided that solidification occurred at a temperature corresponding to the extension of the metastable solidus. However, there is no evidence that equilibrium is maintained at the liquid-solid interface at the high solidification rates involved in splat cooling. We return to this subject later in this chapter.

When the stable phase in an alloy does not form but is replaced by a metastable phase, the resulting metastable equilibrium is thermodynamically indistinguishable from stable equilibrium. Hence, we may construct *metastable phase diagrams*, the most common of which is the $Fe$–$Fe_3C$ diagram. A portion of that diagram is shown in Chap. 5 superimposed on the stable Fe–G diagram. The lines of the metastable diagram lie below those of the stable diagram. A general rule regarding construction of metastable phase diagrams is that when a phase appears in both the stable and metastable diagrams, its field must be larger in the metastable diagram.

Some examples of metastable phases which are found at cooling rates typical of those of castings and ingots include the $Fe_3C$ and $Cd_3Sb_2$ cited earlier. At the high

**278** THERMODYNAMICS OF SOLIDIFICATION

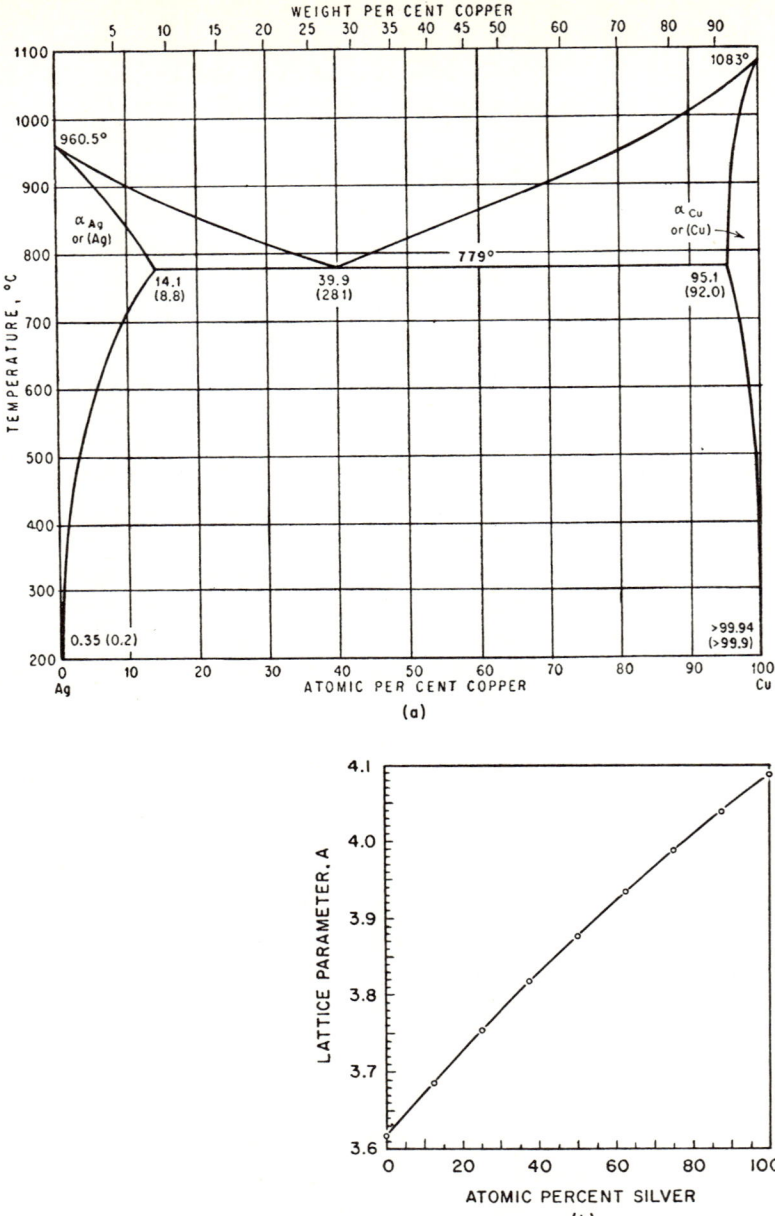

**FIGURE 8-13**
Phase diagram for Ag–Cu system and date of Duwez showing complete solid solubility resulting from splat cooling. (*From Duwez.*[19])

rates of solidification obtained in splat cooling, many dozens of such phases have now been identified. As examples, two different face-centered cubic phases and one gamma brass-type metastable phase have been found in Au–Si alloys.[6,7] A rare, simple cubic-type structure has been obtained in Au–Te alloys,[8] and a new metastable hexagonal close-packed phase has been obtained in iron–carbon alloys.[9]

## COMPOSITION AT THE LIQUID-SOLID INTERFACE

The assumption of interface equilibrium is a valuable and realistic one for describing many solidification processes. The assumption has been used frequently in this text up to this point, and it appears to have broad applicability to metals and many nonmetals for solidification rates encompassed by usual casting and ingot-making processes. In other materials, such as some semiconductors, interface equilibrium is not attained even at low growth rates, as evidenced by the facts that partition ratio is a function of orientation and that facets form. Even in metals, when solidification rate becomes sufficiently rapid (as in splat cooling), significant deviations from interface equilibrium should become possible. Thermodynamics cannot predict what these variations will be, but it can define the domain of possible interface compositions. We shall see this below, following Baker and Cahn,[1] by examining conditions under which binary alloys could solidify at steady state at a given temperature $T$, shown in Fig. 8-14.

Free-energy versus composition curves for the alloys at this temperature are shown in Fig. 8-15. These curves intersect at $C(T_0)$, and the equilibrium solid and liquid compositions are $C_S^*$ and $C_L^*$, respectively. Consider an alloy $C_0$ whose composition lies between $C_L^*$ and $C(T_0)$, that is, in Region I of Fig. 8-14. The possible solid compositions which can form are obtained by drawing a tangent to the liquid free-energy curve at $C_0$ and determining which solid compositions lie below that tangent. Thus, for the alloy chosen, the possible solid compositions lie between $C'_{sm}$ and $C'_{SM}$ (Fig. 8-15). The primes are used hereafter to signify interface compositions that are not necessarily the equilibrium values. Both of these compositions lie below $C_0$, and so steady-state solidification is not possible (since solute cannot be conserved). In a directional solidification experiment, any solid composition between $C'_{sm}$ and $C'_{SM}$ could form but only in nonsteady state. The liquid would become enriched in solute, and the temperature of solidification would drop.

If the initial alloy composition $C_0$ lies between $C(T_0)$ and $C_S^*$, any composition between the limits $C'_{sm}$ and $C'_{SM}$ shown in Fig. 8-16 can form from the liquid. These limits include $C_0$, and so steady-state solidification is now possible. Solidification in this case is occurring in Region II of Fig. 8-14. At steady state, the composition of the solid forming must be exactly $C_0$. Thermodynamics permit, but do not require, the transformation to be diffusionless. If it is not diffusionless, there may be either a

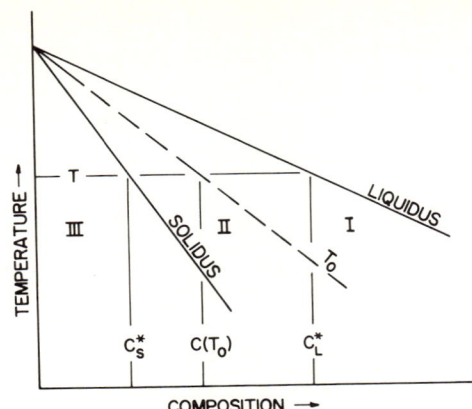

**FIGURE 8-14**
Schematic phase diagram showing three regions of solidification I, II, and III.

solute-enriched or a solute-depleted zone in front of the interface, as sketched in Fig. 8-17. The thermodynamically possible limiting liquid compositions at the interface $C'_{Lm}$ and $C'_{LM}$ are drawn in Fig. 8-18. The limits are imposed by the requirement that the solidification take place as steady state with a decrease in free energy.

For all alloys $C_0$ in Region II, steady-state solidification takes place with a decrease in chemical potential of the solvent $A$, but an increase in chemical potential

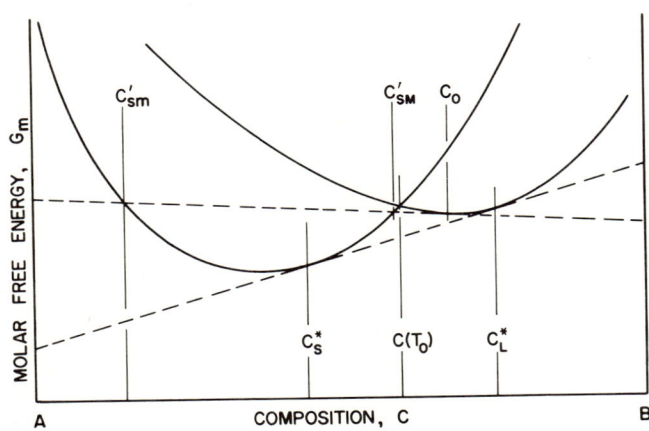

**FIGURE 8-15**
Solidification in Region I. Alloy $C_0$ (at temperature $T$) can lower its free energy by forming solid of any composition between $c'_{sm}$ and $c'_{SM}$. Steady-state solidification for this example is not possible. (*After Baker and Cahn.*[1])

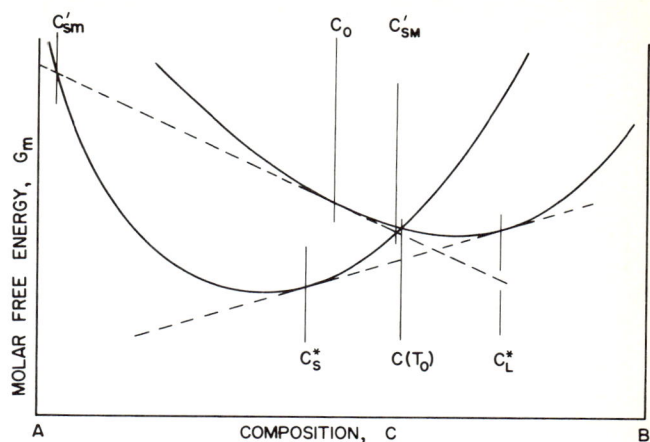

**FIGURE 8-16**
Solidification in Region II. Alloy $C_0$ (at temperature $T$) can lower its free energy by forming solid of any composition between $C'_{sm}$ and $C'_{SM}$. These compositions include $C_0$ and so steady-state solidification is possible. (*After Baker and Cahn.*[1])

of the solute $B$. This can be seen by drawing appropriate tangents to the free-energy curves of Fig. 8-18. The fact that one of the species solidifies with an increase in chemical potential means that it does not solidify completely independently of the other. One species ($B$) either is passively trapped by the advancing interface or is a required participant in an independent solidification reaction involving several species which leads to an overall free-energy decrease. In either case, following Baker and Cahn,

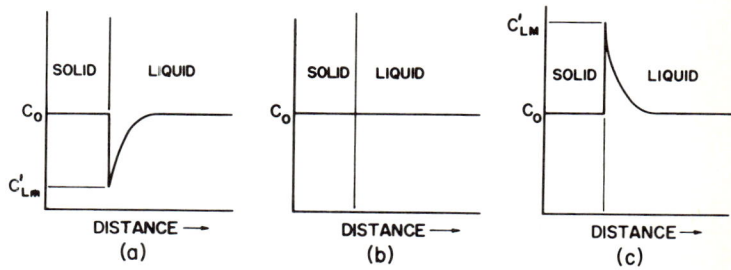

**FIGURE 8-17**
Possible boundary layers for steady-state solidification in Region II.

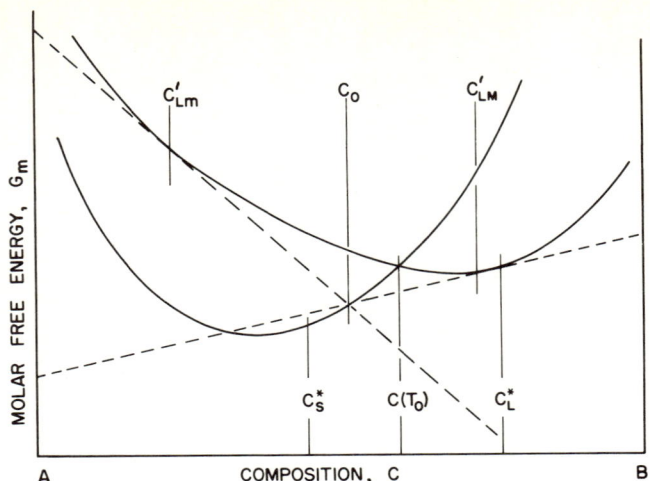

**FIGURE 8-18**
Steady-state solidification in Region II. Alloy $C_0$ at $T$ can form solid of composition $C_0$ provided the liquid composition at the interface lies between $C'_{Lm}$ and $C'_{LM}$. (After Baker and Cahn.[1])

*trapping* of a component is said to occur at an interface when that species experiences an increase in chemical potential there.[1]

Even at solidification rates high enough for solute trapping to occur, there are two factors which might prevent obtaining a homogeneous solid from solidification in Region II. First, there is a possible diffusional instability which prevents the solid composition from returning to $C_0$ if it fluctuates downward. Secondly, as readily seen from Fig. 8-18, the solid which forms at $T$ is unstable and tends to melt. Nonetheless, it appears that in one system at least, solidification does go to completion in this temperature region. In an elegant experiment, Baker and Cahn[10] splat-cooled zinc–cadmium alloys and obtained a fully homogeneous solid of compositions $C_0$ up to 5 wt% Ca. This composition is well above that of the limit of the retrograde solidus, Fig. 8-19, and so solidification could not have occurred below the solidus. It must therefore have occurred in Region II above the solidus and below the $T_0$ temperature.

Similar arguments to the foregoing can be employed to describe steady-state solidification when $C_0 < C_s^*$, that is, in Region III of Fig. 8-14. Unlike solidification in Region II, the resulting solid is stable and the diffusional problem also appears to be stable. There can be either solute depletion or solute enrichment at the interface. Solute trapping occurs at lower values of $C'_L$, and solvent trapping at higher values. There is an intermediate range of $C'_L$, where both components solidify with a decrease in chemical potential.[1]

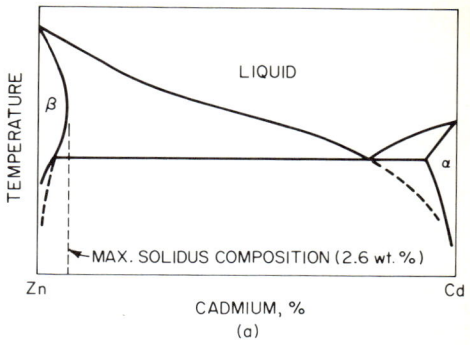

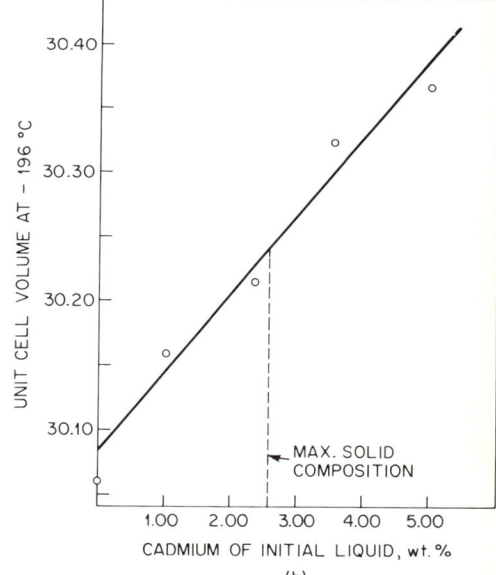

**FIGURE 8-19**
(*a*) Zn–Cd phase diagram with retrograde solidus; (*b*) measurements of Baker and Cahn showing extension of solid solubility beyond limit of retrograde solidus.[10]

At the moment, the only experiments we have which bear on the foregoing are the ones cited, i.e., splat-cooling experiments which show large increases in solubility and therefore indicate that solidification has taken place in either Region II or III, and the result from the system with retrograde solidus which shows for one system that solidification must have occurred in Region II. Of course, there is the familiar case of solidification of an alloy $C_0$ at exactly the solidus, in which case $C'_L \rightarrow C^*_L$. This is the case of equilibrium at the interface in steady-state solidification used as an assumption in earlier chapters.

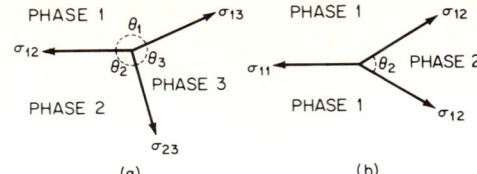

**FIGURE 8-20**
Geometry of interface at three-grain corners.

## THE EQUILIBRIUM SHAPES OF PHASES

Smith[11] first showed that if three phases meet along a common edge, the boundaries will reach local equilibrium at the angles required by vectoral equilibrium of the surface tensions. Assuming isotropic surface tensions and with reference to Fig. 8-20a,

$$\frac{\sigma_{12}}{\sin \theta_3} = \frac{\sigma_{23}}{\sin \theta_1} = \frac{\sigma_{13}}{\sin \theta_2} \quad (8\text{-}27)$$

For the special case where two of the phases are the same, as in Fig. 8-20b,

$$\cos \frac{\theta}{2} = \frac{\sigma_{11}}{2\sigma_{12}} \quad (8\text{-}28)$$

Physical solutions are possible only when $\sigma_{12}$ is greater than $0.5\sigma_{11}$. Above this value, $\theta$ increases progressively from 0 to 180° with increasing ratio of the interface energies.
This angle, the *dihedral angle*, is of great importance in determining equilibrium

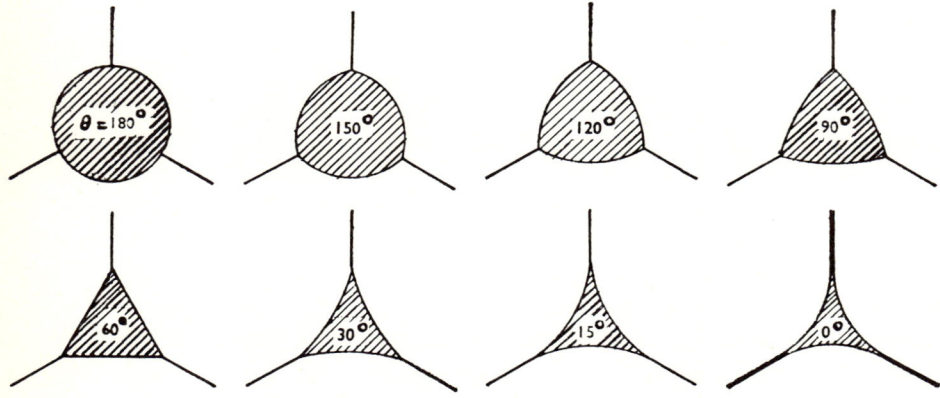

**FIGURE 8-21**
Shape of a minor phase at a grain edge with different dihedral angles. (*From Smith.*[11])

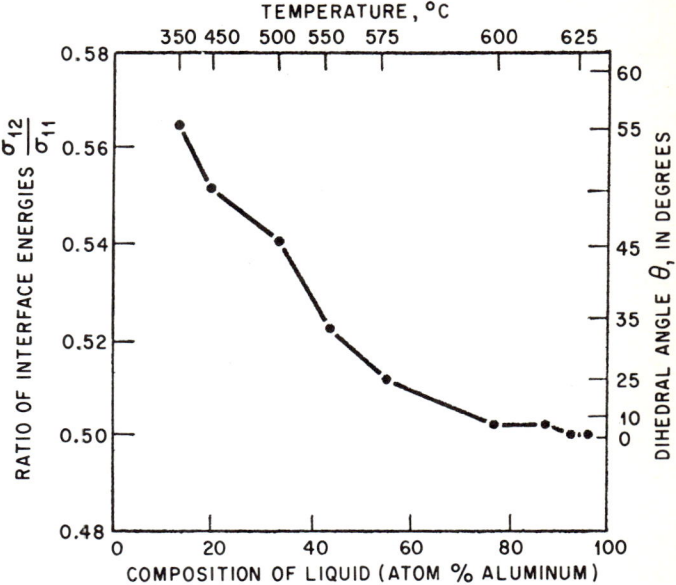

FIGURE 8-22
Variation of interface energy ratio with temperature and composition of liquid phase in Al–Sn alloys. (*From Smith.*[20])

microstructures. Figure 8-21 schematically shows how the equilibrium shape of a minor phase would be expected to depend on the dihedral angle. With high dihedral angles, the phase approaches sphericity; with low angles, it forms a film along a grain boundary. Annealing of metal samples at temperatures where the minor phase is liquid provides a simple way of obtaining structures such as are seen in the sketch. Sometimes, these equilibrium considerations are the dominant aspect controlling solidification structures. An example is the lead droplets which form by monotectic reaction in copper and brasses. In the solidified castings, these have morphologies like those sketched at the upper left of Fig. 8-21 because the dihedral angle between the liquid lead-rich phase and the solid copper-rich phase is high. Oxysulfides in iron are a similar example. More often, solidification structures are determined by the interplay of surface energy, diffusion, and interface kinetics.

Equation (8-28) provides a simple method of experimentally measuring relative liquid-solid surface energies. Apparent dihedral angles are measured on a polished surface of an equilibrium structure. Suitable quantitative metallographic procedures are then used to obtain statistically the true dihedral angle. Equation (8-25) then gives directly relative surface energies. An example is given in Fig. 8-22 for aluminum–tin alloys. Small amounts of the component of lower surface tension (aluminum) markedly lower the surface tension of the mixture. Converse additions have little effect.

A way in which liquid-solid surface energy can be measured directly, at least for pure materials, is by the grain-boundary grooving that occurs when a grain boundary intersects the liquid-solid interface. Assuming equilibrium at the interface and isotropic surface energy, the groove angle is given by Eq. (8-28). Curvature in the neighborhood of the groove is determined by the requirement that

$$T^* = T_M - G\,\Delta X = T_M - \Delta T_r \qquad (8\text{-}29)$$

where $T^*$ is the liquid-solid interface temperature, $G$ is the thermal gradient, and $\Delta X$ is the distance back from the isotherm at $T_M$, the equilibrium melting point of the pure material. Substituting Eq. (8-10) with $\kappa = 1/2r$ for the surface of single radius of curvature,

$$r = -\frac{\sigma T_M V_S}{G\,\Delta H\,\Delta X} \qquad (8\text{-}30)$$

Hence, the curvature depends on the thermal gradient as well as equilibrium properties.

One way to calculate liquid-solid surface energy from grain-boundary grooving is to measure the groove angle and calculate it from Eq. (8-29), knowing the solid-solid surface energy $\sigma_{11}$. Glicksman and Vold[12] made successful measurements in this way on bismuth, using electron microscopy. Low-angle solid grain boundaries were employed so that $\sigma_{11}$ could be calculated from dislocation theory. Results were in substantial agreement with those obtained on nucleation experiments (61 erg/cm$^2$ as compared with 54 erg/cm$^2$ from the earlier nucleation work). Alternatively, measurements can be made of interface curvature in the region of the groove and Eq. (8-30) used to calculate liquid-solid surface energy. Here, the solid-solid surface energy does not enter into calculations. Jones and Chadwick have determined surface energies of several materials in this way.[13] Still another way of measuring liquid-solid surface energies is the indirect way of calculating them from nucleation experiments, as discussed in Chap. 9.

## ANISOTROPY OF LIQUID-SOLID SURFACE ENERGY

The shape of any condensed phase in equilibrium with another is determined by the criterion that total surface free energy be a minimum. For isotropic surface energy, the equilibrium shape is spherical. When $\sigma$ is a function of orientation, the equilibrium shape departs from sphericity in such a fashion as to expose preferentially crystallographic orientations of low energy. The *Wulff theorem* describing the

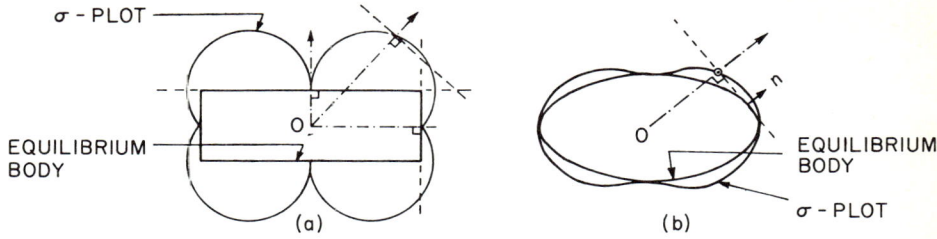

FIGURE 8-23
Wulff construction for two simple equilibrium morphologies. (a) Faceted; (b) nonfaceted. (From Miller and Chadwick.[16])

equilibrium shape of particles with anisotropic surface energy is[14,15]

$$\frac{\sigma_1}{\lambda_1} = \frac{\sigma_2}{\lambda_2} = \cdots \frac{\sigma_i}{\lambda_i} M = \text{const} \qquad (8\text{-}31)$$

where $\lambda_i$ is the distance of the $i$th face of the equilibrium body from the *Wulff center* and $\sigma_i$ is surface energy of the $i$th face. These are equal to a constant, the volume free-energy change on formation of the solid crystal. The Wulff center is the geometric center of nonpolar crystals.

If the variation of interfacial free energy with orientation is known, the equilibrium shape of the crystal can be determined by a simple construction. First, the surface energy is plotted versus orientation on a polar plot ($\sigma$ plot) as in Fig. 8-23. Then, planes are drawn normal to the radius vectors where they intersect its surface. The envelope formed by points which can be reached from the origin without crossing any of these planes is the equilibrium shape. Cusps in the $\sigma$ plot give rise to crystals with sharp facets at equilibrium, as shown in Fig. 8-23a. Smoothly curved $\sigma$ plots yield nonfaceted shapes, an example of which is shown in Fig. 8-23b. Shapes composed of facets with rounded edges are also possible.

Experimentally, it is difficult to equilibrate a small solid particle with a liquid to determine equilibrium shape. The reverse experiment, for alloys, is relatively easy. Small liquid droplets are equilibrated within a solid alloy. The final shape of the liquid droplet must conform to Eq. (8-31); Miller and Chadwick[16] have studied droplet shape in a variety of metallic alloys in this way. In zinc and cadmium alloys, droplet shapes were all elongated parallel the basal planes. Depending on composition, the alloys were either faceted on the (0001) planes (with and without rounding at the edges) or nonfaceted. Droplets equilibrated at higher temperatures showed less faceting and lower aspect ratios. Extrapolation of results suggested that surface tension of pure zinc and cadmium would approach isotropy. Maximum anisotropy at lower temperatures $\gamma_{SL\max}/\gamma_{SL,(0001)}$ was about 2. Particle shape in magnesium alloys studied was spherical, indicating isotropic surface energy.

## REFERENCES

1 BAKER, J. C., and CAHN, J. W.: Thermodynamics of Solidification, "Solidification," p. 23, American Society for Metals, Metals Park, Ohio, 1971.
2 HORVAY, G.: *Proc. 4th Natl. Congr. Appl. Mech.* (*ASME*), 1315 (1962).
3 HUNT, J. D., and JACKSON, K. A.: *J. Appl. Phys.*, **37**:254 (1966).
4 HANSEN, M., and ANDERKO, K.: "Constitution of Binary Alloys," McGraw-Hill Book Company, New York, 1958.
5 DUWEZ, P.: *Trans. ASM*, **60**:607 (1967).
6 ANANTHARAMAN, T. R., LUO, H. L., and KLEMANT, JR., W.: *Nature*, **210**:1040 (1966).
7 PREDECKI, P., GIESSEN, B. C., and GRANT, N. J.: *Trans. AIME*, **233**:1438 (1965).
8 LUO, H. L., and KLEMENT, JR., W.: *J. Chem. Phys.*, **36**:1870 (1962).
9 RUHL, R. C., and COHEN, M.: *Acta Met.*, **15**:1959 (1967).
10 BAKER, J. C., and CAHN, J. W.: *Acta Met.*, **17**:575 (1969).
11 SMITH, C. S.: *Trans. AIME*, **175**:15 (1948).
12 GLICKSMAN, M. E., and VOLD, C. L.: *Acta Met.*, **17**:1 (1969).
13 JONES, D. R. H., and CHADWICK, G. A.: *Phil. Mag.*, **22**:291 (1970).
14 WULFF, G.: *Z. Krist.*, **34**:449 (1901).
15 HERRING, C.: *Phys. Rev.*, **82**:87 (1951).
16 MILLER, W. A., and CHADWICK, G. A.: *Proc. Roy. Soc.* (*London*), *Ser. A*, **312**:257 (1969).
17 THURMOND, C. D.: *J. Phys. Chem.*, **57**:827 (1953).
18 SWALIN, R. A.: "Thermodynamics of Solids," John Wiley & Sons, Inc., New York, 1962.
19 DUWEZ, P.: *Trans. ASM*, **60**:607 (1967).
20 SMITH, C. S.: *J. Inst. Metals*, **76**:728 (1949–1950).

## PROBLEMS

8-1 What is the driving force for solidification of pure nickel at its maximum undercooling (approximately 0.18 of its equilibrium melting temperature)? Compare this with driving force for vapor deposition of nickel on a cold substrate. $\Delta H = -4{,}320$ cal/mol.

8-2 Homogeneous nucleation occurs in pure nickel at atmospheric pressure at an undercooling of about 0.18 of its absolute melting point (1726°K). What pressure is required to homogeneously nucleate nickel at 1726°K? Assume $\Delta H = -4{,}320$ cal/mol, $\Delta V = -0.26$ cm$^3$/mol.

8-3 The equilibrium melting point of pure nickel of zero curvature is 1726°K. By how much will the temperature be lowered if the nickel is a sphere 1 cm radius? 1 μm radius? 0.01 μm radius? $\Delta H = 4{,}320$ cal/mol, $V = 6.6$ cm$^3$/mol, $\sigma = 255$ erg/cm$^2$ (1 cal = 4.184 × 10$^7$ erg).

8-4 A dendritic Ge–1.0 at% Si alloy is held isothermally at 1450°C (see Fig. 8-5). Cylindrical dendrite arms are present ranging in diameter from 2 to 20 μm. The arms have spherical caps. What will be the maximum and minimum compositions of the liquid at these liquid-solid interfaces (spherical and cylindrical) assuming interfacial equilibrium? The molar volume of Ge is 13.64 cm³/mol; assume $\sigma = 181$ cal/cm². Assume molar volume and $\sigma$ are independent of composition.

8-5 Estimate the heat of fusion of pure copper from several of its alloy phase diagrams.

8-6 An Ag–25 at% Cu alloy is splat-cooled so that the final structure is a homogeneously supersaturated solid solution (see Fig. 8-14).
   (a) What is the approximate range of temperatures within which solidification must have occurred?
   (b) Assume the supersaturated phase which forms is metastable at its temperature of formation (with respect to liquid and the same phase of lower solute content). What is the approximate range of temperature within which solidification must now have occurred?

8-7 A low-angle grain boundary lies perpendicular to the liquid-solid interface of a bismuth crystal. The grain-boundary groove angle is 140°; the temperature gradient is 0.10°C/cm. What is the solid-solid interfacial energy of the grain boundary? How deep does the groove penetrate? Give an equation for the shape of the groove boundary. Show this schematically. Assume that the liquid-solid interfacial energy is 54 erg/cm², $T_M = 271°C$, $V_s = 21.3$ cm³/mol, $\Delta H = 2{,}620$ cal/mol.

8-8 Pure aluminum containing a small amount (a few percent or less) of tin is heated to various temperatures in the liquid-solid region, held, and then water-quenched. Show schematically how the interdendritic liquid would appear at junctions of three dendrite arms after equilibrating (a) just above the eutectic temperature and (b) well above the eutectic temperature.

8-9 What increase in pressure is required to change the partition ratio of a dilute nickel–copper alloy by 1 percent? Does the partition ratio increase or decrease? The equilibrium melting point of pure nickel is 1726°K, and the partial molar volume of copper in nickel is 7 cm³/mol.

8-10 How small must the tip radius of a dendrite arm of dilute nickel–copper alloy be to change the partition ratio by 1 percent? Does the partition ratio increase or decrease? Partial molar volume of copper in nickel is 7 cm³/mol. Melting point of pure nickel is 1726°K, liquid-solid surface energy is 255 erg/cm².

# 9
# NUCLEATION AND INTERFACE KINETICS

## HOMOGENEOUS NUCLEATION

When a solid forms within its own melt without aid of foreign materials, it is said to *nucleate homogeneously*. Nucleation in this way requires a large driving force because of the relatively large contribution of surface energy to the total free energy of very small particles. From Eqs. (8-8) and (8-10), the radius $r^*$ of a spherical particle which is just stable at an undercooling $\Delta T$ is given by

$$r^* = -\frac{2\sigma T_M V_s}{\Delta H \, \Delta T} \qquad (9\text{-}1)$$

The particle of radius $r^*$ is termed the *critical nucleus*. Particles larger than this are stable and grow; particles smaller than $r^*$ form spontaneously in the liquid metal both above and below the equilibrium melting point. They do so because in this way they increase the entropy of the system. Following Turnbull,[1] this is seen simply by viewing the liquid as an ideal solution of various size clusters, each containing $i$ atoms or molecules (hereafter termed *atoms*). Let $n$ equal the number of simple single atoms ($i = 1$) per unit volume and $n'_i$ be the number of clusters of $i$ atoms formed from these

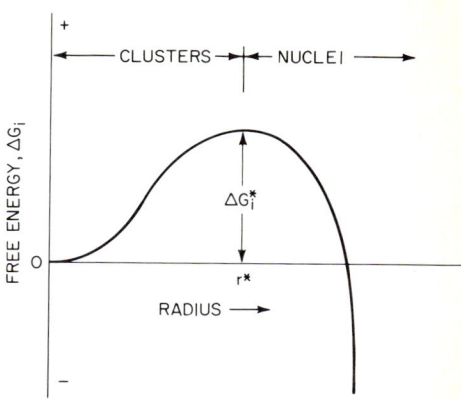

**FIGURE 9-1**
Free-energy barrier to nucleation.

simple atoms. The free-energy change per unit volume of liquid on forming the clusters $\Delta G_c$ is then

$$\Delta G_c = n'_1 \Delta G_i - T \Delta S \qquad (9\text{-}2)$$

where $\Delta G_i$ is the free energy of formation of one cluster of $i$ atoms and $\Delta S$ is the entropy of mixing $n'_i$ clusters, which for an ideal solution is simply

$$\Delta S = -k \left( n'_i \ln \frac{n'_i}{n'_i + n} + n \ln \frac{n}{n'_i + n} \right) \qquad (9\text{-}3)$$

Now, at (metastable) equilibrium, $\partial G_c / \partial n'_i = 0$. At this point, let $n'_i = n_i$ and assume $n \gg n_i$. Then, Eq. (9-3) becomes

$$n'_i = n e^{-(\Delta G^*_i / kT)} \qquad (9\text{-}4)$$

Thus, a liquid at temperature either above or below its melting point will have a distribution of solidlike *clusters* given by Eq. (9-4) at equilibrium. The free energy of formation of each of these clusters $\Delta G_i$ is

$$\Delta G_i = \sigma A + V \frac{\Delta G}{V_s} \qquad (9\text{-}5)$$

where $A$ is the surface area and $V$ is the volume of the cluster; $\Delta G$ is the thermodynamic driving force for solidification as given by Eq. (8-3), and $V_s$ is the molar volume. The form of this curve is as drawn in Fig. 9-1. $\Delta G_i$ increases to a maximum $\Delta G^*_i$ at $r = r^*$, the critical radius; thereafter it decreases rapidly. For a spherical cluster, substituting Eqs. (8-3) and (9-1) in Eq. (9-5) yields

$$\Delta G^*_i = \tfrac{16}{3} \frac{\pi \sigma^3 T_M^2 V_s^2}{\Delta H^2 \, \Delta T^2} \qquad (9\text{-}6)$$

and assuming Eq. (9-4) is valid for $n'_i = n^*_i$,

$$n^*_i = ne^{-(\Delta G_i/kT)} \qquad (9\text{-}7)$$

The thermodynamic barrier to formation of a critical nucleus $\Delta G^*_i$ decreases rapidly with increasing undercooling $\Delta T$. Thus $n^*_i$, the number of such critical nuclei, is a strong function of undercooling. The greater the number of these critical nuclei, the greater the probability that in a given time one will grow and initiate solidification.

We can now derive a rate law for homogeneous nucleation by making two simple assumptions. The first is that the concentration $n^*_i$ of critical nuclei remains that predicted by equilibrium considerations even after nucleation begins. The second is that addition of a single atom to a critical nucleus makes that nucleus supercritical and able to grow rapidly. Then, the rate of formation of nuclei per unit volume $I$ is

$$I = n^*_i \omega^* v_{LS} \qquad (9\text{-}8)$$

where $\omega^*$ is the number of atoms surrounding a critical nucleus and $v_{LS}$ is the frequency with which atoms jump across the liquid-solid interface. For a spherical nucleus, $\omega^*$ is approximately given by

$$\omega^* = \frac{4\pi r^{*2}}{a^2} \qquad (9\text{-}9)$$

The jump of atoms or molecules in the bulk liquid is[2]

$$v_L = \frac{6D_L}{\lambda^2} \qquad (9\text{-}10)$$

where $\lambda$ is jump distance and $D_L$ is liquid diffusion coefficient. The frequency with which each atom strikes the liquid-solid interface is usually taken to be one-sixth its jump frequency in the bulk liquid because it reaches the interface by jumping in only one of six possible directions. Taking $\lambda \simeq a$, we arrive at the expression for $v_{LS}$:

$$v_{LS} = \frac{D_L}{a^2} \qquad (9\text{-}11)$$

Substituting Eqs. (9-6) to (9-9) with the foregoing, we arrive finally at the classical expression for homogeneous nucleation in bulk liquids. This equation is essentially equivalent to that derived by Turnbull and Fisher[3] based on earlier work of nucleation of droplets from the vapor. It is written

$$I = B_1 \frac{D_L}{D_{LM}} \exp\left[-\frac{16\pi\sigma^3 T_M^2 V_s^2}{3 \Delta H^2 \Delta T^2 kT}\right] \qquad (9\text{-}12)$$

where the preexponential term $B_1$ depends on critical nucleus size and surface energy, and $D_{LM}$ is the liquid diffusion coefficient at the equilibrium melting point $T_M$. For liquid metals, $D_L/D_{LM} \simeq 1$ and the preexponential $B_1$ may be taken as a constant of about $10^{33}$ (Holloman and Turnbull[4]).

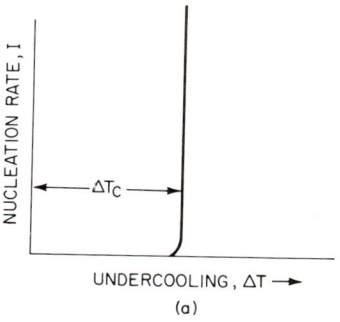
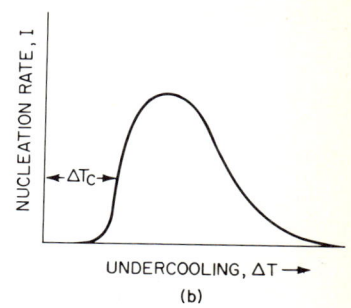

**FIGURE 9-2**
Characteristic shapes for nucleation rate curves. $\Delta T_c$ is the critical undercooling for a sensible nucleation rate. (*a*) Nonviscous liquids such as metals; (*b*) viscous nonmetallics including oxides and organic polymers.

The nucleation rate is so sensitive to the term within the exponential that changes in $B_1$ of several orders of magnitude do not appreciably affect the calculated undercoolings for sensible nucleation. For materials whose liquid diffusivity drops markedly with temperature, e.g., glasses and polymers, the preexponential term $D_L/D_{LM}$ becomes overriding at lower temperatures. As temperature is lowered from the equilibrium transformation temperature, nucleation rate first increases from zero to a maximum at some given undercooling. Then as the diffusivity $D_L$ becomes very small, nucleation rate decreases toward zero again as temperature is further decreased. Figure 9-2 shows schematically the nucleation behavior of a metal as compared with that of a viscous material such as a polymer or oxide glass.

Many refinements and modifications of the classical nucleation theory given here have been proposed, but that described above adequately explains most experimental results thus far obtained. The most fruitful experimental approach employed to confirm the theory is that of solidification studies on finely divided droplets. Two methods have been employed: microscopical observation of small drops and thermal analysis on aggregates of drops separated by thin films. In qualitative agreement with the prediction of Eq. (9-12), these studies show that a given droplet supercools without nucleating until it reaches a given critical temperature $\Delta T_c$ after which nucleation occurs rapidly. The droplet can be held at undercoolings less than $\Delta T_c$ for very long times without nucleating.

Quantitatively, results of Turnbull and coworkers[1,3-4] and of most studies since theirs have shown that for metals the maximum undercooling achieved is about 0.18 of the absolute melting point (Table 9-1). Taking this undercooling as representing the undercooling required for homogeneous nucleation, Turnbull calculated [from Eq. (9-12)] the liquid-solid surface energies $\sigma$. These surface energies are of the right order of magnitude and are self-consistent, as shown by the fact that they plot linearly against heat of fusion. This latter correlation is in agreement with expectation from simple theory.[5,6] Until recently, there were no independent measurements of liquid-solid surface energies to compare with those calculated from nucleation experiments. Now a few such measurements are available, as discussed in Chap. 8, and they agree rather well with those listed in Table 9-1. It is perhaps surprising the agreement is as good as it is since the comparison is between the surface energy of tiny nuclei and that of the bulk material.

Experiments which have been performed on metal alloys show that homogeneous nucleation occurs at approximately the same undercooling, calculated from the liquidus, as in the pure metal.[7] Figure 9-3 shows one example. In usual commercial practice heterogeneous nuclei are present in ample quantities so that the large undercoolings required to attain homogeneous nucleation are never achieved. However, techniques have been developed by a number of investigators which do permit obtaining large undercoolings in bulk metals and alloys.[8-10] One of the simplest and most effective techniques is to encase the molten metal charge in a viscous glass. The glass prevents the metal from being nucleated by crucible walls and perhaps dissolves impurities which act as nucleating agents. Undercoolings approaching those required for homogeneous nucleation have been obtained in this way in a number of metals in

Table 9-1 SUMMARY OF RESULTS ON SUPERCOOLING OF LIQUID METALS†

| Metal | Maximum supercooling, °C | Surface energy, erg/cm$^{-2}$ |
|---|---|---|
| Aluminum | 195 | 121 |
| Manganese | 308 | 206 |
| Iron | 295 | 204 |
| Cobalt | 330 | 234 |
| Nickel | 319 | 255 |
| Copper | 236 | 177 |
| Palladium | 332 | 209 |
| Silver | 227 | 126 |
| Platinum | 370 | 240 |
| Gold | 230 | 132 |
| Lead | 80 | 33 |
| Gallium | 76 | 56 |
| Germanium | 227 | 181 |
| Tin | 118 | 59 |
| Antimony | 135 | 101 |
| Mercury | 77 | 28 |
| Bismuth | 90 | 54 |

† After Hollomon and Turnbull.[4]

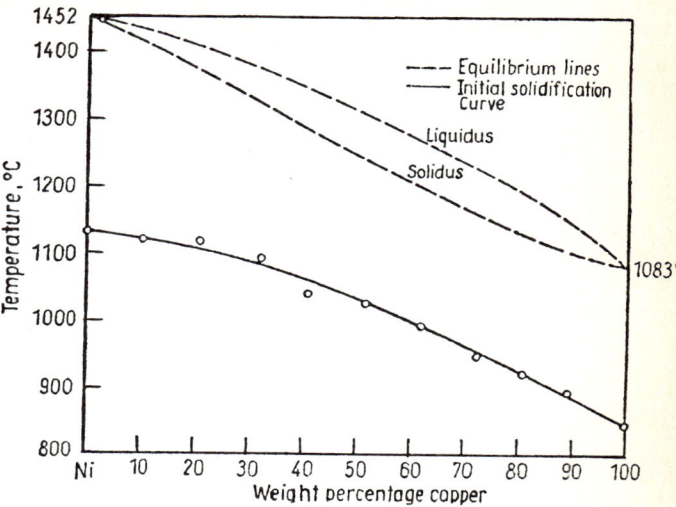

FIGURE 9-3
Solidification temperatures of copper–nickel alloy droplets as a function of composition. (*From Cech and Turnbull.*[7])

sizes up to several pounds. This technique would have great commercial interest if sufficient undercooling could be achieved so that the metal did not recalesce to its melting point. If such undercooling could be achieved in alloys, then large bodies of liquid could be solidified adiabatically in an exceedingly short time and hence with very fine structure. The necessary condition is that the undercooling be greater than the heat of fusion divided by the specific heat of the solid. Unfortunately, so far as is now known, the homogeneous nucleation temperature is such that this required undercooling is not obtainable in metallic alloys of commercial interest.

Still another technique of obtaining undercooling in liquid metallic alloys is by levitation melting of droplets, usually the order of a few grams. This technique has been used in performing measurements on the density of undercooled liquids[10] and on the metastable extensions of liquidus lines.[11]

## HETEROGENEOUS NUCLEATION AND GRAIN REFINEMENT

As is well known, metals and most other liquids (except those that readily form glasses) rarely undercool by more than a few degrees before beginning to crystallize. The crystallization begins on impurity particles, i.e., nucleating agents or mold walls,

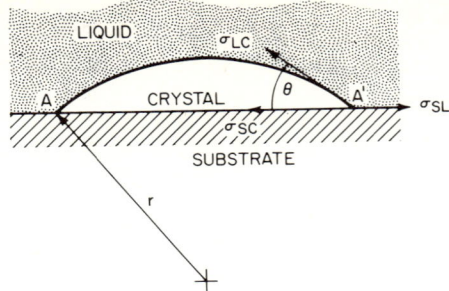

**FIGURE 9-4**
Formation of a cluster on a substrate.
(*From Chalmers.*[5])

and by so doing avoids the very large thermodynamic barrier to homogeneous nucleation.

Imagine a cluster as in Fig. 9-4 at equilibrium on a flat substrate. Equilibrium pertains when

$$\sigma_{sL} - \sigma_{sc} = \sigma_{Lc} \cos \theta \qquad (9\text{-}13)$$

where $\sigma_{sL}$, $\sigma_{sc}$, $\sigma_{Lc}$ are the surface energies of substrate-liquid, substrate-cluster, and liquid-cluster interfaces, respectively. The equilibrium form of the cluster (assuming isotropic surface energy) is a spherical cap of radius $r$. A *critical-size* cluster is just that where $r = r^*$, as given by Eq. (9-1); and it can be shown that the free energy to form a cap of this size on the substrate $\Delta G_{ci}^*$ is less by a factor $f(\theta)$ than that required to form a homogeneous nucleus

$$\Delta G_{ci}^* = \Delta G_i^* f(\theta) \qquad (9\text{-}14)$$

where $f(\theta)$ goes to zero as the angle $\theta$ decreases to zero. Thus, as *wetting* between crystal and substrate improves, the nucleation barrier decreases, ultimately vanishing altogether. $f(\theta)$ is given by[3]

$$f(\theta) = \tfrac{1}{4}(2 + \cos \theta)(1 - \cos \theta)^2 \qquad (9\text{-}15)$$

Following an analysis similar to that given for homogeneous nucleation, take $n_{ci}^*$ as the number of critical-size clusters per unit volume of liquid. Then, if $n_s'$ is the number of surface atoms of the substrate per unit volume of liquid,

$$n_{ci}^* = n_s' e^{-(\Delta G_{ci}^*/kT)} \qquad (9\text{-}16)$$

The rate expression for homogeneous nucleation is, by direct analogy with Eq. (9-8),

$$I_c = n_{ci}^* \omega_{si}^* \nu_{Ls} \qquad (9\text{-}17)$$

where $I_c$ is the rate of heterogeneous nucleation per unit volume of liquid, $n_{ci}^*$ is the number of critical clusters per unit volume of liquid, $\omega_{ci}^*$ is the number of surface

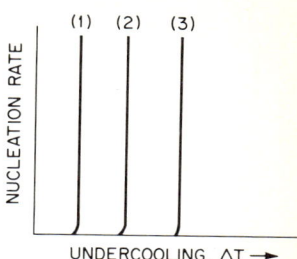

FIGURE 9-5
Nucleation-rate curves for a nonviscous liquid (e.g., a metal). Curve 3 for homogeneous nucleation; curves 1 and 2 for heterogeneous nucleation.

atoms surrounding a critical cluster, and $v_{Ls}$ is the jump frequency. Substituting in Eq. (9-17), as was done to obtain the rate expression for homogeneous nucleation, we obtain the analogous expression for heterogeneous nucleation

$$I_c = B'_1 \frac{D_L}{D_{LM}} \exp\left[-\frac{16\pi\sigma^3 T_M^2 V_s^2}{3 \Delta H^2 \Delta T^2 kT} f(\theta)\right] \qquad (9\text{-}18)$$

where the preexponential $B'_1$ depends on critical nucleus size, surface energy, and $n'_s$, the number of surface atoms on the substrate per unit volume of liquid. $B'_1$ differs little from $B_1$ except for the factor $n'_s/n$, where $n$ is the number of atoms in the liquid per unit volume. Since, for metals, $B_1$ is about $10^{33}$, $D_L/D_{LM} \simeq 1$, we may write Eq. (9-7) for metals as[3]

$$I_p = \frac{n'_s}{n} 10^{33} \exp\left[-\frac{16\pi\sigma^3 T_M^2 V_s^2}{3 \Delta H^2 \Delta T^2 kT} f(\theta)\right] \qquad (9\text{-}19)$$

where $f(\theta)$ is given by Eq. (9-15).

The form of this expression for heterogeneous nucleation is identical to that for homogeneous nucleation. The important difference is that the free-energy barrier is reduced by an amount depending on contact angle $\theta$. It disappears altogether when $\theta$ goes to zero, i.e., when the growing solid "wets" the substrate. Note also that nucleation rate should depend on the total surface area of heterogeneous nuclei present since $n'_s$ is directly proportional to this quantity. Figure 9-5 shows the expected curves of nucleation rate versus undercooling for a metal containing inoculants of varying degrees of effectiveness.

As with homogeneous nucleation, many refinements have been made in details of heterogeneous nucleation theory. One factor that has been considered by a number of workers which is probably of considerable importance is surface geometry. Surface roughness and local pits or grooves could alter nucleation behavior appreciably. As one example, it is possible that solid is retained in cavities even at temperatures above its equilibrium melting point.[4] If solid survives in this way in, say, a

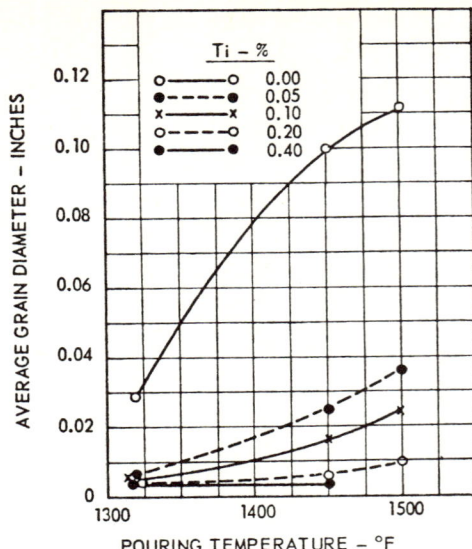

FIGURE 9-6
Effect of pouring temperature and percent titanium on grain size of cast aluminum–copper alloy. (*From Sicha and Boehm.*[13])

cylindrical cavity of radius $a$, then it will act as a critical nucleus and grow at an undercooling $\Delta T$ for which the critical radius for nucleation is $a$. As another example, in cavities whose cross section is a simple reentrant angle, nucleation is given by expressions identical in form to Eq. (9-19). The quantitative difference is that a sensible nucleation rate can be obtained at much lower undercooling than that given by Eq. (9-19). In the limit, for cavities of sufficient size and sufficiently sharp angle, nucleation occurs at zero undercooling.

In commercial practice, *inoculating agents* are added to many molten alloys to produce fine-grained materials. These include titanium and boron for aluminum alloys, impure ferrosilicon for cast iron (to nucleate graphite), carbon for certain magnesium alloys, and zirconium for others.[12-15] Cobalt, zinc, and other materials sprayed on mold walls are effective in nucleating ferrous metals.[16] Unfortunately, the powerful nucleating effect obtained by this surface treatment does not persist in ferrous metals if the agents are added before pouring. In each of the cases above, it is presumed that the addition agent forms a compound with some component of the melt and that this compound then acts as a heterogeneous catalyst according to the classical theory outlined. The behavior of the agents is qualitatively what one would expect from theory. Undercooling before initial nucleation is reduced by the grain refiners. The effect of the grain refiners diminishes with temperature and time of holding the metal molten—even though the overall chemistry does not change measurably; Fig. 9-6 is an example. This *fading effect* of the inoculant is explained by assuming the heterogeneous nuclei change their chemistry or surface characteristics or coalesce in the melt with time.

It is instructive to consider in more detail the action of one grain refiner, zirconium, in magnesium and its alloys. This element refines by a peritectic reaction in which the nucleating agent (metallic zirconium) forms in finely divided form during the reaction. The parent phase (magnesium) then nucleates and grows around the zirconium. The lattice parameter of zirconium is close to that of magnesium, suggesting a low contact angle between magnesium and zirconium, as required if the zirconium is to act as an effective refiner. This example is simpler and better understood than most. It is also of particular interest since zirconium in magnesium is the most powerful grain-refining agent in commercial use. It is so effective that it prevents development of even the tiny amount of undercooling that is necessary for dendrites to form. Thus, the grains grow, not dendritically, but spheroidally, as shown in Chap. 5. The final grain size is then determined, not by nucleation potency, but by the extent to which grain coarsening takes place during solidification, analogous to the dendrite coarsening discussed earlier.

Based on classical heterogeneous nucleation theory, the general characteristics of a good grain refiner can be stated simply. The refiner should be one that produces a small contact angle between the nucleating particle and the growing solid. This implies, from Eq. (9-13), high surface energy between particle and melt $\sigma_{Lp}$ and low surface energy between solid and particle $\sigma_{sp}$. The quantity $\sigma_{sp}$ should decrease with decreasing lattice mismatch between particle and solid and with increasing *chemical affinity* between particle and solid. In addition to these criteria, a successful nucleating agent should be as stable as possible in the molten metal, possess a maximum of surface area, and have optimum surface character (perhaps be rough or pitted).

Of the foregoing, little is usually known about the liquid-solid surface energy of nucleating particles, their specific surface area, or their surface character. Thus, in attempting to explain behavior of known nucleants, attention has been focused primarily on the presumed particle-solid surface energy $\sigma_{sp}$, particularly on the lattice disregistry that should influence $\sigma_{sp}$. A number of workers have found that undercooling required to initiate heterogeneous nucleation increases with increasing lattice disregistry between nucleant and solidifying particle.[17-20] In a particularly interesting study, Glicksman and Childs[18] showed that yttrium is an effective catalyst for tin, and that its effectiveness (as measured by amount of undercooling required for nucleation) depends on crystal orientation. It is more effective when the prismatic plane is exposed to the melt than when the basal plane is exposed. The lattice spacing on the prismatic plane is nearly exactly that along the *a* axis of tin, while no such favorable orientation relationship exists with the basal plane.

Qualitative agreement of experiment with these simple ideas on lattice mismatch is not always good. As an example, Glicksman and Childs[18] also showed that platinum is a more effective nucleating agent for tin than is silver even though silver has a lattice parameter that more closely matches that of tin. Here, apparently, chemical or physical characteristics of the surface are more important than lattice disregistry in determining nucleation potency. From a practical point of view, the effectiveness of an inoculant is to be measured, not in how much it reduces undercooling for nucleation, but in how effectively it refines the grain. Using this as a measure, the effectiveness of the various commercial grain refiners can all be rationalized on the basis of lattice mismatch ideas. Results of studies on grain refiners not used commercially have been similarly interpreted.[21-23]

As with the studies above-mentioned on the effects of nucleants on undercooling for nucleation, the predictive power of this approach has been very low indeed. It is possible that factors such as surface area or surface character of the nucleant have an overriding importance in practice. It is also possible that other factors not envisioned by classical nucleation theory have an overriding influence in determining the important final end point—the grain size of the casting. If, for example, growth is extremely rapid compared with nucleation, then one particle can produce only one grain regardless of its surface area. In this simple example, the number of nucleating particles becomes an important independent variable distinct from that of total interface area between nucleating particles and the melt. Still another factor not considered in the classical nucleation theory is grain refinement by *grain multiplication*, discussed in Chap. 5. Perhaps some of our commercial grain refiners act in part by enhancing this grain multiplication.

Another way of stimulating nucleation in undercooled liquids has aroused much interest in past years, i.e., the introduction of vibration to an undercooled liquid. As examples, Walker[8] and Hunt and Jackson[24] have shown that a pulse of sufficient intensity causes nucleation to begin in undercooled liquid nickel and in water. Cavitation appears to be necessary for the vibration to be effective. The most generally accepted explanation is that the pressure resulting from the collapse of a void formed by cavitation is very high, perhaps tens of thousands of atmospheres. This pressure changes the melting point of the liquid by very large amounts. Thus, the pressure (or subsequent rarefaction) wave could activate nuclei that would otherwise be inactive at the temperature of the experiment. Many attempts have been made to apply these ideas to the grain refinement of commercial castings and ingots. Vibration has been applied by mechanical, sonic, and ultrasonic means, generally with some success. However, it is now understood that when vibration is introduced during solidification, the ensuing convection can also cause grain refinement by a grain-multiplication mechanism. In retrospect, the successes of many of the earlier experiments in achieving grain refinement by vibration were due to grain multiplication, not to enhanced heterogeneous nucleation.

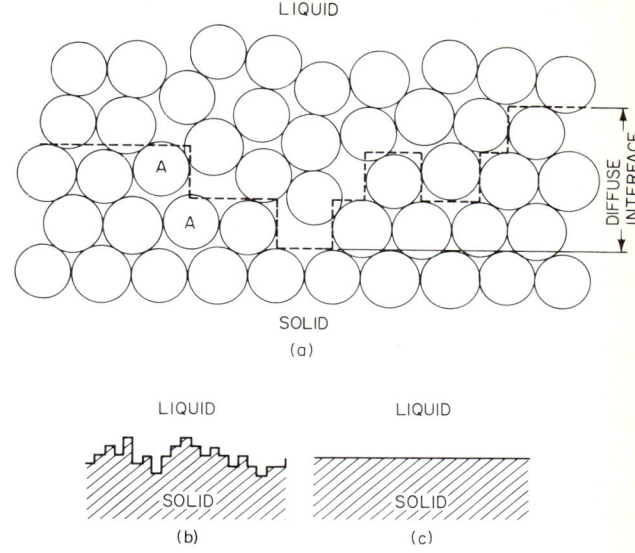

**FIGURE 9-7**
Two types of interfaces. (*a*) Atom packing in a diffuse interface; (*b*) schematic representation of a diffuse interface; (*c*) atomically flat interface.

# GROWTH

The ease with which atoms can attach themselves to a growing solid interface depends on interface structure. We can imagine two different types of interface structures, as sketched in Fig. 9-7. In the first, the transition from liquid to solid takes place over a number of atomic layers which comprise a *diffuse interface*, Fig. 9-7a. Within the diffuse interface, ordering of the atoms increases gradually with distance toward the fully crystalline side until essentially all the atoms are in their appropriate lattice sites and all heat of fusion has been removed. The dotted line in Fig. 9-7a is a schematic demarcation between atoms that are "part of the solid" and those that are "part of the liquid." This schematic device is used in Fig. 9-7b and subsequently; but the reader should be aware that it is an oversimplification. The atoms themselves in the diffuse layer are not so sure whether they are part of the liquid or part of the solid.[25] Note that the dotted line is drawn such that some of the atoms above it have close packing characteristics of the solid; some atoms below (those marked *A*) are slightly out of their crystalline lattice sites. The important characteristic of the diffuse interface is that, with increasing distance into the interface, the thermodynamic

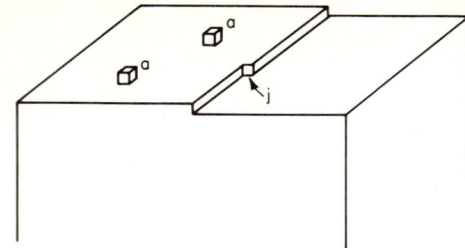

**FIGURE 9-8**
Atomically flat interface with a step, a jog in the step (at $j$) and two adsorbed atoms (at $a$).

properties of the layers of atoms within the zone vary continuously from those of the liquid to those of the solid.

The second type of interface is sketched in Fig. 9-7c, the atomically flat, close-packed interface. Here, a transition from liquid to solid is envisioned to take place across a single atomic layer. Our ideas concerning the flat interface in solidification have developed largely from earlier work on vapor deposition, where it can be demonstrated such interfaces do exist. In solidification, they probably do not, and the interfaces here must be at least a few layers thick. However, as Cahn has shown, the classical laws developed for atomically flat interfaces can be extended with little difficulty to the slightly diffuse interfaces expected for many liquid-solid transitions.

Diffuse interfaces grow much more easily than those with an atomically flat interface, and we shall first attempt to see the reason for this with a simple quasi-chemical model in line with ideas developed in vapor deposition. To do this, we must imagine an ideally flat, close-packed interface, as sketched in Fig. 9-8, except that there is a *step* in the interface and a jog in the step. If an atom deposits on the solid at point $a$ on the close-packed face, it is joined to the three nearest neighbors and will have lost three-sixths, or one-half, of its heat of fusion. However, if it deposits at the jog $j$, it will have three nearest neighbors in the close-packed plane below and three nearest neighbors in its own plane. It will now have lost six-sixths, or all, of its heat of fusion. Thus, the driving force for deposition of an atom at $j$ is twice that at position $a$, and any atom that deposits at point $a$ can lower its energy further by migrating and depositing at $j$. Jogs $j$ are preferred sites for growth, and in an *ideally diffuse* interface a high proportion of interface sites are jogs such as this.

We expect ideally diffuse interfaces to grow more easily than flat interfaces, and we also expect them to grow in a different way. In the ideally diffuse interface, most lattice sites are favorable for deposition; and so as atoms strike the interface it moves forward more or less uniformly and growth is said to be *continuous*. If, on the other hand, the interface is atomically flat, then forward growth occurs preferentially at steps which sweep laterally across the interface, as shown schematically in Fig. 9-9a.

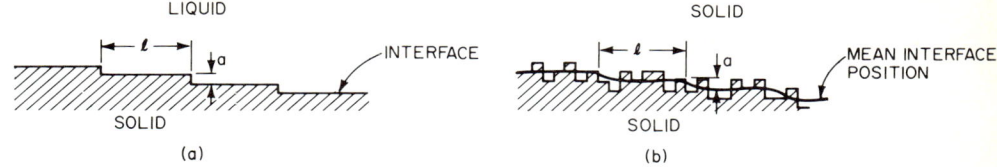

**FIGURE 9-9**
Steps in (a) an atomically flat interface and (b) a diffuse interface. Growth is by lateral movement of these steps.

Solidification is then said to proceed by *lateral* growth. Between the *ideally diffuse* interface and the atomically flat stepped interface lies the diffuse interface shown schematically in Fig. 9-9b. Here, the liquid-solid transition occurs over a number of layers, but growth occurs by lateral propagation of rather diffuse steps.

A number of attempts have been made to predict the degree of diffuseness of liquid-solid interfaces.[25-27] We can see the problem qualitatively from a quasi-chemical viewpoint by viewing a flat interface with a step and jog, and a small number of *adsorbed* atoms in positions as sketched in Fig. 9-8. If the interface is in its equilibrium configuration, there will then be a number of atoms at positions $a$ in equilibrium with the jog, given at the equilibrium melting point $T_M$ approximately by[28]

$$\frac{n_a}{n_s} = e^{\Delta H_{aj}/kT_M} \qquad (9\text{-}20)$$

where $n_a$ is the number of adsorbed atoms $a$ per unit area, $n_s$ is the number of surface atoms per unit area, and $\Delta H_{aj}$ is the energy difference per atom between an atom at $j$ and one at $a$. Since the energy difference between $j$ and $a$ is one-half the heat of fusion,

$$\frac{n_a}{n_s} = e^{\Delta H_m/2kT_M} \qquad (9\text{-}21)$$

where $\Delta H_m$ (a negative quantity) is the molecular (or atomic) heat of fusion and $k$ is Boltzman's constant. Equation (9-21) shows that the population of adsorbed atoms is strongly dependent on $\Delta H_m/T_M$, the *entropy of fusion*.

When the adsorbed atom population becomes large, the atoms begin to interact and Eq. (9-21) is no longer valid. The simplest treatment considering this interaction is that of Jackson,[27] who concludes that 50 percent of the surface sites in a single layer of a close-packed interface will be filled when $\Delta H_m/T_M = -4k$, that is, when $\Delta H/T_M = -4R$. This he takes to correspond to an ideally diffuse interface which can move forward by continuous growth. Close-packed interfaces for which the entropy

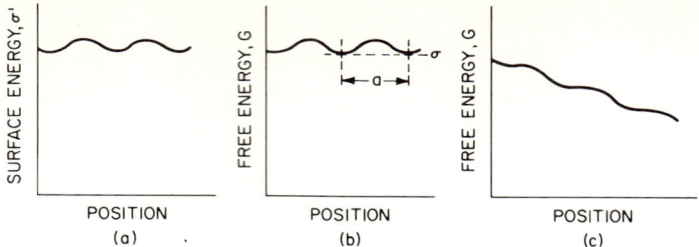

**FIGURE 9-10**
Surface energy and free energy of a solidifying material as a function of interface position. (*a*) Surface energy; (*b*) free energy at low driving force; (*c*) free energy at high driving force. (*From Cahn.*[25])

of fusion is greater than $4R$ are assumed to be flat and require growth by a lateral mechanism. Non-close-packed interfaces are diffuse at higher entropies of fusion than close-packed interfaces.

The quasichemical model discussed above assumes that a line of demarcation can be drawn (such as the dashed line of Fig. 9-7*a*) that clearly separates "solid" atoms from "liquid" atoms. Cahn's more recent treatment shows that such a line is an idealization at best, and we would do better to picture the diffuse interface as a region in which gradual organization of atoms takes place, with the atoms moving more and more into their equilibrium positions with depth in the interface. This treatment shows that the liquid-solid transition must, in general, take place over at least a few atomic layers, as sketched in Fig. 9-7*a*. That is, it must always be *diffuse*. To advance this diffuse interface below the melting point by an interplanar distance (i.e., for solidification to occur) may or may not require that the interface pass through a configuration of higher surface energy. If it does not, then *continuous growth* results. If it does, then growth must be by lateral propagation of steps.

Imagine a diffuse interface in its equilibrium configuration. Its surface energy $\sigma'$ is at a minimum, denoted by $\sigma$ in Fig. 9-10*a*. As new atoms are added, the surface energy $\sigma'$ increases to a maximum and falls again to $\sigma$ when the new layer of atoms is completed and the interface is again in its equilibrium configuration.

In the presence of a driving force, the variation per unit area of total free energy of the system $\delta G$ due to uniform motion of the interface an amount $\delta x$ is then given by

$$\delta G = \left( \frac{\Delta G}{V_s} + \frac{d\sigma'}{dx} \right) \delta x \qquad (9\text{-}22)$$

where $\Delta G$ is the thermodynamic driving force for solidification, as given by Eq. (8-3), and $V_s$ is the molar volume. When there is a thermodynamic barrier to growth (that is, when $\partial G/\partial x > 0$, as in Fig. 9-10b), a lateral growth mechanism is required; when $\partial G/\partial x > 0$, as in Fig. 9-10c, growth is continuous. From Eq. (9-22), the critical driving force necessary for continuous growth is

$$\Delta G = -V_s \frac{d\sigma'}{dx} \qquad (9\text{-}23)$$

and Cahn[25] shows this to be

$$\Delta G = -\frac{\pi \sigma g V_s}{a} \qquad (9\text{-}24)$$

where $a$ is interplanar spacing and $g$ is a *diffuseness parameter* which depends on the number of atoms comprising the transition from the solid to liquid at the melting temperature. For sharp interfaces, $g$ is the order of 1 and rapidly becomes very much less than 1 as the interface becomes diffuse. Unfortunately, $g$ depends so sensitively on the number of layers comprising the transition that reliable theoretical estimates of $g$ cannot yet be made. However, estimates of $g$ can be made for real materials by indirect experiment and we return to this later.

An important result of Cahn's analysis, embodied in Eq. (9-24), is that whether or not a given material solidifies by a lateral growth mechanism depends both on the diffuseness of the interface and on the thermodynamic driving force. No matter how diffuse an interface, there should always be a driving force (undercooling) below which lateral growth is required. Of course, for very diffuse interfaces, $g$ may be so small that it would be difficult experimentally to observe lateral growth.

## CONTINUOUS GROWTH

Turnbull[29] has described kinetics of continuous growth by using classical rate theory. An activation energy $\Delta G_b$ is envisioned for transport of atoms (or molecules) from the liquid to the solid, Fig. 9-11. The frequency with which atoms surmount the barrier will be approximately the vibration frequency $v_0$ times the number of atoms at any given moment whose free energy is just that of the barrier $\Delta G_b$ above the free energy of the liquid. Assuming an equilibrium concentration of such atoms, it will be $e^{-(\Delta G_b/kT)}$. Thus, the frequency $v_{LS}$ with which atoms will pass from the liquid to the solid will be

$$v_{LS} = v_0 e^{-(\Delta G_b/kT)} \qquad (9\text{-}25)$$

Similarly, atoms will jump in the reverse direction, from solid to liquid. If the solid-liquid interface is below the equilibrium melting point, the barrier in this direction will be greater by $\Delta G_m$, the free-energy change per molecule (or atom) for the

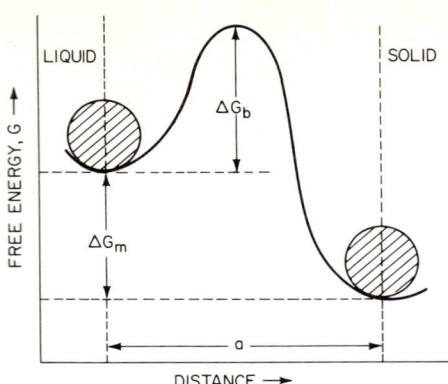

**FIGURE 9-11**
Schematic free-energy diagram for a solid-liquid interface.

liquid-solid transition. Thus, this reverse reaction $v_{SL}$ is given by

$$v_{SL} = v_0 e^{-[(\Delta G_b + \Delta G_m)/kT]} \qquad (9\text{-}26)$$

and the net jump frequency across the interface $v_{net}$ is

$$v_{net} = v_{LS} - v_{SL} = v_{LS}(1 - e^{-(\Delta G_m/kT)}) \qquad (9\text{-}27)$$

Now, substitute Eq. (8-3) (on a per molecule basis) in (9-27) and assume the exponent is small. Also, denote the undercooling as $\Delta T_k$ to signify the undercooling referred to is kinetic undercooling to drive the interface and take it as positive for $T < T_M$. Then, for $T \simeq T_M$,

$$v_{net} = -v_{LS} \frac{\Delta H_m \, \Delta T_k}{k T_M^2} \qquad (9\text{-}28)$$

Assuming all sites on the interface are favorable for growth, the rate of this continuous growth $R$ is $a v_{net}$, where $a$ is the amount the interface advances when a molecule is added. Hence,

$$R = -a v_{LS} \frac{\Delta H_m \, \Delta T_k}{k T_M^2} \qquad (9\text{-}29)$$

The widely used kinetic law for continuous growth is now obtained by substituting Eq. (9-11) in the above equation:

$$R = -\frac{D_L \, \Delta H_m \, \Delta T_k}{a k T_M^2} \qquad (9\text{-}30)$$

This law is sometimes written

$$R = B_2 \frac{D_L}{D_{LM}} \Delta T_k \qquad (9\text{-}31)$$

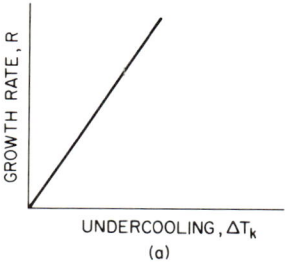 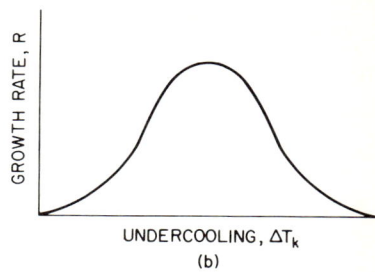

FIGURE 9-12
Characteristic growth rate curves for (a) nonviscous liquids, such as metal, and (b) oxide or organic polymers.

where $B_2$ is a constant defined by Eq. (9-30). Growth rate varies linearly with undercooling, provided the diffusion coefficient $D_L$ does not change much with undercooling from its value $D_{LM}$ at the equilibrium melting point. A schematic example is shown in Fig. 9-12a. When $D_L$ varies strongly with temperature, as in oxide glasses and polymers, growth rate increases to a maximum at some undercooling $\Delta T_k$ and then decreases at higher undercoolings, Fig. 9-12b.

Cahn, Hillig, and Sears[30] have suggested that the assumption of $\lambda \simeq a$ in Eq. (9-11) is quite inaccurate and $\lambda$ may be as much as an order of magnitude less than $a$. Also, while $v_{LS}$ should be about $\frac{1}{6}v_L$ for symmetric molecules, it should be less for unsymmetric molecules which must rotate before they fit in place in the solid. Thus, they suggest Eq. (9-30) should be corrected by a factor $\beta$ to read

$$R = -\frac{\beta D_L \Delta H_m \Delta T_k}{akT_M^2} \quad (9\text{-}32a)$$

where

$$\beta = \left(\frac{a}{\lambda}\right)^2 \frac{6v_{LS}}{v_L} \quad (9\text{-}32b)$$

There is great difficulty in verifying the linear growth law for materials with high solute diffusivities, such as liquid metals. For a typical metal, the kinetic coefficient relating $R$ and $\Delta T_k$ in Eq. (9-31) is the order of 100 cm/(°C)(s). Thus, for a growth rate conveniently obtained in a laboratory crystal-growing experiment, such as 0.01 cm/s, the interface undercooling $\Delta T_k$ is the order of $10^{-4}$°C, a very difficult quantity to measure indeed! Many laboratory attempts have been made to measure the relation between $R$ and $\Delta T_k$ in metals, examples being those on tin.[31,32] However, the very great experimental difficulties make the numerical results obtained to date open to question. The one unambiguous result of these and other experiments is that for such materials the kinetic undercooling at low growth rates is very small indeed.

## LATERAL GROWTH

Consider now a sharp interface with a step as pictured in Fig. 9-8, but with many jogs in the step so that any atom which strikes the step deposits on it and moves that portion of the step forward one atomic distance. Growth of the interface takes place by steps such as this sweeping across it. The velocity of movement of this straight step $R_\infty$ would be given exactly by Eq. (9-32a) if the only atoms to reach the step were those that impinged directly on it from the liquid metal. However, there is the possibility of surface diffusion bringing additional atoms to the step. Following Hillig and Turnbull,[33] suppose that atoms which strike the interface at the atomic positions on the two sides of the step also migrate to the step. Then, the growth rate of the step of infinite radius of curvature is three times Eq. (9-32a), or

$$R_\infty = -\frac{3\beta D_L \Delta H_m \Delta T_k}{ak T_M^2} \qquad (9\text{-}33)$$

Steps in a diffuse interface are expected to grow much faster than is predicted by Eq. (9-33). Since the steps themselves are diffuse, as sketched in Fig. 9-9b, many more atoms strike the step per unit time than in the case of the sharp step. Following Cahn et al.,[30] step width is proportional to $g^{-1/2}$ and so for the diffuse interface Eq. (9-33) becomes

$$R_\infty = -\frac{\beta D_L \Delta H_m \Delta T_k}{ak T_M^2}(2 + g^{-1/2}) \qquad (9\text{-}34)$$

where $g$ varies from unity for the sharp step to a very small number for very diffuse interfaces.

If a step is not flat but has a curvature $r$, the undercooling available to drive the step is no longer the total undercooling of the interface $\Delta T_k$ but is less by the melting-point depression $\Delta T_{rs}$, corresponding to the step curvature $r$. That is,

$$R_\perp = R_\infty \left(1 - \frac{\Delta T_{rs}}{\Delta T_k}\right) \qquad (9\text{-}35)$$

where $R_\perp$ is the velocity of a step of radius curvature $r$. From equation (8-10) $\Delta T_{rs}$ is proportional to $1/r$. The radius of a step that would be at equilibrium at the total undercooling $\Delta T_k$ is of critical radius $r^*$, as given by Eq. (9-1). Hence, $\Delta T_k$ is proportional to $1/r^*$ and Eq. (9-35) can be written

$$R_\perp = R_\infty \left(1 - \frac{r^*}{r}\right) \qquad (9\text{-}36)$$

Now, imagine a number of steps on a crystal surface separated by a length $l$ and all moving past a given point with velocity $R_\perp$ as sketched in Fig. 9-9. $R_\perp/l$ steps move past the point per unit time. The point moves ahead one step height $a$ as each plane passes, and so the rate of forward advance is

$$R = \frac{R_\perp a}{l} \qquad (9\text{-}37)$$

where $R_\perp/l$ is the frequency of passage of steps past a given point. Equation (9-37) is the general equation relating the rate of solidification to the rate of step movement $R_\perp$. We need to turn now to considerations of the source of steps to determine the number available and hence their spacing $l$.

## GROWTH BY TWO-DIMENSIONAL NUCLEATION

One source of steps on low index faces is two-dimensional nucleation. The classical theory for this type of growth was developed by many workers, including Volmer[34] and Becker and Doring.[35] Physically and mathematically, the problem of nucleus formation is closely analogous to the three-dimensional nucleation problem described earlier. The free-energy barrier $\Delta G_{gi}^*$ required to form a critical two-dimensional cylindrical nucleus of step height $a$ is given by an equation of the form of Eq. (9-5) where the surface area $A$ is that of the new cylindrical surface. Hence

$$\Delta G_{Gi}^* = 2\pi r^* a \sigma + \pi r^{*2} a \frac{\Delta G}{V_s} \qquad (9\text{-}38)$$

and, from Eq. (8-10), the undercooling $\Delta T$ for which the critical nucleus of single radius of curvature $r^*$ will neither melt nor freeze is

$$\Delta T = -\frac{\sigma T_M V_m}{\Delta H_m r^*} \qquad (9\text{-}39)$$

where $V_m$ is the volume of a molecule (or atom). At the undercooling $\Delta T$, a distribution of surface clusters of different sizes will form, as in the nucleation examples discussed earlier and as sketched in Fig. 9-13. Following steps analogous to those in developing Eq. (9-7), the resulting comparable expression for the number of critical two-dimensional nuclei per unit area $n_{gi}^*$ is

$$n_{gi}^* = n_s e^{-(\Delta G_{gi}^*/kT)} \qquad (9\text{-}40)$$

Now, assume that during growth the concentration $n_{gi}^*$ remains as predicted by Eq. (9-40), and that addition of a single atom makes the critical nucleus unstable and therefore able to grow rapidly. The rate of formation of new nuclei is then given simply as

$$I_{2d} = n_{gi}^* \omega^* \nu_{LS} \qquad (9\text{-}41)$$

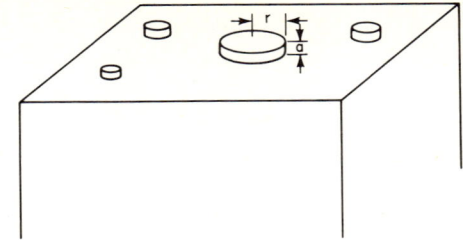

**FIGURE 9-13**
Two-dimensional clusters on a flat close-packed surface.

where $I_{2d}$ is the nucleation rate (per unit area), $n_A^*$ is the number of critical nuclei per unit area, and $\omega^*$ is the number of atoms sufficiently close to the edge of the critical nucleus so that they reach it when jumping from liquid to solid; $\nu_{LS}$ is the jump frequency from liquid to solid. We assume that the growth rate of each nucleus that forms is sufficiently rapid relative to the crystal size so that it spreads to form a full atomic plane before the next nucleus forms. Then the rate of passage of steps $R_\perp/l$ is exactly the rate of nucleation times the surface area of the growing face $A$. Thus Eq. (9-37) becomes

$$R = AI_{2d}a \qquad (9\text{-}42)$$

where $A$ is the surface area of the growing plane.

To be consistent with the derivation of Eq. (9-34), we now take $\omega^*$ to be $2 + g^{-1/2}$ times the number of atoms that would surround a sharply stepped nucleus of critical size. For a cylindrical nucleus this is

$$\omega^* = \frac{2\pi r^*}{a}(2 + g^{-1/2}) \qquad (9\text{-}43)$$

Substituting the various necessary relations in Eq. (9-42), the equation for rate of growth by two-dimensional nucleation can now be written as

$$R = B_3 \frac{D_L}{D_{LM}} e^{(\pi\sigma^2 aV_s/k\,\Delta H\,\Delta T_k)} \qquad (9\text{-}44)$$

where the preexponential $B_3$ depends on the variables $\Delta T_k$, $g$, and $A$. $D_{LM}$ is the liquid diffusion coefficient at the equilibrium melting point, and so $D_L/D_{LM}$ represents a correction to the growth rate from varying diffusivity. This expression has the form

$$R = B_3 \frac{D_L}{D_{LM}} e^{-(B_3'/\Delta T_k)} \qquad (9\text{-}45)$$

where both $B_3$ and $B_3'$ are constants. As in the case of homogeneous nucleation, quite large variations in $B_3$ should have little effect on overall growth rate.

Equation (9-45) is identical to that derived by many workers, although the constants $B_3$ and $B_3'$ differ depending on details of the analysis. The predicted

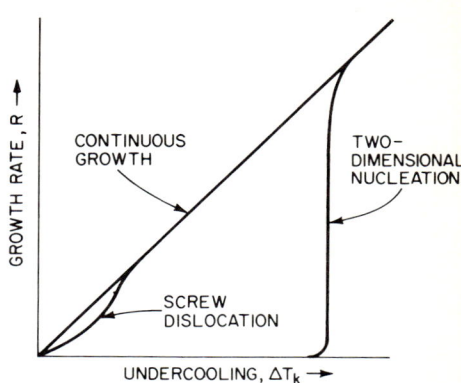

FIGURE 9-14
Growth rate versus interface undercooling according to the three classical laws.

variation of growth rate with temperature is shown in Fig. 9-14. Typically, for sharp interfaces ($g \simeq 1$) very large calculated values of undercooling $\Delta T_k$ are required to achieve an observable growth rate. The predicted undercoolings required for growth are so much larger than have ever been observed that this mechanism of growth has not been viewed with much enthusiasm in recent years. However, the introduction of the diffuseness parameter $g$ by Cahn and coworkers[25,30] results in a growth law of identical form but in which required undercooling for observable growth rate can be very small indeed, depending on the diffuseness parameter $g$.

At high undercoolings, the rate of two-dimensional nucleation becomes so high that many nuclei form for each plane that grows. When the spacing between steps becomes the order of the interatomic spacing, the actual growth rate deviates from that given by Eq. (9-45); it approaches the continuous growth law as an upper limit, as shown in Fig. 9-14. In the case of a diffuse interface growing by two-dimensional nucleation, a deviation from Eq. (9-45) in the opposite direction occurs at intermediate undercooling. This deviation results when the critical radius becomes less than the step width.[30]

Workers have generally attempted to compare theory with experiment only for low undercoolings, where Eq. (9-45) should be valid. Among experiments performed are those of Hillig,[36] who showed that perfect (dislocation-free) ice crystals do not grow at measurable rates at less than about 0.03°C undercooling; at greater undercoolings, growth rate was described by Eq. (9-45), as shown in Fig. 9-15. Similar results were obtained on ice by Sperry[37] and on tri-α-naphthyl benzene by Magill and Plazek.[38] In the latter case, 2°C undercooling was required to initiate growth. Alfintsev and Ovsienko[39] found that gallium did not grow at observable rates when $\Delta T_k$ was less than about 0.5°C. In all these studies, the materials solidified with sharp facets. In most, there was strong evidence that when dislocations were introduced into

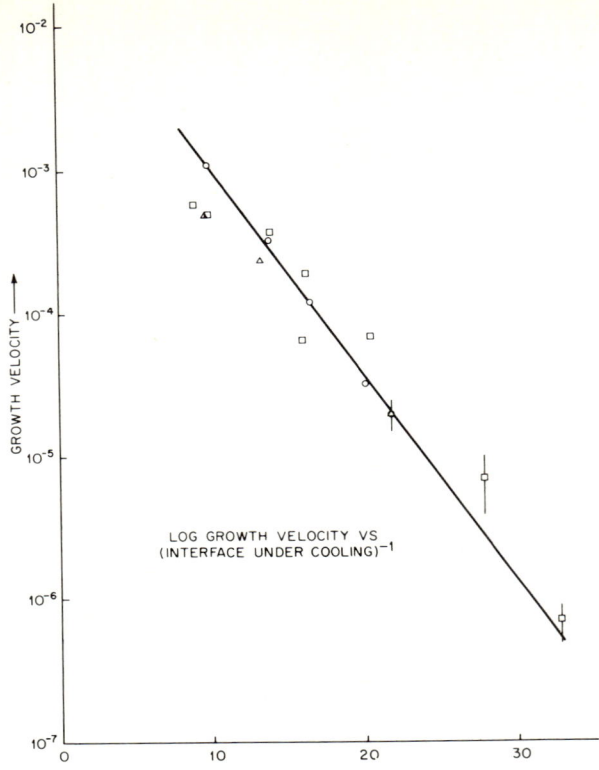

**FIGURE 9-15**
Growth kinetics of perfect ice crystal along $c$ direction. (Log growth velocity versus reciprocal of interface undercooling.) (*From Hillig.*[36])

the crystal, growth rate increased rapidly, suggesting a change in mechanism to one involving growth by dislocations. All of these observations provide strong and probably conclusive proof of the general applicability to dislocation-free crystals of the concept of growth by two-dimensional nucleation.

## GROWTH BY SCREW DISLOCATIONS

Suppose, now, that a crystal is not perfect but has a screw dislocation emerging at its liquid-solid interface. The close-packed face can no longer be flat but must have a step in it, as shown schematically in Fig. 9-16. Frank[40] first pointed out that such a step obviates the need for two-dimensional nucleation and permits growth at much lower undercoolings. The step is self-perpetuating; no matter how many layers of

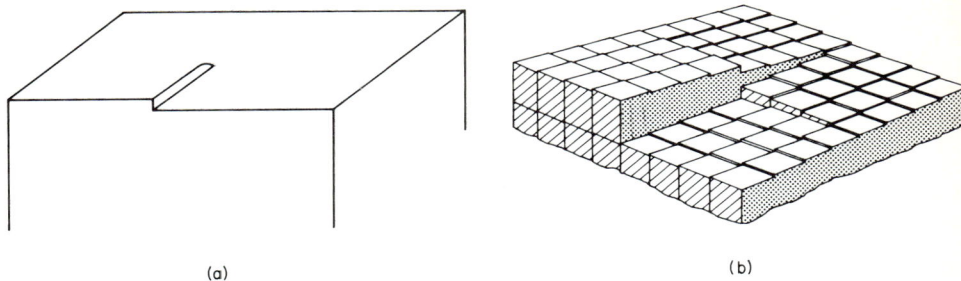

(a)  (b)

FIGURE 9-16
Screw dislocation emerging at liquid-solid interface. (*a*) Simple line sketch; (*b*) block model.

atoms are deposited on the face, the step will persist, as can be seen by imagining atoms deposited on the model diagram of Fig. 9-16. For a single dislocation, the crystal grows up a spiral staircase by continuously depositing atoms at the exposed step. The step thereby rotates continuously about the point where the dislocation emerges. Since one end of the step is fixed at the dislocation line, the step rapidly winds itself up as sketched in Fig. 9-17. At the center of the spiral, it reaches a minimum radius of curvature $r^*$, which is exactly that radius given by Eq. (9-1), the critical radius for two-dimensional nucleation. At this curvature, the step edge is just in equilibrium with the surrounding melt and neither advances nor retreats. Further out along the spiral, the curvature is less and the step advances at a greater rate. At steady state, the *wound-up* spiral would appear to be rotating at constant angular velocity so that step velocity at each point $R_\perp$ would be constant with time.

Equation (9-36) gives the dependency of step velocity on step radius, and the problem of growth by the screw dislocation mechanism is now completely specified by Eqs. (9-34), (9-36), and (9-37), with the boundary condition that $r = r^*$ at the point of emergence of the screw dislocation. These equations are not readily solved

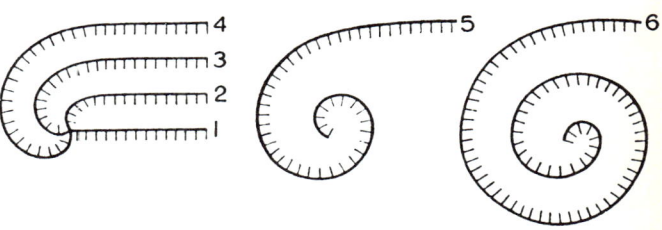

FIGURE 9-17
Development of spiral step structure. (*After Christian.*[28])

directly, but Burton, Cabrera, and Frank[26] have shown that very nearly the correct final result is obtained if the growth spiral is assumed to be archimedean, according to the relation

$$r = 2r^*\theta \qquad (9\text{-}46)$$

where $\theta$ is angle. For the archimedean spiral, there is one step for each $2\pi$ radius and so the length between steps is

$$l = 4\pi r^* \qquad (9\text{-}47)$$

As can be seen by differentiating Eq. (9-46) with respect to time, the rate of advance of steps $dr/dt$ is independent of position and therefore equal to $R_\infty$. This is not, of course, consistent with Eq. (9-36) but gives a final result close to that calculated by a more rigorous and much more cumbersome method. From Eqs. (9-37) and (9-47),

$$R = \frac{R_\infty a}{4\pi r^*} \qquad (9\text{-}48)$$

and combining this with Eqs. (9-34) and (9-39) yields the dislocation growth law for growth from the melt in the form given originally by Hillig and Turnbull[33] modified by $\beta$ and the diffuseness parameter $g$:

$$R = \frac{(1 + 2g^{1/2})\beta}{g} \frac{D_L \Delta H_m^2 \Delta T_k^2}{4\pi\sigma T^3 k V_s} \qquad (9\text{-}49)$$

For sufficiently small values of $\Delta T_k$,

$$R = B_4 \frac{D_L}{D_{LM}} (\Delta T_k)^2 \qquad (9\text{-}50)$$

where $D_{LM}$ is liquid diffusivity at the equilibrium melting point.

Thus, we expect the square dependence of growth rate on undercooling shown in Figs. 9-14 and 9-18. Growth approaches the curve for continuous growth as an upper limit. For the case of sharp steps ($g \simeq 1$), growth rate deviates from that given by Eq. (9-49) when the steps become so closely spaced that they compete with one another for available atoms from the melt. For $g < 1$, deviation from Eq. (9-49) comes at lower undercooling and the deviation is such that the curve for continuous growth is approached more rapidly. This can be seen schematically in Fig. 9-19, in which $R/\Delta T_k(D_{LM}/D_L)$ is plotted versus $\Delta T_k$. At low undercooling (*classical regime*), the screw dislocation law [Eq. (9-50)] is obeyed; the resulting curve is a straight line which passes through the origin. At high undercoolings (*continuous regime*), Eq. (9-31) is obeyed and the curve is horizontal. Between the two regimes is a transitional

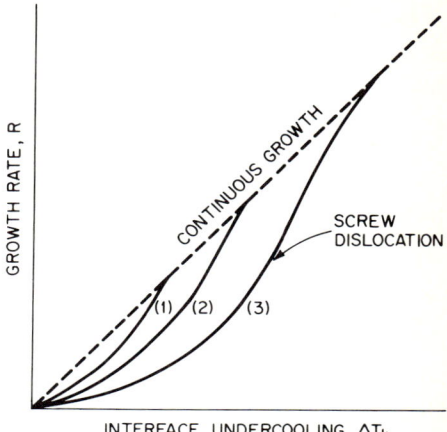

FIGURE 9-18
Growth-rate curves for growth by screw dislocation mechanism. Curve 3 for classical growth of an atomically flat interface ($g = 1$); curves 1 and 2 for diffuse interface ($g < 1$).

regime, which is shown by Cahn et al.[25,30] to lie between about $\Delta T^*$ and $\pi \Delta T^*$, where

$$\Delta T^* = -\frac{\sigma g V_s T_M}{a \, \Delta H} \qquad (9\text{-}51)$$

Thus, the undercooling necessary for continuous growth is seen to depend on the diffuseness parameter $g$.

The foregoing quantitative treatment has dealt with only a single dislocation emerging as the growing face. However, real crystals, especially metals, may have as many as $10^8$ lines per square centimeter. The simplest case of multiple dislocations is shown in Fig. 9-20, where growth is proceeding from two dislocations of opposite sign spaced apart not less than $2r^*$. Here, we obtain closed terraces on a flat cone rather than the continuous terrace of the isolated dislocation. Overall growth rate is unchanged, however, and Eq. (9-49) is equally valid for this case. Similarly, it can be shown that the equation remains valid no matter how many dislocations are present provided there are equal numbers of opposite dislocations separated by greater than $r^*$. Pairs of opposite dislocations less than $r^*$ apart are inoperative. Deviations from Eq. (9-49) result when there is an unbalanced number of opposite dislocations, but these deviations do not alter the form of the equation. They result in a growth law that is just Eq. (9-49) multiplied by a constant factor that lies between unity and the number of unbalanced dislocations.[26]

Many growth cones (spirals), such as those sketched in Figs. 9-17 and 9-20, have now been observed experimentally on materials grown from the vapor, from

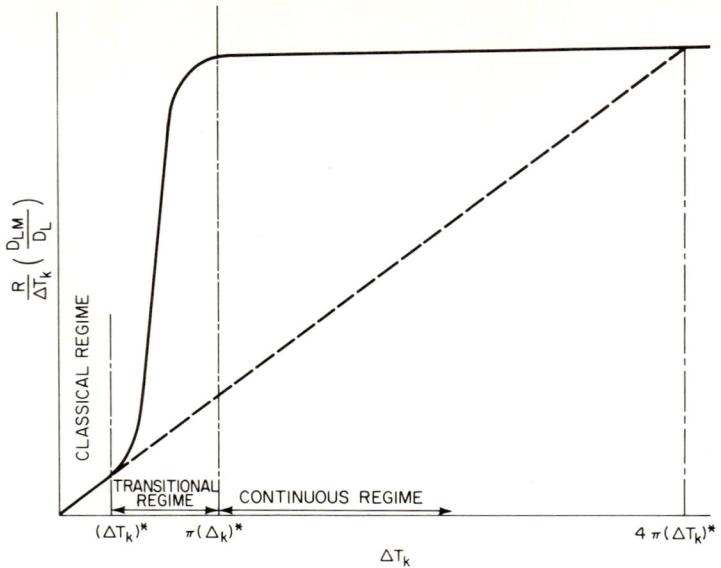

**FIGURE 9-19**
Predicted growth rate curve for surface with an emergent dislocation. Ordinate is interface velocity divided by undercooling and corrected for temperature dependence of the diffusion coefficient. (*After Cahn et al.*[30])

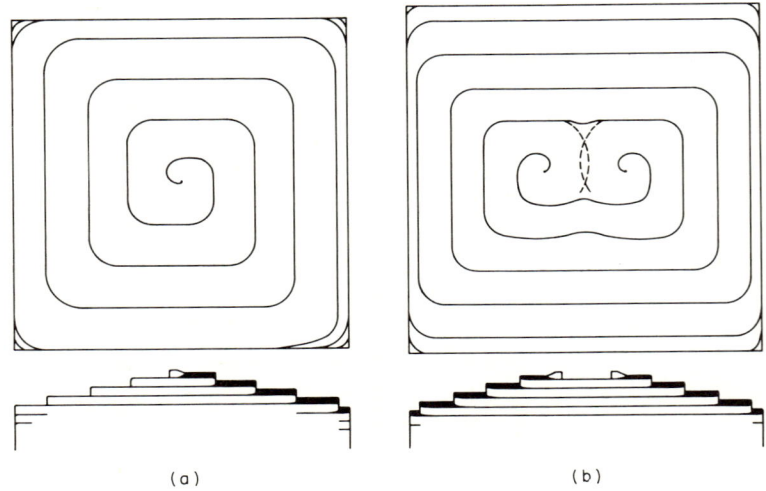

**FIGURE 9-20**
Growth pyramid due to (*a*) a single dislocation and (*b*) a pair of dislocations. (*After Burton, Cabrera, and Frank.*[26])

### GROWTH BY SCREW DISLOCATIONS 317

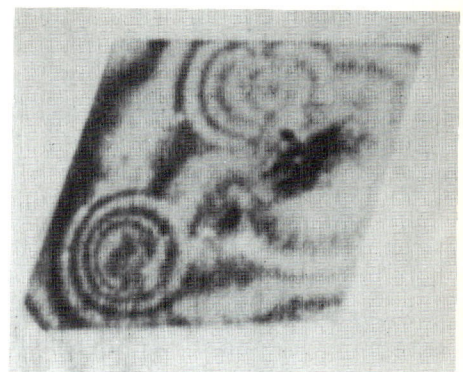

**FIGURE 9-21**
Spiral growth pattern on a single crystal of salol grown from its melt (the crystal is 1 mm across). (*From Chadwick.*[51])

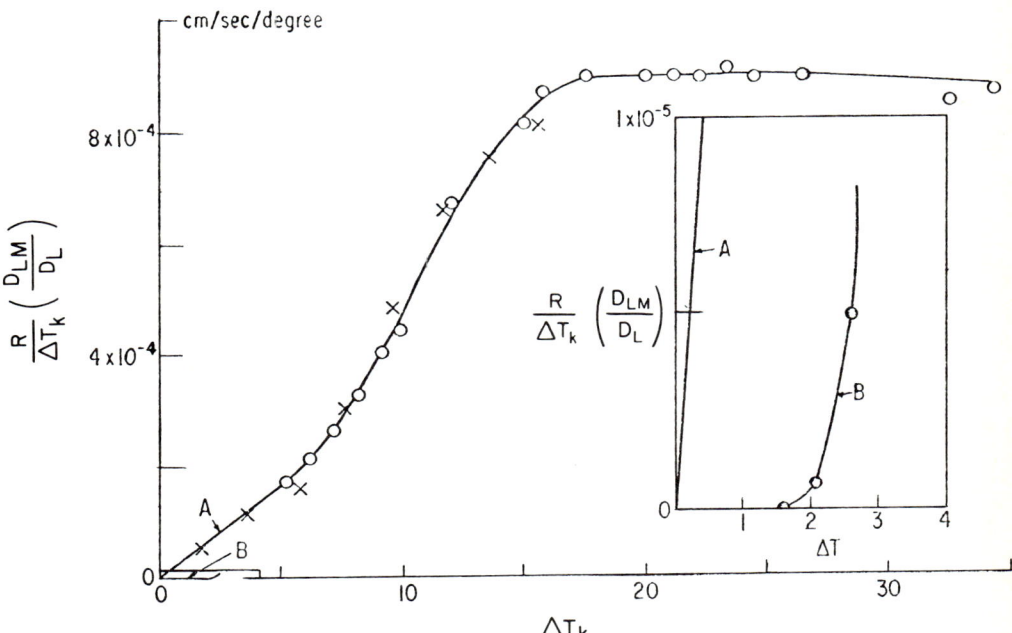

**FIGURE 9-22**
Growth rate curves for salol (*from Cahn et al.*[30]). (*Curve A from data of Pollatschek*[52] *and of Newmann and Micus; curve B from Danilov and Malkin.*[53])

solution, electrolytically, and from the melt.[41,42] The step height in some instances is of unit molecular height, as assumed herein, and in other cases it is much larger. The larger steps presumably result from step *bunching* as a result of impurities. Figure 9-21 shows cones on the surface of a solid formed from its melt. For small undercoolings, a parabolic relation between $R$ and $\Delta T_k$ (at small $\Delta T_k$) has been observed for a number of materials grown from their melt. This is in agreement with prediction of Eq. (9-49) and provides further confirmation of the screw dislocation mechanism of growth. Materials in which this behavior has been observed include salol, potassium di-silicate, and sodium di-silicate, as reviewed by Cahn et al.[30] and by Jackson et al.[43] Data on salol compiled from several investigators are summarized in Fig. 9-22. Curve $A$ is comparable in form to that of Fig. 9-19, suggesting growth by a screw dislocation mechanism at low undercoolings and continuous growth at high undercoolings. Curve $B$ is for growth on what appears to be a perfect crystal surface by two-dimensional nucleation.

## GROWTH BY PROPAGATION OF TWIN PLANES

There is still another source of steps at liquid-solid interfaces: the reentrant angle resulting from the emergence of a twin plane at a crystal surface. Figure 9-23 shows a germanium crystal with a single twin plane bounded entirely by flat, close-packed {111} facets. There are now three directions which have a reentrant angle and from which, therefore, growth can proceed rapidly: the $[2\bar{1}\bar{1}]$ direction (shown) and the $[\bar{1}2\bar{1}]$ and $[\bar{1}\bar{1}2]$ directions, which lie at 60 and 120° from the first. The problem is that rapid growth of these three planes causes them to disappear, leaving a flat crystal of triangular cross section with its edge bounded by a ridge structure with no grooves; growth therefore ceases.

If an additional twin plane is introduced parallel to the first, the structure shown in Fig. 9-24 results, again bounded by flat, close-packed {111} planes. There are now favored reentrant growth sites at six $\langle 211 \rangle$ directions, and rapid growth takes place in these directions. Most important, these favored growth sites will not disappear. Growth can continue indefinitely in two or more of the six directions, with little growth in the thickness direction. This mechanism leads to the dendritic *ribbon* crystals of germanium first described by Billig.[44] The same mechanism was thereafter shown to be operative in other materials, including InSb and Si.[45] The requirement that there be at least two twin planes was first demonstrated by Wagner[46] and by Hamilton and Seidensticker.[47] When more than two twin planes are present, growth in other directions than the $\langle 211 \rangle$ becomes possible. A quantitative growth law for growth by the twin-plane mechanism has not been formulated.

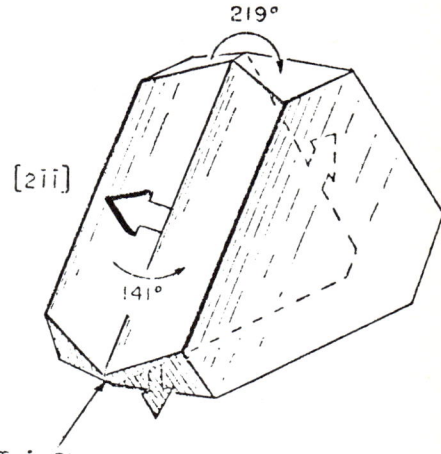

**FIGURE 9-23**
Twinned germanium crystal with reentrant angle in three growth directions. (*From Hamilton and Seidensticker.*[47])

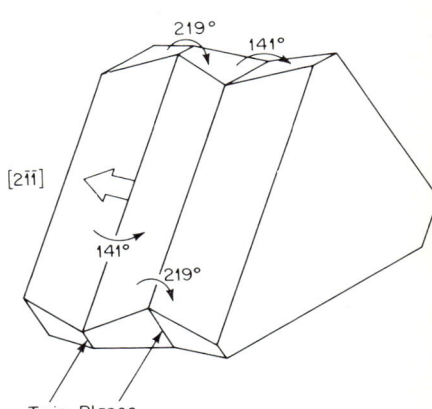

**FIGURE 9-24**
Germanium crystal with two twin planes has six favored reentrant sites 60° apart. (*From Hamilton and Seidensticker.*[47])

## GROWTH MORPHOLOGIES

The important practical consequence of the interface kinetics discussed earlier in this chapter is that these kinetics are important in determining solidification morphologies. A consequence that is immediately apparent is the facets that develop on low index faces in materials which grow by a lateral spreading mechanism. Higher index faces grow more rapidly and so disappear, as sketched in Fig. 9-25, leaving the low index faces behind.

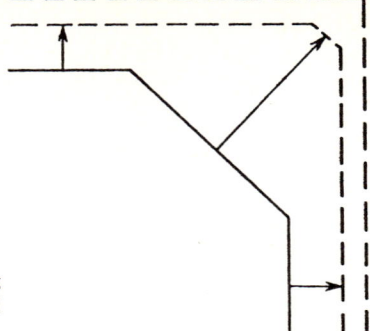

FIGURE 9-25
Sketch showing how rapidly growing crystal faces disappear from the crystal form. (*From Christian.*[28])

We can examine the stability of a growing flat facet of such a crystal in a simple way. Suppose a cubic crystal as sketched in Fig. 9-26 is growing in an undercooled pure melt. The liquid is undercooled $\Delta T$ below its melting point, and the crystal grows by dissipation of heat into the melt. The temperature distribution outward from the center of a flat face is as in Fig. 9-26a; the interface is at temperature $T_i$, undercooled an amount $\Delta T_k$ below the equilibrium melting point. A similar plot can be drawn for a region of the same face near an edge or corner, as is done in Fig. 9-26 b The difference here is that heat diffuses more readily because of the *corner effect* (divergence of heat flux) and so the kinetic undercooling $\Delta T_k$ must be larger than at the center of a face. The growth rate of the face is exactly that which would result if the kinetic undercooling over the entire face were equivalent to that of the maximum (near the corners). Steps originate near the corners and spread across the face at a velocity $R_\perp$, which must decrease as the local undercooling $\Delta T_k$ decreases. The flat (or nearly flat) face is maintained by a gradual crowding of the steps at the center of the face, as shown schematically in Fig. 9-27. The interface advances everywhere with the same velocity $R$ as long as

$$\frac{R_\perp}{l} = \text{const} \qquad (9\text{-}52)$$

where $l$ is step spacing. The macroscopically flat facet is retained as long as the angle $\phi$ is small, where

$$\phi = \frac{a}{l} = \frac{R}{R_\perp} \qquad (9\text{-}53)$$

Exactly the same argument as the foregoing could be repeated for the case of a faceted crystal growing in an alloy melt. In this case, the diffusivity of heat is so much greater than the diffusivity of solute that we need generally consider only the latter.

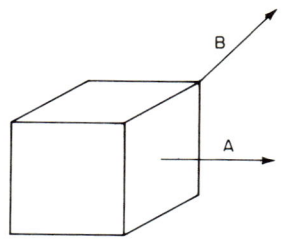

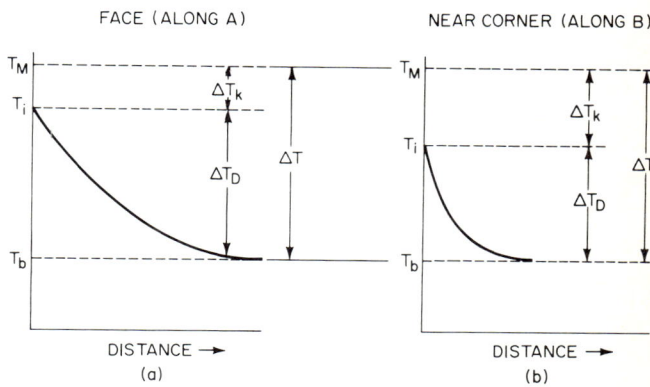

**FIGURE 9-26**
Growth of a cubic crystal in a pure melt at undercooling $\Delta T$ (or an alloy melt of uniform temperature) at constitutional undercooling $\Delta T$.

Assuming a bath of uniform temperature, Fig. 9-26 again applies, but $\Delta T$ now represents *constitutional supercooling*, as defined in Chap. 5. The slopes of the curves near the interface are the *gradient of constitutional supercooling* at the liquid-solid interface. Clearly, in the case of both the pure material and the alloy, the growth kinetics of the faceted crystal act as a stabilizing influence, preventing breakdown of the plane front interface. In Chap. 3 it was seen that a gradient of constitutional supercooling of very nearly zero was all that was required to break down the plane front interface for metals where $\Delta T_k \simeq 0$ and growth is continuous. For materials

**FIGURE 9-27**
Steps on a faceted surface separated by a distance $l$. Surface remains macroscopically faceted as long as $\phi$ is small. Step spacing decreases from left to right.

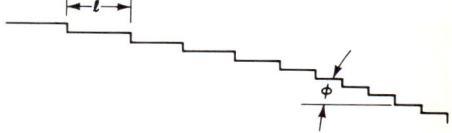

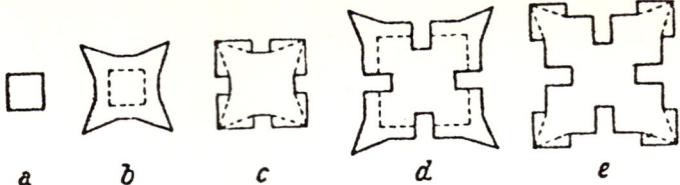

**FIGURE 9-28**
Development of a KCl dendrite. (*From Papapetrou.*[54])

where the kinetic limitation to growth is larger, large gradients of constitutional supercooling are required to break down the plane interface.

When, during growth, the diffusional undercooling at the center of a face becomes so large that $\Delta T_k$ here approaches zero, steps cease to propagate in this area and the flat facet breaks down. One way in which this breakdown can occur is for the corners to grow outward to form faceted dendrites, as in the classical sketch of Papapetrou of the formation of a KCl dendrite (Fig. 9-28). Contrast this with Papapetrou's observations of the formation of an NH$_4$Cl dendrite (Fig. 9-29). Here, no facets are present and the dendrite forms as in metal systems. Some other forms that can be taken by the faceted dendrites are shown in Figs. 9-30 and 9-31. If the edges, instead of the corners, grow preferentially, then *hopper crystals* form, as in bismuth (sketched in Fig. 9-32).

Keith and Padden[48] and Cahn[14] propose relations for the stability of a growing faceted crystal, both of which can be written for small undercoolings approximately as

$$\frac{Rr}{D_L} \leq -\frac{\Delta T}{m_L C_0 (1 - k)} \tag{9-54}$$

where $r$ is the approximate radius of the faceted crystal. With an equal sign, Eq. (9-54) is simply the equation for diffusion limited growth of a sphere at small undercoolings. Approximately the same equation applies for a chunky faceted crystal. If growth rate

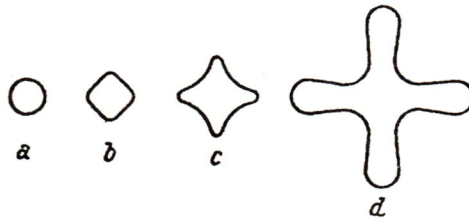

**FIGURE 9-29**
Development of a NH$_4$Cl dendrite.
(*From Papapetrou.*[54])

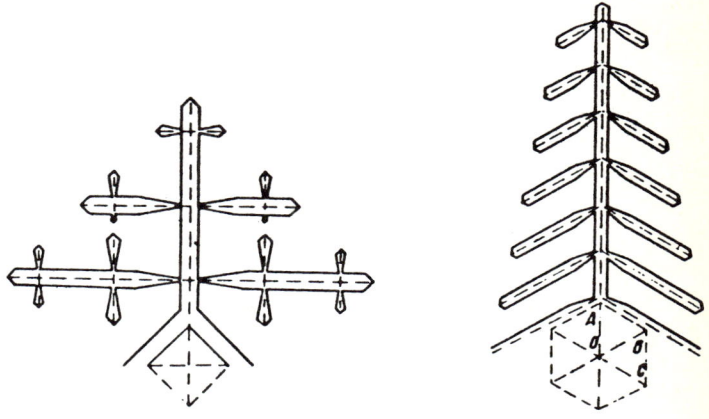

FIGURE 9-30
Faceted dendrites of materials of two different crystal structures. (*From Saratovkin.*[55])

$R$ is less than this equality, overall growth rate is determined by layer spreading and we expect no breakdown. As it approaches the equality, growth rate becomes limited by diffusion and we now expect the *corner effect* to become important, resulting in dendrites or hopper crystals.

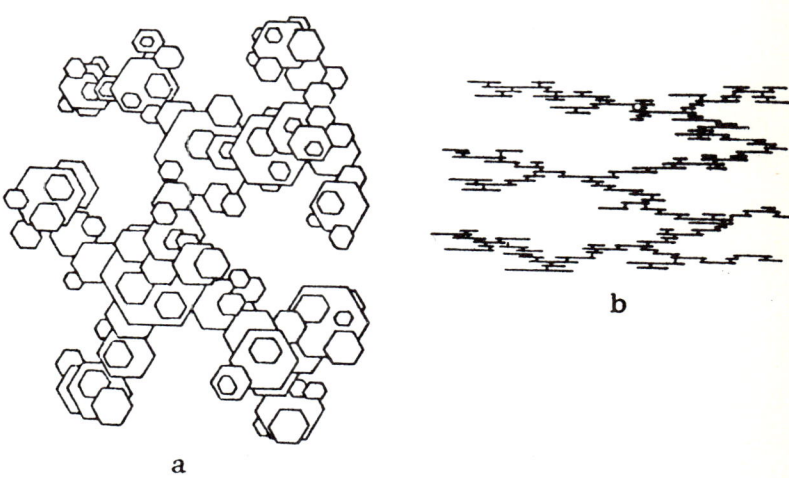

FIGURE 9-31
Foliated dendrite. (*a*) In plan; (*b*) in section. (*From Saratovkin.*[55])

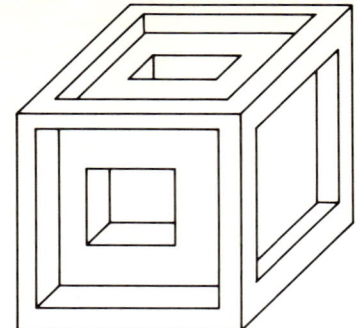

FIGURE 9-32
Idealized representation of development of hopper crystals.

As discussed earlier, various attempts have been made to predict which materials will grow by layer spreading (and therefore develop facets) and which materials will not. Jackson's criterion has met with substantial success in this regard. This criterion states that pure materials which have an entropy of melting greater than $4R$ should solidify with facets and those which have a lower entropy of melting should not.[27] In agreement with this criterion, it is found that materials which usually solidify without facets include most metals and some nonmetals, including carbon tetrabromide, cyclohexane, and succinonitrile. These latter materials have been used in many studies of solidification to simulate the behavior of metals. The advantage is that their transparency permits direct observation of growth morphology. Also in agreement with Jackson's criterion, pure materials which usually solidify with facets include the metalloids (Bi, Ga, Sb, and As), semiconductors (Si and Ge), and most oxide and organic nonmetals. One limitation of the Jackson criterion is that it cannot explain the observed effects of undercooling on morphology. An example is the work of Glicksman and Schaeffer,[50] who showed that when white phosphorus was solidified within 1°C of its equilibrium melting point, it possessed a faceted morphology. At melt undercoolings somewhere between 1 and 9°C undercooling, the morphology changed to nonfaceted. These results are qualitatively interpretable by Cahn's analaysis.

## REFERENCES

*1* TURNBULL, D.: "Solid State Physics," vol. 3, p. 225, Academic Press, Inc., New York, 1956.
*2* SHEWMON, P. G.: "Diffusion in Solids," McGraw-Hill Book Company, New York, 1963.
*3* TURNBULL, D., and FISHER, J. C.: *J. Chem. Phys.*, **17**:71 (1949).
*4* HOLLOMAN, J. H., and TURNBULL, D.: *Progr. Met. Phys.*, **4**:333 (1953).

5 CHALMERS, B.: "Principles of Solidification," John Wiley & Sons, Inc., New York, 1964.
6 TURNBULL, D.: in T. J. Hughel (ed.), "Liquids: Structure, Properties, Solid Interactions," Elsevier Publishing Company, New York, Amsterdam, 1965.
7 CECH, R. E., and TURNBULL, D.: *Trans. AIME*, **191**:242 (1951).
8 WALKER, J.: in G. R. St. Pierre (ed.), "Physical Chemistry of Process Metallurgy II," p. 845, Interscience Publishers, Inc., New York, 1961.
9 KATTAMIS, T. Z., and FLEMINGS, M. C.: *Trans. AIME*, **236**:1523 (1966).
10 SHIRAISHI, S. Y., and WARD, R. G.: *Can. Met. Quart.*, **3**(1):117 (1964).
11 YARWOOD, J., FLEMINGS, M. C., and ELLIOTT, J. F.: *Met. Trans.*, **2**:2573 (1971).
12 GLASSON, E. L., and EMLEY, E. F.: "The Solidification of Metals," p. 1, Iron and Steel Institute Publ. No. 110, London, 1968.
13 SICHA, W. E., and BOEHM, R. C.: *Trans. AFS*, **56**:398 (1948).
14 EMLEY, E. F.: "Principles of Magnesium Technology," Pergamon Press, Oxford, New York, 1966.
15 HUGHES, I. C. H.: "The Solidification of Metals," Iron and Steel Institute Publ. No. 110, p. 184, 1968.
16 REYNOLDS, J. A., and TOTTLE, C. R.: *J. Inst. Metals*, **80**:1328 (1951).
17 TURNBULL, D., and VONNEGUT, B.: *Ind. Eng. Chem.*, **44**:1292 (1952).
18 GLICKSMAN, M. E., and CHILDS, W. J.: *Acta Met.*, **10**:925 (1962).
19 CROSLEY, P. B., DOUGLAS, A. W., and MONDOLFO, L. F.: in "Solidification of Metals," Iron and Steel Institute Publ. No. 110, p. 10, 1968.
20 BRAMFITT, B. L.: *Met. Trans.*, **1**:1987 (1970).
21 CIBULA, A.: *J. Inst. Metals*, **82**:513 (1954).
22 DENNISON, J. P., and TULL, E. V.: *J. Inst. Metals*, **85**:8 (1956).
23 WALLACE, J. F.: *J. Metals*, **15**:372 (1963).
24 HUNT, J. D., and JACKSON, K. A.: *J. Appl. Phys.*, **37**:254 (1966).
25 CAHN, J. W.: *Acta Met.*, **8**:554 (1960).
26 BURTON, W. K., CABRERA, N., and FRANK, F. C.: *Phil. Trans.*, **A243**:299 (1950).
27 JACKSON, K. A.: in R. H. Doremus et al. (eds.), "Growth and Perfection of Crystals," p. 319, John Wiley & Sons, Inc., New York, 1958.
28 CHRISTIAN, J. W.: "The Theory of Transformations in Metals and Alloys," Pergamon Press, Oxford, New York, 1965.
29 TURNBULL, D.: "Thermodynamics in Metallurgy," American Society for Metals, Metals Park, Ohio, 1949.
30 CAHN, J. W., HILLIG, W. B., and SEARS, G. W.: *Acta Met.*, **12**:1421 (1964).
31 KRAMER, J. J., and TILLER, W. A.: *J. Chem. Phys.*, **42**:257 (1965).
32 RIGNEY, D. A., and BLAKELY, J. N.: *Acta Met.*, **14**:1375 (1966).
33 HILLIG, W. B., and TURNBULL, D.: *J. Chem. Phys.*, **24**:914 (1956).
34 VOLMER, M.: "Kinetik de Phasenbildung," Steinkopff, Dresden and Leipzig, 1939.
35 BECKER, R., and DORING, W.: *Ann. Phys.*, **24**:719 (1935).
36 HILLIG, W. B.: in R. H. Doremus, B. W. Roberts, and D. Turnbull (eds.), "Growth and Perfection of Crystals," p. 350, John Wiley & Sons, Inc., New York, 1958.
37 SPERRY, P. R.: Sc.D. thesis, M.I.T., Cambridge, Mass., 1965.

38 MAGILL, J. H., and PLAZEK, D. J.: *J. Chem. Phys.*, **46**:3757 (1967).
39 ALFINSTSEV, G. A., and OVSIENKO, D. E.: *Soviet Phys. Tech. Phys. (English Transl.)*, **9**:489 (1964).
40 FRANK, F. C.: *Discussions Faraday Soc.*, **5**:48 (1949).
41 FRANK, F. C.: *Advan. Phys.*, **1**:91 (1953).
42 VERMA, A. R.: "Crystal Growth and Dislocations," Academic Press, New York, 1953.
43 JACKSON, K. A., UHLMANN, D. R., and HUNT, J. D.: *J. Crystal Growth*, **1**:1 (1967).
44 BILLIG, E.: *Proc. Roy. Soc.*, **229**:346 (1955).
45 ALBAN, N., and OWEN, A. E.: *J. Phys. Chem. Solids*, **24**:899 (1963).
46 WAGNER, R. S.: *Acta Met.*, **8**:57 (1960).
47 HAMILTON, D. R., and SEIDENSTICKER, R. G.: *J. Appl. Phys.*, **31**:1165 (1960).
48 KEITH, H. D., and PADDEN, JR., F. J.: *J. Appl. Phys.*, **34**:2409 (1963).
49 CAHN, J. W.: in H. S. Peiser (ed.), "Crystal Growth," p. 681, Pergamon Press, New York, 1967.
50 GLICKSMAN, M. E., and SCHAEFFER, R. J.: *J. Crystal Growth*, **67**:297 (1967).
51 CHADWICK, G.: *Sheffield Univ. Met. Soc. J.*, **9**:15 (1970).
52 POLLATSCHEK, H.: *Z. Physik. Chem. (Frankfurt)*, **142A**:289 (1929).
53 DANILOV, V. I., and MALKIN, V. I.: *Zh. Fiz. Khim*, **27**:1837 (1954).
54 PAPAPETROU, A.: *Z. Krist.*, **A92**:89 (1935).
55 SARATOVKIN, D. D.: "Dendritic Crystallization," Consultants Bureau Transl., New York, 1959.

# PROBLEMS

9-1 The maximum undercooling observed in liquid nickel is 319°C. Assuming homogeneous nucleation occurs at this temperature, calculate the liquid-solid surface energy of nickel. Assume $T_M = 1453°C$, $V_s = 6.6$ cm$^3$/mol, $\Delta H = -4{,}320$ cal/mol.

9-2 A substrate is introduced to a melt of liquid cobalt which is such that when solid cobalt forms, it does so on the substrate with a 30° contact angle. What is the undercooling you would expect to achieve in liquid cobalt with this substrate present?

9-3 Describe five different procedures which you might follow to obtain a fine-grained solidification structure in an aluminum alloy. Explain how each of your methods probably grain-refines.

9-4 Vibration has been shown to produce a relatively fine grain size in Al–4.5% Cu alloy (approximately 0.02 cm). Equivalent grain refinement can be obtained by gentle stirring during solidification. What can you conclude and what can you not conclude from these experiments?

9-5 What is the approximate kinetic undercooling of the liquid-solid interface in a pure nickel single crystal growing with a plane front $10^{-3}$ cm/s? (Assume continuous growth.) Explain why the solidification rate of metals is seldom limited by interface kinetics. Assume $D_L = 5 \times 10^{-5}$ cm$^2$/s, $V_s = 6.6$ cm$^3$/mol, $\Delta H = -4{,}320$ cal/mol.

*9-6* Suppose nickel were to solidify with an atomically flat interface. What would be the growth rate of a planar close-packed interface at 1°C undercooling, assuming growth by two-dimensional nucleation? Assume $\sigma = 255$ erg/cm$^2$ and other necessary data as given in the above problems.

*9-7* Repeat Prob. 9-6 assuming growth by a screw dislocation mechanism.

*9-8* How small must Cahn's diffuseness parameter be to achieve continuous growth in nickel at 0.1°C undercooling?

*9-9* Give some examples of experimental evidence we have of growth by the screw dislocation mechanism.

*9-10* By a simple sketch show how a reentrant twin can accelerate the growth rate of a liquid-solid interface.

*9-11* What factors might favor formation of hopper-type crystals rather than dendrites when the plane front of a faceting alloy breaks down?

*9-12* A small crystal is growing from Bi–0.1% Sn alloy at 1°C total undercooling. At a crystal size of 20 μm, what is the maximum rate at which it can grow with faceted morphology? Assume $D_L = 5 \times 10^{-5}$ cm$^2$/s, $m_L = 1.63$°C/percent, $k = 0.37$.

*9-13* By deriving Eq. (9-12), show what variables comprise the factor $B_1$.

*9-14* Write the three growth laws in their simplest form for continuous growth, growth by two-dimensional nucleation, and growth by the screw dislocation mechanism. Describe briefly two different types of experiments you would do to try to determine which law applies to a given material.

# 10
## PROCESSING AND PROPERTIES

### HOMOGENIZATION

As discussed in Chap. 5, even single-phase alloys possess microsegregation after dendritic solidification. This microsegregation (*coring*) can be reduced by a high-temperature treatment termed *homogenization*. A useful parameter for discussion of homogenization of microsegregation is the *index of residual microsegregation* $\delta_i$:

$$\delta_i = \frac{C_M - C_m}{C_M{}^0 - C_m{}^0} \quad (10\text{-}1)$$

where $C_M$ = maximum solute concentration of element $i$ (in interdendritic spaces) at time $t$

$C_m$ = minimum solute concentration of element $i$ (in center of dendrite arms) at time $t$

$C_M{}^0$ = maximum initial solute concentration of element $i$

$C_m{}^0$ = minimum initial solute concentration of element $i$

$\delta_i$ = index of residual microsegregation of element $i$

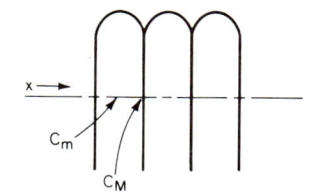

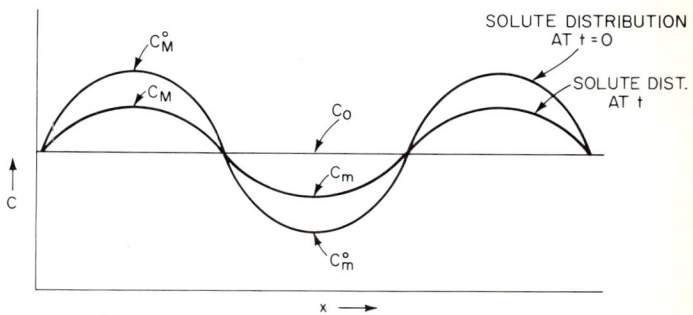

**FIGURE 10-1**
Simple model for homogenization. *Top*: platelike dendrite arms; *bottom*: solute distribution across dendrite arms.

Before any homogenization has taken place, $\delta_i = 1$; after complete homogenization, $\delta_i = 0$. The value of $\delta_i$ after a given homogenization treatment depends on the simple dimensionless group of variables $D_s t/l^2$, where $D_s$ is the diffusion coefficient in the solid at the temperature of homogenization of element $i$ (assumed here a constant), $t$ is the homogenization time, and $l$ is a characteristic diffusion distance the order of the dendrite arm spacing. We can readily see this dependency for a very simple dendrite model. Imagine, as in Fig. 10-1, dendrite arms are simple plates and concentration across the arms is sinusoidal (with maxima at interdendritic regions and minima at the center of dendrite arms). The initial solute distribution is then

$$\frac{C^0 - C_0}{C_M^0 - C_0} = \sin \frac{\pi x}{l_0} \qquad (10\text{-}2)$$

where  $C^0$ = concentration at $x$ at $t = 0$

$C_0$ = mean alloy composition

$l_0$ = one-half the dendrite arm spacing

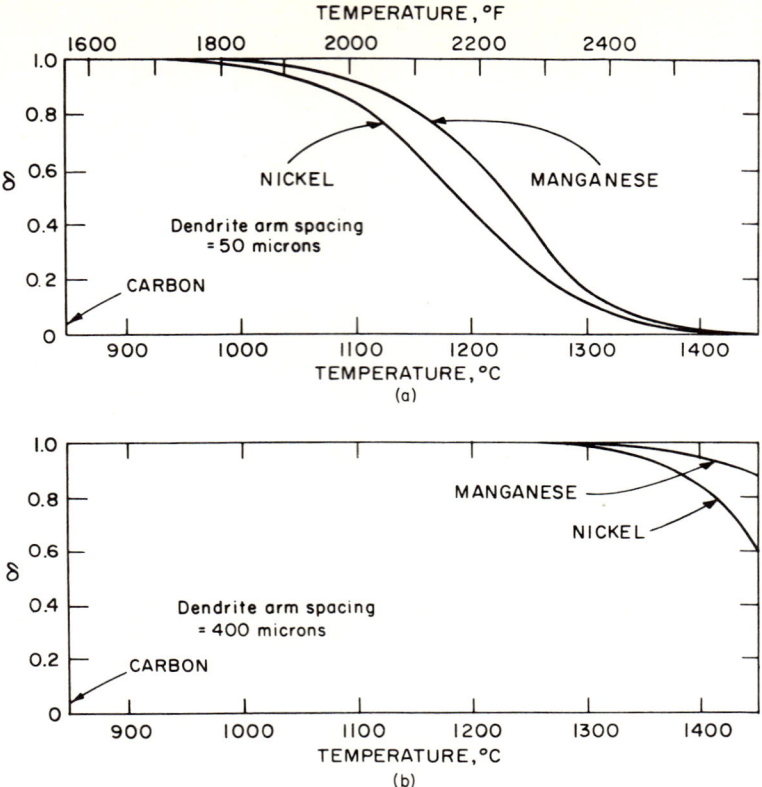

**FIGURE 10-2**
Homogenization of a low-alloy steel. Index of residual microsegregation $\delta$ is shown for 1-h treatments at various temperatures. (a) 50 μm dendrite arm spacing; (b) 400 μm dendrite arm spacing. (*From Kattamis and Flemings.*[1])

The solution of Fick's second law for the change in the concentration profile with time yields

$$\frac{C - C_0}{C_M{}^0 - C_0} = \sin\frac{\pi x}{l} e^{-\pi^2(D_s t / l_0^2)} \qquad (10\text{-}3)$$

at $x = l_0/2$, $C = C_M$. Substituting these values in Eq. (10-1) yields

$$\delta_i = e^{-\pi^2(D_s t / l_0^2)} \qquad (10\text{-}4)$$

Equation (10-4) is useful for approximate prediction of times and temperatures required to homogenize a given cast structure. More accurate prediction requires that the initial dendrite geometry and solute distribution be described. Numerical analysis can then be employed to determine exactly the time to achieve a given amount of homogenization. Results of such calculations for a low-alloy steel are shown in Fig. 10-2. Figure 10-2a is for metal of fine dendrite arm spacing (50 μm) that is achieved only in chilled castings within about ¼ in from the chill or in thin-section sand castings. Even with this fine spacing, 1-h homogenization treatments under about 1150°C effect little reduction in segregation, and higher temperatures (or longer times) are required to effectively eliminate the segregation. The calculation for the dendrite arm spacing of 400 μm relates to material 4 to 5 in away from the chilled mold wall in a heavy casting or ingot. In this case, microsegregation is not appreciably reduced by 1-h treatments even as high as 1350°C. In both cases, calculations for carbon show that complete homogenization of this element is achieved below 900°C.[1]

Usual homogenization treatments for steel castings are carried out below 1100°C and so remove little segregation of elements other than carbon. However, vacuum heat-treating furnaces are now beginning to be employed by foundries producing high-quality castings; these permit achieving temperatures in excess of 1350°C. For longer times and moderate dendrite arm spacings, very large improvements in properties of castings result.

## SOLUTION TREATMENT

In accordance with industrial terminology, we reserve the term *homogenization* for treatments designed primarily to even out concentration gradients within a single phase. *Solution* treatments are those treatments designed to dissolve one or more nonequilibrium second phases, and these treatments are discussed below.

The simplest model for considering solution kinetics is that of Singh and Flemings[2] for a binary alloy containing nonequilibrium eutectic. In this model, dendrite arms are again assumed platelike and solute distribution within them sinusoidal, as sketched in Fig. 10-3. In addition, eutectic is assumed divorced, and so the interdendritic region consists of plates of second phase with uniform composition $C_\beta$. The amount of second phase is small so that motion of the $\alpha$-$\beta$ boundary can be neglected. Dissolution is limited by diffusion in the $\alpha$ phase. The solution to the diffusion equation is similar in form to that for homogenization of single-phase alloys. It is

$$\frac{g+a}{g_0+a} = e^{-(\pi^2 D_s t/4 l_0^2)} \quad (10\text{-}5)$$

where

$$a = \frac{C_M - C_0}{C_\beta}$$

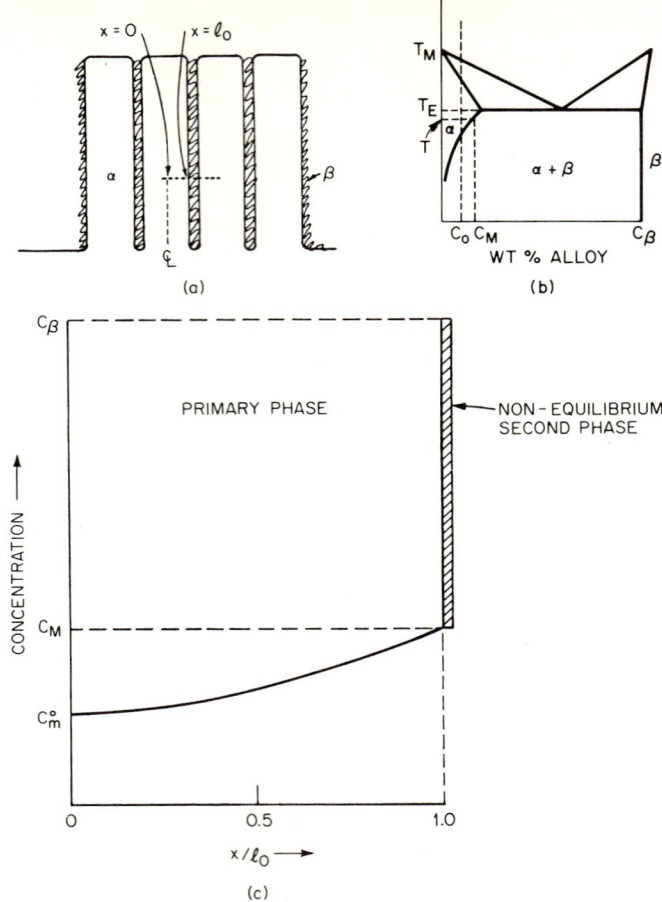

**FIGURE 10-3**
Simple model for solution treatment. (*a*) Platelike dendrite arms with second phase (shaded); (*b*) phase diagram showing temperature of solution treatment $T$; (*c*) initial solute distribution.

$g$ and $g_0$ are volume fraction of eutectic at times $t$ and $t_0$, respectively; $C_M$ is maximum solute content of the primary phase in units of moles per cubic centimeter; and $C_0$ is overall alloy composition. When heat treatment is at a temperature very close to the solvus, $C_0 = C_M$ and Eq. (10-5) is simply

$$\frac{g}{g_0} = e^{-(\pi^2 D_s t / 4 l_0^2)} \qquad (10\text{-}6)$$

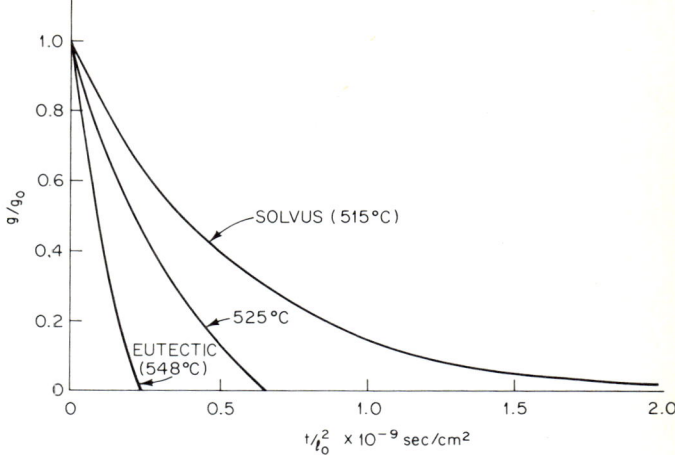

**FIGURE 10-4**
Calculated rate of solution of Al–4.5% Cu alloy.

Figure 10-4 shows Eq. (10-5) plotted for Al–4.5% Cu alloy for several different temperatures using values of the diffusion coefficient from the work of Murphy.[3] Solution temperature, time, and segregate spacing (dendrite arm spacing) are the important variables determining the effectiveness of a solution treatment. A moderate size sand casting, with dendrite arm spacing of 200 μm and given a standard commercial solution treatment (10 h at 515°C), will not have the volume fraction of eutectic present reduced very much by the heat treatment. According to Fig. 10-4, approximately 40 h would be required to eliminate the second phase. This agrees with observations in practice which show that considerable second phase is left in sand castings after usual solution treatments. On the other hand, modern *premium-quality* aluminum foundries apply substantial chilling to their castings to ensure their dendrite arm spacing stays below about 100 μm. Furthermore, they employ high-purity materials and carefully controlled equipment so that they can solutionize within 10 to 20°C of the melting point (eutectic temperature) of the alloy being solution treated. A 10-h solution treatment at these temperatures is now more than ample to dissolve all the second phase.

Figure 10-5, from Singh et al.,[4] shows excellent agreement between the simple theory of Eq. (10-5) and experiment except for long solution times. Dissolution of the last traces of segregate requires longer times than predicted by the analysis. The reason for this is that more segregate exists between primary dendrite arms than between secondary dendrite arms, and the last remnant of solute must diffuse over larger distances than simply the secondary dendrite arm spacings.

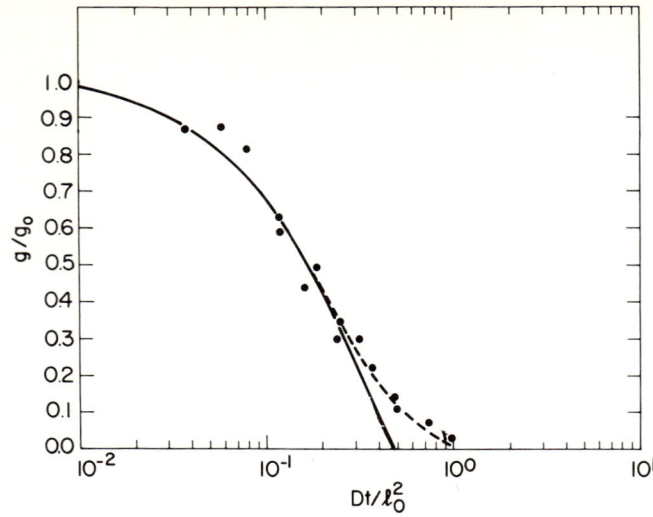

**FIGURE 10-5**
Solution kinetics of Al–4.5% Cu alloy at 545°C. Dashed line is experimental; solid line is calculated from Eq. (10-5). (*From Singh, Bardes, and Flemings.*[4])

The models discussed above for both homogenization and solution treatment have involved a number of simplifying assumptions, including simple segregate geometry and that bulk diffusion is the rate limiting step. These assumptions seem valid for approximate description of homogenization and solution treatment of usual cast structures. A number of more detailed models are available, however, and have recently been reviewed.[5,6]

High-temperature homogenization and solution treatments affect cast structures in ways other than reducing concentration gradients within the dendrite arms, or removing the second phase. Nonmetallic insoluble inclusions tend to *ripen* so that their total number is decreased and their size increased. An example is Type II manganese sulfides in low-alloy steel. As an extreme example, Gnanamuthu[7] has shown that after a 40-h homogenization at 1315°C, the number of such inclusions decreased from $1.6 \times 10^6$ to $5 \times 10^4$ per square centimeter of polished surface. The driving force for the change is reduction of surface area between the sulfide particles and the matrix. Homogenization also results in some reduction of microporosity in cast alloys, with the pores that remain tending to spherodize. It has been suggested that moderate pressures (a few atmospheres or less) should aid in removal of porosity during heat treatment. The much larger pressures obtained in isostatic hot pressing completely eliminate detectable microporosity that is not surface-connected.[8,9]

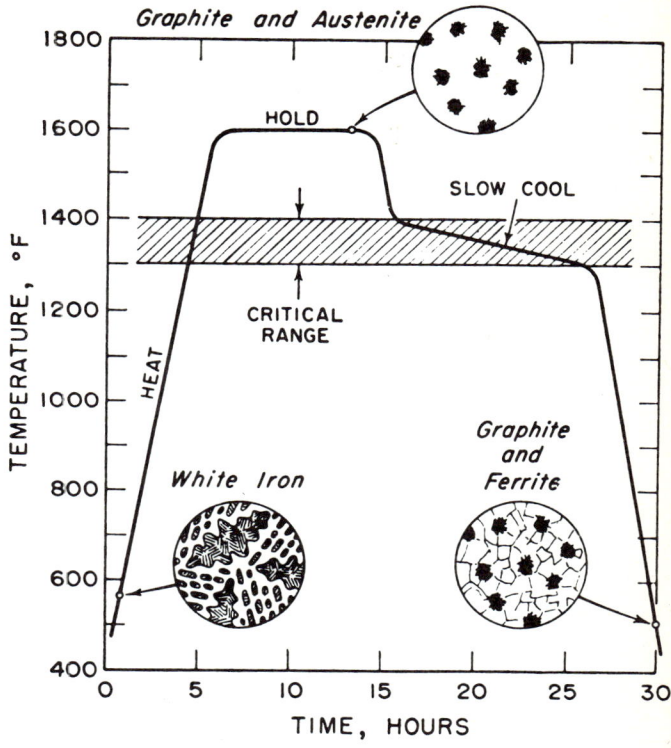

FIGURE 10-6
Typical annealing cycle and structure of ferritic malleable cast iron. (*From Taylor, Flemings, and Wulff.*[10])

## MALLEABLE IRON

If a white cast iron such as the simple Fe–3% C alloy discussed in Chap. 6 is held for a sufficiently long time at a temperature above the eutectoid, the metastable iron carbide present decomposes to form graphite *temper nodules* in an austenite matrix. If, then, cooling to room temperature is sufficiently slow, the austenite transforms to ferrite, with the excess carbon precipitating on the temper nodules. The result is a *ferritic malleable iron*. The malleablization process is shown schematically in Fig. 10-6, and an actual microstructure is shown in Fig. 10-7. This material possesses many of the advantages of gray cast iron and has much greater ductility (20 percent or more compared with about 1 percent for ferritic gray iron). In addition, high yield and tensile strengths can be obtained in malleable iron by subsequent heat treatments to produce *pearlitic malleable iron* or *quenched and tempered malleable iron*.[10,11]

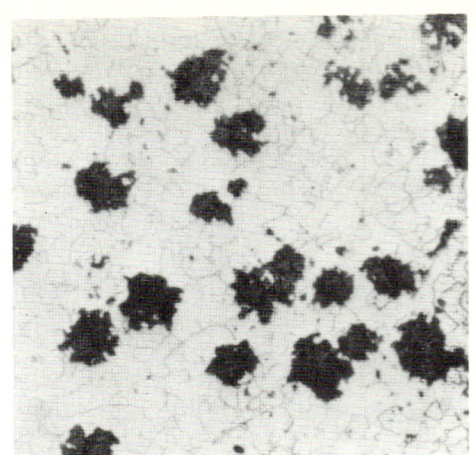

FIGURE 10-7
Ferritic malleable iron. (Magnification × 100.) (*From Ref.* 31.)

The process of forming graphite from ledeburite above the eutectoid temperature is termed *first-stage graphitization*. This process involves simultaneous solution of $Fe_3C$, diffusion of C through the austenite, and reprecipitation of the carbon as graphitic temper nodules. Assuming this process is limited by diffusion of carbon through the austenite, Fig. 10-8 describes the kinetics of the process.

Briefly, the solubility of carbon in austenite at the $\gamma$–$Fe_3C$ interface $C_{\gamma 2}$ is higher than that at the austenite–graphite interface $C_{\gamma 1}$. If the spacing between a cementite particle and adjacent graphite particle is $l_0$, carbon diffuses from the cementite to the graphite down the gradient $(C_{\gamma 2} - C_{\gamma 1})/l_0$. An approximate relation describing this process quantitatively may now be written, assuming equilibrium at the particle-matrix interfaces, one-dimensional diffusion, and neglecting curvature effects. From Fick's first law, the rate of carbon diffusion is

$$j = -D_s \frac{C_{\gamma 2} - C_{\gamma 1}}{l_0} \qquad (10\text{-}7)$$

where  $j$ = solute flux, mol/(cm²)(s)

$C_{\gamma 1}, C_{\gamma 2}$ = solubilities of carbon in austenite in equilibrium with graphite and $Fe_3C$, respectively, mol/cm³

$l_0$ = spacing between $Fe_3C$ graphite precipitates

The solute flux comes entirely from dissolution of a volume fraction $g_c$ of $Fe_3C$ with composition $C_c$. Thus, the total transport of solute across a face of a cross-

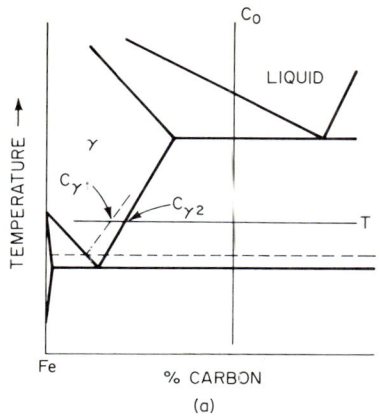 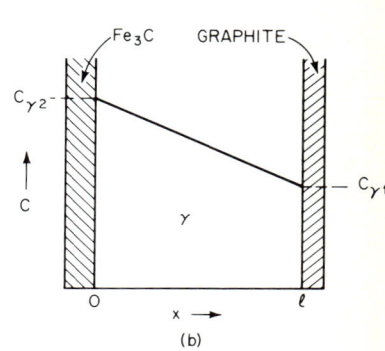

**FIGURE 10-8**
Malleablization of white iron of composition $C_0$ at temperature $T$. (*a*) Iron-carbon phase diagram. Solid lines are for the metastable and dashed lines for the stable (Fe-G) equilibrium; (*b*) carbon diffuses through austenite from $C_{\gamma 2}$ (at $Fe_3C$ interface) to $C_{\gamma 1}$ (at graphite interface).

sectional area $A$ of a small volume $Al_0$ is

$$jA = Al_0 \frac{d}{dt} C_c g_c \qquad (10\text{-}8)$$

Combining Eqs. (10-7) and (10-8) and integrating from $g_c = g_c^{\,0}$ at $t = 0$ yields

$$g_c = g_c^{\,0} - \frac{D_s}{l_0^{\,2}} \frac{C_{\gamma 2} - C_{\gamma 1}}{C_c} \qquad (10\text{-}9)$$

And, for complete elimination of the ledeburite, at $t = t_{\text{crit}}$, $g_c = 0$; thus

$$t_{\text{crit}} = g_c^{\,0} \frac{C_c}{C_{\gamma 2} - C_{\gamma 1}} \frac{l_0^{\,2}}{D_s} \qquad (10\text{-}10)$$

Equations (10-9) and (10-10), although they are certainly simplifications of the real case, illustrate the important variables involved in malleablizing. Extent of malleablization, like other heat-treating processes examined, depends on the value of the parameter $D_s t/l_0^{\,2}$. Quite large values of $D_s t/l_0^{\,2}$ are needed to obtain significant solute redistribution because of the small driving force $C_{\gamma 2} - C_{\gamma 1}$. One way to increase the malleablization rate is to reduce $l_0$, and this can be done by obtaining a finer ledeburitic structure through a rapid cooling rate. It can also be done by controlling melt chemistry in such a way that a larger number of temper carbon nodules nucleate and grow during malleablization. Another and highly effective way to increase the malleablization rate is to add an element such as silicon, which separates the metastable limit of austenite solubility further from the stable limit, thus increasing $C_{\gamma 2} - C_{\gamma 1}$. There is a limit, however, to the amount of silicon that can be added to malleable iron. Above this limit, the iron solidifies according to the stable rather than the metastable diagram; the limit, of course, depends on casting section size and other elements present.

## FERRITIC, PEARLITIC, AND MARTENSITIC IRONS

The three major types of graphitic irons (gray, malleable, ductile) are produced commercially with matrices of ferrite, pearlite, or tempered martensite. For each material, the ferritic iron is more machinable and ductile while the martensitic iron is hardest and most wear-resistant. By careful control of chemistry and cooling conditions in the mold, gray iron and ductile iron can be cast directly with either fully ferritic or fully pearlitic matrices. Often, however, these materials, like malleable iron, are heat-treated to obtain optimum and uniform properties. Figure 10-9 shows typical properties for as-cast and heat-treated ductile iron. A similar range of properties is obtained in as-cast and heat-treated malleable iron. Comparable hardness levels are obtained by similarly heat-treating gray cast iron, although tensile properties are much lower due to the embrittling effect of the graphite flakes.[12]

To produce a ferritic iron by heat treatment, a cast iron is first heated to the $\gamma + G$ region and then slowly cooled through the eutectoid region, the aim being to diffuse all the excess carbon from the matrix to existing graphite nodules with no formation of cementite. Figure 10-10 describes schematically the kinetics of the ferritizing anneal, taking the simple case of isothermal transformation at temperature $T$ between the metastable and stable transformation temperatures. As drawn, and for the simplification of unidirectional solute flow, the reader should recognize this as the mass flow analog of a heat-flow problem solved in Chap. 1. The heat-flow problem is that of unidirectional solidification of a pure metal poured at its melting point against a metal mold at constant temperature with no interface resistance. The solution may therefore be written down directly as

$$g_\alpha = 2\gamma \frac{\sqrt{D_s t}}{l_0} \qquad (10\text{-}11)$$

where $g_\alpha$ is fraction of austenite transformed to ferrite and $\gamma$ is given by the relation

$$\gamma e^{\gamma^2} \operatorname{erf} \gamma = \frac{1}{\sqrt{\pi}} \frac{C_\gamma - C_{\alpha 2}}{C_{\alpha 2} - C_{\alpha 1}} \qquad (10\text{-}12)$$

| Structure | Thermal Treatment | Typical Mechanical Properties | | | |
|---|---|---|---|---|---|
| | | Tensile Strength, psi | Yield Strength, psi | Elongation, % | BHN |
| 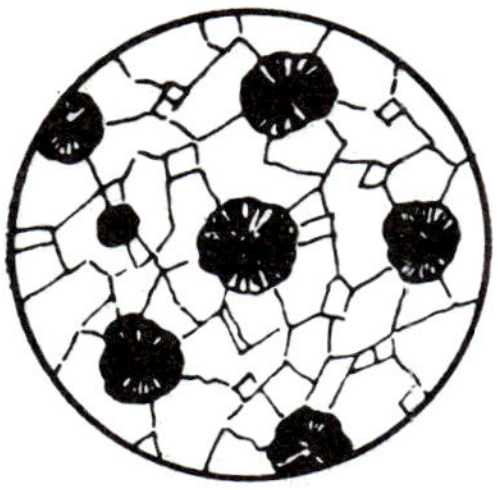 Graphite + ferrite | As cast (or annealed) | 70,000 | 50,000 | 20 | 170 |
|  Graphite + pearlite | Normalized and tempered (or as cast) | 110,000 | 80,000 | 6 | 270 |
| 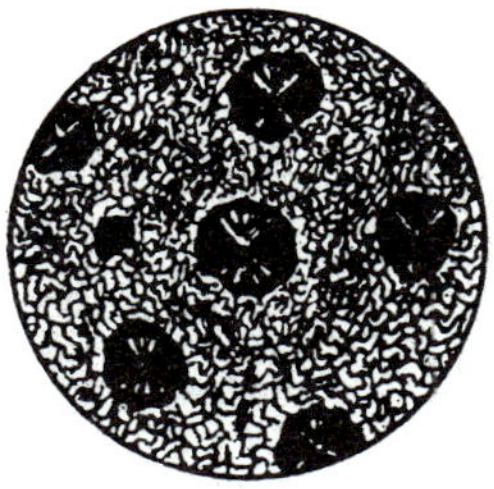 Graphite + tempered martensite | Quenched and tempered | 140,000 | 110,000 | 5 | 310 |

FIGURE 10-9
Structure and properties of ductile cast iron. (*From Taylor, Flemings, and Wulff*.[10])

If the ferritizing anneal is not allowed to go to completion but the casting is cooled rather rapidly at the time sketched in Fig. 10-10, the austenite remaining is transformed to pearlite while rings around each of the graphite particles are ferrite. In the case of ductile iron, this is termed *bullseye iron*, Fig. 10-11. If cooling through the eutectoid temperature region is too rapid, pearlite forms and extremely long times of isothermal holding just under the eutectoid are required to eliminate the $Fe_3C$. The time to achieve a fully ferritic structure is increased by 100 or more. A problem

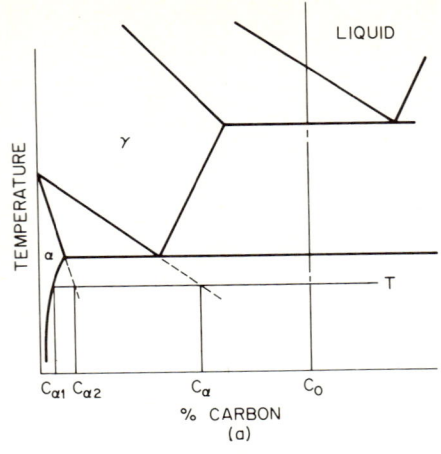

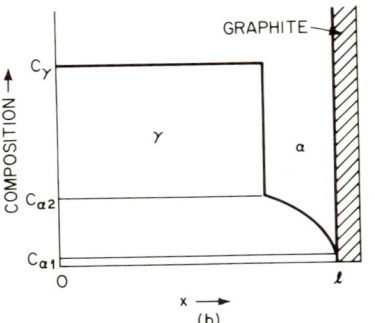

**FIGURE 10-10**
Ferritizing anneal at temperature $T$ of a cast iron of composition $C_0$ which initially has an austenitic matrix. (*a*) Iron-carbon phase diagram. Solid lines are for the stable (Fe-G) equilibrium, and dashed lines are metastable extensions of phase boundaries; (*b*) carbon diffuses from the austenite through the ferrite to the graphite.

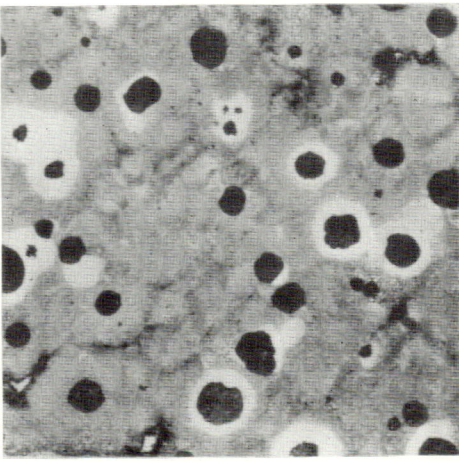

**FIGURE 10-11**
Bullseye ductile iron. (Magnification × 100.) (*From Ref.* 31.)

at the end of this chapter (Prob. 10-6) shows why this is so by deriving an expression for this case similar to those given as Eqs. (10-10) and (10-11). Similar equations can also be derived for the processes of austenitizing a pearlitic and a ferritic iron. Both of these processes, particularly austenitizing a pearlitic iron, take place very much more rapidly than does a ferritizing anneal.

After austenitizing, if a pearlitic structure is desired rather than a ferritic one, the cooling rate is increased so that no ferrite forms. Still more rapid quenching yields a fully austenitic matrix. In general, the matrix of any graphitic cast iron can be brought to and retained at about 0.8 percent carbon. At this carbon level, any of the various structures of carbon steel can be obtained, including pearlite, martensite, and tempered martensite. As in the case of steel, elements such as nickel and chromium can be added to increase *hardenability*. Tempering is as in steel, with the tempering temperature being adjusted to obtain the desired balance between hardness and ductility. If a soft intermediate or final structure is desired, heat treatment can be such that the carbon is rejected from the matrix to the graphite on cooling. The resulting matrix is then very low-carbon iron and is soft and ductile.

## MECHANICAL PROPERTIES OF EQUIAXED CAST STRUCTURES

Room-temperature mechanical properties of cast structures are generally found to increase with decreasing grain size. An example is shown in Fig. 10-12 for Al–4.5% Cu alloy, from the work of Sicha and Boehm.[13] Note that by comparison with many wrought structures, these grain sizes are relatively coarse, the minimum being about 0.01 cm. In the common commercial casting alloys only zirconium-refined magnesium alloys have a finer grain size, as discussed in Chap. 5. In nongrain-refined aluminum, or in ferrous- or copper-base-alloy castings, grain sizes are much larger than this, often being in the neighborhood of 1 cm in diameter.

The effects of grain size on properties appear to result primarily from changes in distribution of porosity, inclusions, and microsegregate. Often, these heterogeneities are particularly severe at grain boundaries, and the preferred fracture paths that are present in coarse-grained structures result in lower mechanical properties. A much more effective way to improve properties in cast metals than by decreasing grain size is to directly reduce the heterogeneities. In combination with heat treatment of the cast structure, an effective way to do this is by *chilling* to promote rapid and directional solidification. As an example, Fig. 10-13 shows properties from a plate casting that was chilled at one end. The properties near the chill are better than those far away because near the chill there is lower microporosity, finer distribution of undissolved second phases, and lower amounts of such phases (after heat treatment).

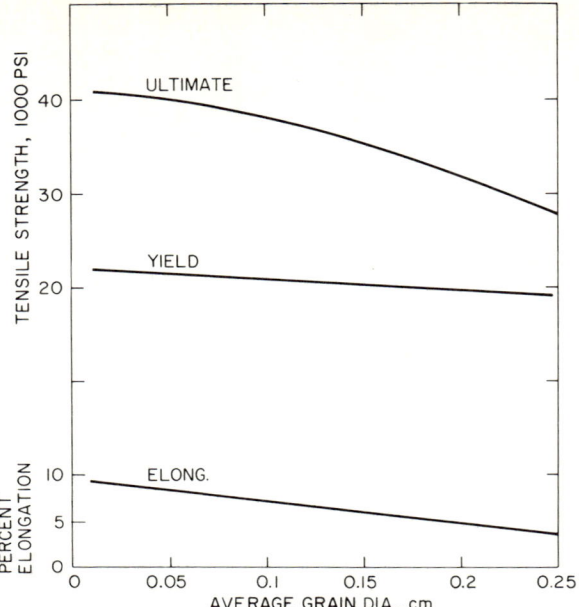

**FIGURE 10-12**
Effect of grain size on mechanical properties of cast Al–4.5% Cu alloy. (*From data of Sicha and Boehm.*[13])

Grain-size differences contribute little, if anything, to the variation in properties seen. The grain size is 0.016 cm near the chill and 0.019 cm at the end opposite the chill.

The usual commercial solution-treatment temperature for this alloy is 515°C. Solution treatment at a higher temperature (543°C) improves the properties of the portion of the plate far removed from the chill but has little influence on those properties of the portion near the chill. This is because dendrite arm spacing near the chill was sufficiently fine so that nearly complete solutionizing was obtained at the lower temperature. Far away from the chill, more intensive solutionizing was necessary to remove the last remaining soluble second phase. Even with complete removal of the second phase, the more slowly solidified material (of coarser dendrite arm spacing) has lower properties than the material of finer spacing. This is presumed to be caused by the coarse distribution of insoluble inclusions such as $Cu_2FeAl_7$ in the material with larger dendrite arm spacing, coarser porosity, and larger amounts of porosity.

In recent years, it has come to be recognized that dendrite arm spacing of cast structures usually correlates far better with mechanical properties than does grain size. Increasing solidification rate always reduces dendrite arm spacing, and a mechanical

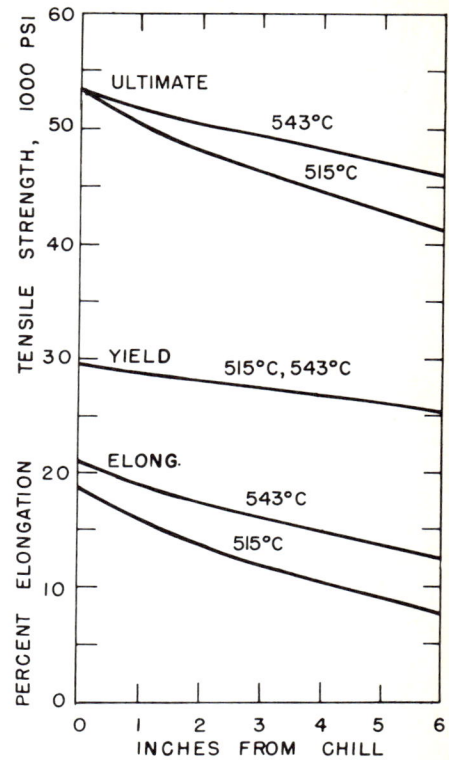

FIGURE 10-13
Mechanical properties in a plate casting of Al–4.5% Cu alloy. Test bars were solution treated at the temperature shown and aged. (*Data from Passmore et al.*[32])

property improvement usually accompanies this reduction. Many examples are to be found in the literature, and the correlation is sufficiently strong so that dendrite-arm-spacing measurements have become an important metallurgical and quality-control tool in foundries producing high-strength castings. Special foundry techniques have been developed for achieving fine dendrite arm spacings in sand and investment castings to obtain special "premium" mechanical properties. These techniques involve the careful placement of chills in the mold to achieve fine dendrite arm spacing while also achieving soundness through directional solidification.

The tensile properties that are influenced to greatest degree in cast alloys by chilling are ductility and tensile strength. Other properties such as fatigue strength are also improved; however, as in the example of Fig. 10-13, yield strength (after heat treatment) is not significantly altered. New higher-strength premium-quality ferrous and nonferrous alloys involve both modified alloy analysis and solidification control, usually chilling. The alloy modification raises yield strength, and chilling provides the alloy with sufficient ductility to be an acceptable engineering material.

The form, distribution, and amount of insoluble inclusions is also of importance in determining properties of cast structures. Rapid solidification results in finer inclusions, but small chemical variations can be of great importance also. The manganese sulfide inclusions in steel discussed in Chap. 6 exhibit very different morphologies, depending on subtle differences in the chemistry of the steel. When inclusions are of the eutectic film type (Type II), properties suffer greatly. In one study of 150 low-alloy steels, only 5 percent of test bars cast with Type II sulfides had impact values over 40 ft-lb. For Types I and III the comparable percentages were 94 and 80, respectively.[14]

In alloys containing a substantial quantity of eutectic, the eutectic structure itself is of importance in determining mechanical properties. Fine eutectic *cell size* in cast iron is desirable, and the spheroidal graphite of ductile iron is much to be preferred over flakes for room-temperature strength and ductility. The nonfaceted *modified* eutectic structure of silicon-rich aluminum–silicon alloys is generally found to produce better properties than the unmodified structure.

## PROPERTIES OF COLUMNAR STRUCTURES

When castings or ingots are unidirectionally solidified without heat input to the liquid, columnar dendritic structures result, as discussed in Chap. 5. Quite remarkable property improvements have been found in a number of alloys solidified in this way. As an example, Fig. 10-14 shows reduction in area of several low-alloy steel castings that were heat-treated to 160,000 psi yield strength. The casting that was unidirectionally solidified shows the highest ductility, and this ductility is independent of test-bar orientation with respect to the columnar structure. Of course, in low-alloy steel, the solid-state transformations result in an actual grain size much smaller than cast columnar grains. The higher properties of the columnar structure result because this structure favorably affects distribution of microsegregation, inclusion size and distribution, and microporosity.

Improved high-temperature homogenization improves the mechanical properties of columnar structures as it does for equiaxed structures. Figure 10-15, from the work of Quigley and Ahearn,[15] shows the increase in ductility of low-alloy steel that results from the high-temperature heat treatment. The improved properties presumably result from a combination of reduced microsegregation and from changes in porosity and inclusions, as discussed earlier.

The improved properties of unidirectionally solidified structures have been exploited in several applications, most notably in turbine blades and buckets of high-temperature alloys by workers at Pratt and Whitney. Directional solidification is used

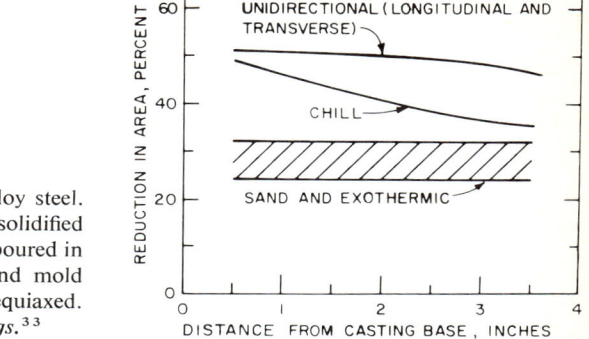

FIGURE 10-14
Reduction in area of a low-alloy steel. Structure of unidirectionally solidified casting was columnar. Those poured in the chill mold, sand mold, and mold exothermic material were equiaxed. (*From data of Polich and Flemings.*[33])

to produce columnar structures such as the two shown in Fig. 5-26. In the second of the two structures, the entire casting is a single-oriented dendrite (*monocrystal*) obtained by a competitive growth process. As shown by Versnyder,[16] these different structures influence significantly the creep and stress-rupture properties of the superalloys. Some comparisons are shown in Fig. 10-16 and Table 10-1 for Mar-M200, a

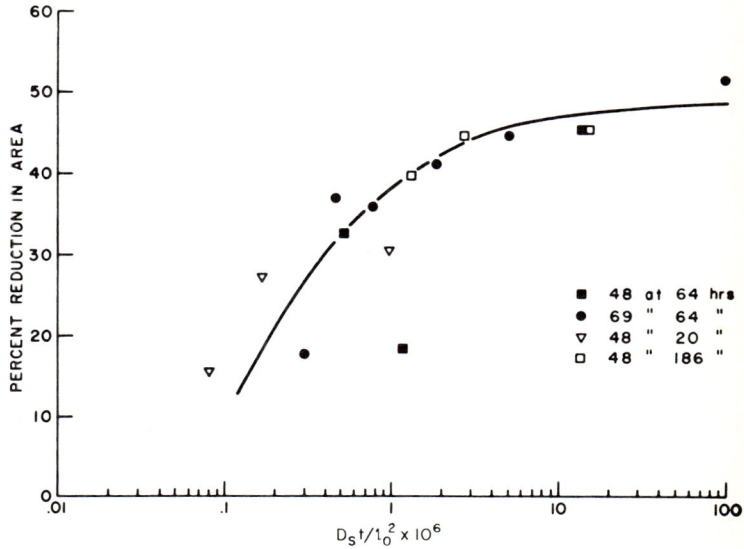

FIGURE 10-15
Reduction in area of a columnar low-alloy cast steel versus dimensionless homogenization time. (*From Quigley and Ahearn.*[15])

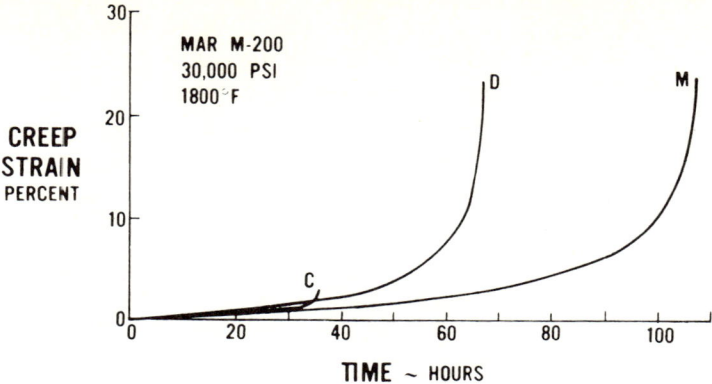

FIGURE 10-16
Comparison of the creep properties at 1800°F of conventional (C), directional (D), and monocrystal (M) Mar-M200. (*From Versnyder and Shank.*[16])

nickel-base superalloy containing Cr, Co, W, and other elements. The higher properties of the columnar and monocrystal material result from a number of factors. Two of these are improved distribution of inclusions and reduced microporosity. Another is the alignment of as-cast grain boundaries with direction of stress (or their elimination in the case of the monocrystal). The preferred crystallographic texture of the columnar material also has an effect. Finally, the distribution of microsegregation in the columnar material is more uniform than that in the conventionally cast material,

Table 10-1  CREEP AND STRESS-RUPTURE PROPERTIES OF CONVENTIONALLY CAST, DIRECTIONALLY SOLIDIFIED, AND MONOCRYSTAL MAR-M200[†]

|  | 1400°F/100 ksi | | | 1800°F/30 ksi | | |
|---|---|---|---|---|---|---|
|  | Rupture life, h | Elongation | Minimum creep rate, in/in/h | Rupture life, h | Elongation | Minimum creep rate, in/in/h |
| Conventionally Cast | 4.9 | 0.45 | $70.0 \times 10^{-5}$ | 35.6 | 2.6 | $23.8 \times 10^{-5}$ |
| Directionally Solidified | 366.0 | 12.6 | $14.5 \times 10^{-5}$ | 67.0 | 23.6 | $25.6 \times 10^{-5}$ |
| Monocrystal | 1914.0 | 14.5 | $2.2 \times 10^{-5}$ | 107.0 | 23.6 | $16.1 \times 10^{-5}$ |

† From F. L. Versnyder.[16]

and this makes it possible to homogenize the columnar material at higher temperature.

Another important application of columnar cast structures is for permanent magnets, primarily of the Alnico type. Magnetic properties depend on crystallographic orientation, and better magnets can be made with oriented columnar structures than with equiaxed structures. Table 10-2 shows how great this improvement is. The *maximum energy product*, an important magnetic measure, is doubled by columnar crystallization of Alnico 8. These magnets develop their full strength by heat treatment in a magnetic field to produce ferromagnetic single-domain precipitates. The strength of the columnar-grain magnets is improved because they have precipitates that are more nearly aligned with the field and more parallel one to another than those in equiaxed structures.[17]

## ALIGNED COMPOSITES

Growth of aligned composites of eutectic and off-eutectic composition has been discussed in Chap. 4. A wide variety of potential applications can be envisioned for these structures, resulting from the fine microstructure and its directionality. Many applications have been proposed in the areas of optics and electromagnetics.[18,19] One eutectic now used commercially is the InSb–NiSb eutectic in magnetoresistance devices. With the aligned eutectic one obtains 60 percent of the theoretical magnetoresistance effect but only 6 percent with the unaligned eutectic. Use of the aligned structure has permitted realization of a number of ideas which have appeared in the

Table 10-2 EFFECT OF CAST STRUCTURE ON PROPERTIES OF TWO ALNICO ALLOYS†

| Alloy | Composition | Coercive force $H_c$, Oe | Residual induction $B_r$, G | Maximum energy product $(BH)_{max}$, kG-Oe |
|---|---|---|---|---|
| Alnico 5 (equiaxed) | 24Co, 14Ni, 8Al, 3Cu, Fe | 630 | 12,250 | 5.0 |
| Alnico 7 (columnar) | 24Co, 14Ni, 8Al, 3Cu, Fe | 760 | 12,800 | 7.3 |
| Alnico 8 (equiaxed) | 35Co, 15Ni, 8Al, 4Cu, 5Ti, Fe | 1,250 | 9,000 | 5.0 |
| Alnico 9 (columnar) | 35Co, 15Ni, 8Al, 4Cu, 5Ti, Fe | 1,500 | 11,000 | 10.0 |

† From Bower, Granger, and Keverian.[17]

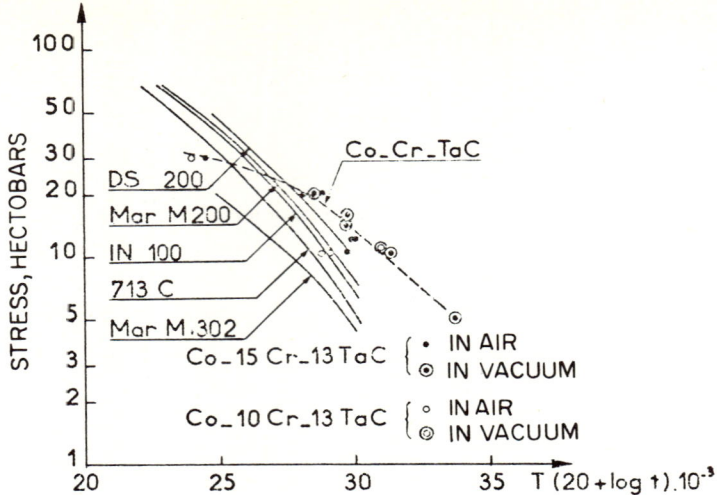

**FIGURE 10-17**
Larson-Miller curves for high-temperature alloys. Stress for rupture is plotted against a function of temperature $T$ (°K) and time $t$ (hours). (*From Bibring et al.*[21])

literature over the last century in the areas of magnetic field measurement, modulation of small dc currents and voltages, and others.

Much attention has been given to the mechanical properties of aligned eutectic or eutecticlike composites. These offer an inherent advantage over other types of aligned composites. Since they are grown in situ, the problem of achieving adequate bonding between the phases does not arise, nor does the problem of achieving low reactivity between the phases. Fiber strengthening at room temperature has been reported for a number of alloys, but the strengthening achieved so far is not sufficient to make such materials competitive with other cast or wrought alloys.

High-temperature properties, however, are another matter. Excellent high-temperature properties have been obtained by Thomson et al.[20] in an aligned composite consisting of carbides of composition $(Cr,Co)_7C_3$ in a matrix of cobalt with chromium in solid solution. Bibring et al.[21] have also obtained attractive properties in a cobalt–chromium alloy strengthened with TaC. Data from this latter investigation are given in Fig. 10-17. At high temperatures and long times these alloys are markedly superior to any existing metallic high-temperature alloys, including the Mar-M200 alloy discussed in the previous section and the same alloy when directionally solidified (DS200). Directionally solidified composites such as this are certain to receive much attention in research and development laboratories.

FIGURE 10-18
Bands in wrought low-alloy steel, Fe–1.5% Cr–1.0% C. (Magnification × 75.)
(*From Flemings et al.*[22])

## EFFECTS OF WORKING

Hot and cold working alters cast structures in a number of important ways. In some materials the original cast grain structure is replaced by a different structure as a result of recrystallization either from working or from solid-state transformation. In others, such as many aluminum alloys, the original grain structure may persist in the wrought structure although a fine subgrain structure may also be present. Modest amounts of hot working close and presumably weld microporosity. Microsegregation is seldom completely eliminated by hot working of commercial alloys. Figure 10-18 shows segregation in a commercial steel after it was reduced by hot working by a factor of nearly 3,000:1; the segregation ratio in this material was reduced by only a factor of 2 as a result of all this work.[22]

When nonequilibrium second phases are present in the cast structure, as in most light alloys, the extent to which working promotes solutionizing depends on how the working influences the second phase. As simple examples, suppose the second phase is present as discrete small particles which are so hard and strong that they neither deform nor break during working. Now, as long as volume diffusion controls the rate of solution treatment, the working will have no effect on this rate since diffusion distances are unchanged; this is shown schematically in Fig. 10-19a. On the other hand, suppose the particles are soft, so that they deform, or brittle, so that they

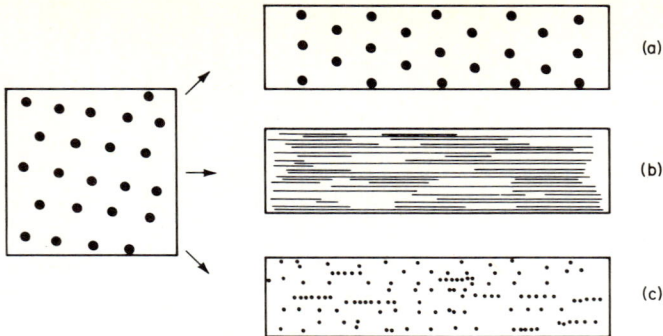

**FIGURE 10-19**
Schematic illustration of effect of working on second-phase particles. (*a*) Particles and particle spacing are not changed by working; (*b*) particles deform plastically; and (*c*) particles fracture.

break during working. Now working alters diffusion distance, as shown in Fig. 10-19*b* and *c*, and so can be expected to alter the rate of solutionizing. The way the included second phase deforms (or fractures) during working depends sensitively on temperature and rolling conditions, and so it is no wonder that conflicting results are sometimes obtained as to the effect of working on homogenization even for the same alloy. Figure 10-20 shows the effect of one working practice on rate of solutionizing of a high-strength Al–Zn–Mg–Cu alloy. Here the second phase deformed and fractured during working, and so the rate of solutionizing depended sensitively on the amount of reduction. A similar working practice on a different alloy (Al–4.5% Cu) had little effect on the rate of solutionizing because the second phase particles were not significantly deformed by the working.[2,24]

Nonmetallic inclusions in steel behave during working as do the nonequilibrium second phases observed above. Softer inclusions deform plastically, whereas harder ones either break up or remain intact during working. Sometimes hard inclusions, such as $Al_2O_3$, separate from a softer phase such as an alumino-silicate phase during working. How an inclusion behaves during working depends on its deformability with respect to the surrounding matrix, and so it is dependent both on deformation rate and temperature as well as inclusion composition. Optimum working practice (with regard to inclusions) would appear to be those practices that break down large inclusions into less harmful dispersions by promoting their fracture.[24-26]

Much can be found in the literature on the effects of ingot structure on the properties of wrought material. Second-phase particles, whether retained nonequilibrium alloy second phases or nonmetallic inclusions, are particularly injurious.

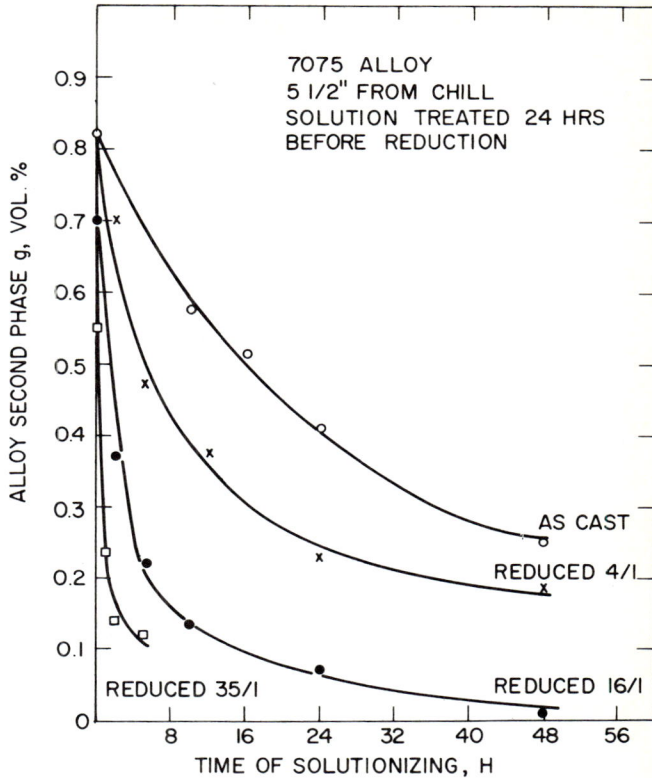

FIGURE 10-20
Undissolved second phase versus solution time for a high-strength aluminum alloy.[22]

Large increases in properties, especially transverse properties, have been obtained by lowering amounts of these second phases through high-temperature-solution treatments and improved alloy purity; examples are the work of Mulherin and Rosenthal[27] and Jatczak et al.[28]

Microsegregation in cast ingots, sometimes compounded by porosity and coarse grain structure, makes initial breakdown of highly alloyed materials very difficult indeed. Because of this, compositions of many high-strength ferrous and non-ferrous alloys represent compromises between those that could give optimum properties when worked and those that can be economically worked. Reduced grain size and dendrite arm spacing promote ingot workability, and so in recent years much attention has been given to control of these structural features in all ingot-making processes. A major advantage of several of the newer processes such as consumable electrode ingot-making and electroslag remelting is that these permit obtaining finer dendrite arm spacing and grain size than the older processes.

One interesting, if still expensive, way to reduce the size and amounts of second phases in wrought alloys is to make these alloys from atomized droplets of liquid metal. The droplets are made by various means, often by striking a stream of liquid metal with one or more gas jets. The droplets so formed solidify rapidly and so usually have a dendrite arm spacing under 10 $\mu$m. They can now be compacted and sintered and subsequently extruded or rolled. The resulting wrought material is free of any macrosegregation; moreover, any microsegregation or included second phase is on a very fine scale. Workability of the compacted billet is often much superior to that of a conventionally cast ingot, and the final structure can be fully homogenized without difficulty. This result has been used in aluminum-base alloys to develop new high-strength alloys. Hardening elements (Zn, Cu, Mg) have been added to aluminum in amounts above the limit at which the material can be worked if processed by usual means. However, when the material is fabricated by the powder-metallurgy route, working is readily accomplished. Ultimate tensile strengths as high as 125,000 psi have been obtained in wrought aluminum by this method (compared with about 95,000 psi for the usual commercial high-strength material).[29,30]

Processing of atomized powders is also being used to produce limited quantities of specialty tool steel where alloy carbides are never dissolved and so are much finer in material produced this way than by usual methods. Other applications are anticipated, particularly in difficult-to-forge high-temperature alloys. One alloy of current industrial interest made in this way is IN 100 (a nickel-base superalloy containing Co, Mo, Ti, Al, and other elements). In the cast form this alloy has an ultimate tensile strength of 140,000 psi, and, with a ductility of 1 percent, it is difficult or impossible to forge. Compacted powders can be worked and have a tensile strength of 220,000 psi with up to 18 percent elongation. This material with its fine dendrite structure and grain size is ideal for use at temperatures below about 1400°F (760°F). For higher-temperature applications, castings, especially columnar castings, are preferred.

# REFERENCES

*1* KATTAMIS, T. Z., and FLEMINGS, M. C.: *Trans. AIME*, **233**:992 (1965).
*2* SINGH, S. N., and FLEMINGS, M. C.: *Trans. AIME*, **245**:1803 (1969).
*3* MURPHY, J. B.: *Acta Met.*, **9**:563 (1961).
*4* SINGH, S. N., BARDES, B. P., and FLEMINGS, M. C.: *Met. Trans.*, **1**:1383 (1970).
*5* PURDY, G. R., and KIRKALDY, J. S.: *Met. Trans.*, **2**:371 (1971).
*6* AARON, H. B., and KOTLER, G. R.: *Met. Trans.*, **2**:393 (1971).

7 GNANAMUTHU, D., Met. E. Thesis, Dept. of Metallurgy and Material Science, Mass. Inst. of Tech., Cambridge, Mass., 1972.
8 BASARAN, M., MEHRABIAN, R., and FLEMINGS, M. C.: (to be published).
9 COBLE, R. C., and FLEMINGS, M. C.: *Trans. Met. Soc.*,
10 TAYLOR, H. F., FLEMINGS, M. C., and WULFF, J.: "Foundry Engineering," John Wiley & Sons, Inc., New York, 1959.
11 "Malleable Iron Castings," Malleable Founders' Society, Cleveland, Ohio, 1960.
12 WALTON, C. F. (ed.): "The Gray Iron Castings Handbook," Gray Iron Foundries Society, Cleveland, Ohio, 1958.
13 SICHA, W. E., and BOEHM, R. C.: *Trans. AFS*, **56**:398 (1948).
14 SIMS, C. E., and BRIGGS, C. W.: *Trans. 25th Intern. Foundry Congr.*, Brussells (1958).
15 QUIGLEY, F. C., and AHEARN, P. J.: *Trans. AFS*, **72**:813 (1964).
16 VERSNYDER, F. L., and SHANK, M. E.: *Mater. Sci. Eng.*, **6**:213 (1970).
17 BOWER, T. F., GRANGER, D. A., and KEVERIAN, J.: in "Solidification," p.385, American Society for Metals, Metals Park, Ohio, 1971.
18 GALASSO, F. S.: *J. Metals*, **19**:17 (June 1967).
19 WEISS, H.: *Met. Trans.*, **2**:1513 (1971).
20 THOMPSON, E. R., KOSS, D. A., and CHESNUTT, J. C.: *Met. Trans.*, **1**:2807 (1970).
21 BIBRING, H., TROTHER, J. P., RABINOVITCH, M., and SEIBEL, G.: *Mem. Sci. Rev. Met.*, **68**:23 (1971).
22 FLEMINGS, M. C., POIRIER, D. R., BARONE, R. V., and BRODY, H. D.: *J. Iron Steel Inst.*, **208**:371 (1970).
23 RETI, A. M., and FLEMINGS, M. C.: *Met. Trans.*, **3**:1869 (1972).
24 KIESSLING, R.: *J. Metals*, **21**:48 (1969).
25 CHAO, H., and VAN VLACK, L. H.: *Trans. AIME*, **233**:1227 (1965).
26 CHAO, H., and VAN VLACK, L. H.: *Trans. ASM*, **58**:335 (1965).
27 MULHERIN, J. H., and ROSENTHAL, H.: *Met. Trans.*, **2**:427 (1971).
28 JATCZAK, C. F., GIRADI, D. J., and ROWLAND, E. S.: *Trans. ASM*, **48**:279 (1956).
29 HAAR, A. P.: Development of Aluminum Base Alloys, pts. 1 to 3, Alcoa Res. Labs., New Kensington, Penna., Repts. to Frankford Arsenal, 1961–1965.
30 BOWER, T. F., SINGH, S. N., and FLEMINGS, M. C.: *Met. Trans.*, **1**:191 (1970).
31 "Typical Microstructures of Cast Metals," 2d ed., Institute of British Foundrymen, London, 1966.
32 PASSMORE, E. M., FLEMINGS, M. C., and TAYLOR, H. F.: *Trans. AFS*, **66**:96 (1958).
33 POLICH, R. F., and FLEMINGS, M. C.: *Trans. AFS*, **73**:28 (1965).

# PROBLEMS

*10-1* A low-alloy steel ingot contains 2.0 percent Ni and 0.4 percent C. It is slowly solidified so that dendrite arm spacing is 500 $\mu$m. How long a time will be required to reduce the coring of carbon by 90 percent? To reduce the coring of nickel by the same amount? Assume sinusoidal variation of composition across dendrite arms of both carbon and nickel. The diffusion coefficient of carbon in austenite at 1200°C is $2.23 \times 10^{-6}$ cm²/s; the diffusion coefficient of nickel is $7.03 \times 10^{-11}$ cm²/s.

**10-2** Using Fig. 5-16a, what is the dendrite arm spacing in the Al–4.5% Cu ingot of Fig. 1-16 at 10 cm from the chill? At 1 cm? What solution-treatment time is required to eliminate all the second phase at these locations at just under the eutectic temperature (545°C)? Would it be practical to solution treat the casting above the eutectic temperature?

**10-3** What would be the required time to eliminate 90 percent of the second phase in an Al–5.0% Cu (Al–2.2 at%) alloy at the solvus temperature 530°C if dendrite arm spacing is 100 μm? The diffusion coefficient of copper in aluminum is given by

$$D_s = 0.29 e^{-(31,120/RT)} \text{ cm}^2/\text{s}$$

where $R = 1.99$ cal/(mol)(°K)

**10-4** Starting with a cast structure of ferritic ductile iron, how would you obtain the following structures?
(a) Graphite spheres embedded in a pearlitic matrix
(b) Graphite spheres surrounded by ferrite and embedded in a pearlitic matrix
(c) Graphite spheres embedded in a tempered martensite matrix
(d) Graphite spheres surrounded by pearlite and embedded in a ferritic matrix
(e) Flake graphite embedded in a pearlite matrix

**10-5** (a) Assuming diffusion control, make a very rough estimate of the time required at just above the eutectoid temperature to malleablize the white cast-iron structure of Fig. 6-8 to obtain a structure of graphite in austenite which would appear like the structure of Fig. 1-7 after ferritizing anneal. Assume the alloy is 2.5 percent carbon and 0 percent Si.
(b) What is done in practice to speed up this malleablization time? Explain the mechanisms.
(c) Why does dissolution of ledeburite in graphitization require much more time than solution of pearlite (at just above the eutectoid)?
*Note:* The diffusion coefficient of carbon in austenite is given by

$$D_\gamma = 0.21 e^{-(33,800/RT)} \text{ cm}^2/\text{s}$$

**10-6** Explain why the times required for second-stage malleablization (at temperatures just below the eutectoid) are greatly increased if cementite forms. Do this by demonstrating that an equation similar to Eq. (10-11) applies but that the driving force is much lower.

**10-7** It is possible to solidify the same alloy with an equiaxed dendritic structure, columnar structure, or with a plane front. Pick an alloy that would be single-phase after equilibrium solidification and suggest applications for each type of structure.

**10-8** Repeat Prob. 10-7 for an alloy that would consist of two or more phases after equilibrium solidification.

*10-9*      The casting for which properties are shown in Fig. 10-13 was a plate ¾ in thick chilled at one end and risered at the other. Suppose you wished to produce this plate with minimum mechanical properties of 50,000 psi tensile strength, 25,000 psi yield strength, and 15 percent elongation. How would you alter the chilling and risering arrangement?

*10-10*    How could you raise the yield strength of the plate casting of Prob. 10-9 if you were willing to sacrifice some ductility?

*10-11*    Derive an expression for effect of working on solution kinetics of a binary alloy by relating amount of reduction to amount of residual second phase. Assume second phase is originally present as thin plates, uniformly spaced, and reduction is uniform throughout the sample.

# APPENDIX A
## TABULATION OF ERROR FUNCTIONS

| N | erf N | N | erf N | N | erf N | N | erf N |
|---|---|---|---|---|---|---|---|
| 0.00 | 0.00000 | 0.26 | 0.2869 | 0.52 | 0.5379 | 0.78 | 0.7300 |
| 0.01 | 0.01128 | 0.27 | 0.2974 | 0.53 | 0.5465 | 0.79 | 0.7361 |
| 0.02 | 0.02256 | 0.28 | 0.3079 | 0.54 | 0.5549 | 0.80 | 0.7421 |
| 0.03 | 0.03384 | 0.29 | 0.3183 | 0.55 | 0.5633 | 0.81 | 0.7480 |
| 0.04 | 0.04511 | 0.30 | 0.3286 | 0.56 | 0.5716 | 0.82 | 0.7538 |
| 0.05 | 0.05637 | 0.31 | 0.3389 | 0.57 | 0.5798 | 0.83 | 0.7595 |
| 0.06 | 0.06762 | 0.32 | 0.3491 | 0.58 | 0.5879 | 0.84 | 0.7651 |
| 0.07 | 0.07886 | 0.33 | 0.3593 | 0.59 | 0.5959 | 0.85 | 0.7707 |
| 0.08 | 0.09008 | 0.34 | 0.3694 | 0.60 | 0.6039 | 0.86 | 0.7761 |
| 0.09 | 0.1013 | 0.35 | 0.3794 | 0.61 | 0.6117 | 0.87 | 0.7814 |
| 0.10 | 0.1125 | 0.36 | 0.3893 | 0.62 | 0.6194 | 0.88 | 0.7867 |
| 0.11 | 0.1236 | 0.37 | 0.3992 | 0.63 | 0.6270 | 0.89 | 0.7918 |
| 0.12 | 0.1348 | 0.38 | 0.4090 | 0.64 | 0.6346 | 0.90 | 0.7969 |
| 0.13 | 0.1459 | 0.39 | 0.4187 | 0.65 | 0.6420 | 0.91 | 0.8019 |
| 0.14 | 0.1569 | 0.40 | 0.4284 | 0.66 | 0.6994 | 0.92 | 0.8068 |
| 0.15 | 0.1680 | 0.41 | 0.4380 | 0.67 | 0.6566 | 0.93 | 0.8116 |
| 0.16 | 0.1790 | 0.42 | 0.4475 | 0.68 | 0.6638 | 0.94 | 0.8163 |
| 0.17 | 0.1900 | 0.43 | 0.4569 | 0.69 | 0.6708 | 0.95 | 0.8209 |
| 0.18 | 0.2009 | 0.44 | 0.4662 | 0.70 | 0.6778 | 0.96 | 0.8254 |
| 0.19 | 0.2118 | 0.45 | 0.4755 | 0.71 | 0.6847 | 0.97 | 0.8299 |
| 0.20 | 0.2227 | 0.46 | 0.4847 | 0.72 | 0.6914 | 0.98 | 0.8342 |
| 0.21 | 0.2335 | 0.47 | 0.4937 | 0.73 | 0.6981 | 0.99 | 0.8385 |
| 0.22 | 0.2443 | 0.48 | 0.5027 | 0.74 | 0.7047 | 1.00 | 0.8427 |
| 0.23 | 0.2550 | 0.49 | 0.5117 | 0.75 | 0.7112 | | |
| 0.24 | 0.2657 | 0.50 | 0.5205 | 0.76 | 0.7175 | | |
| 0.25 | 0.2763 | 0.51 | 0.5292 | 0.77 | 0.7238 | | |

| N | 1−erf N | N | 1−erf N |
|---|---|---|---|
| 1.0 | 0.1573 | 1.5 | 0.03389 |
| 1.1 | 0.1198 | 1.6 | 0.02365 |
| 1.2 | 0.08969 | 1.7 | 0.01621 |
| 1.3 | 0.06599 | 1.8 | 0.01091 |
| 1.4 | 0.04771 | 1.9 | 0.007210 |
| | | 2.0 | 0.004678 |

Notes:

$\text{erf } N = (2/\sqrt{\pi}) \int_0^N e^{-u^2} du$.

When $N < 0.2$, $\text{erf } N \simeq (2N/\sqrt{\pi})$.
When $N > 2.0$, $1 - \text{erf } N \simeq (e^{-N^2}/\sqrt{\pi})$.

# APPENDIX B
## TABLES OF APPROXIMATE THERMAL DATA

### MOLD AND METAL CONSTANTS

| Material | Specific heat, cal/(g)(°C) | Density, (g)/cm³ | Thermal conductivity, cal/(cm)(°C)(s) |
|---|---|---|---|
| Sand | 0.27 | 1.5 | $14.5 \times 10^{-4}$ |
| Plaster | 0.20 | 1.1 | $8.3 \times 10^{-4}$ |
| Mullite investment | 0.18 | 1.6 | $9.1 \times 10^{-4}$ |
| Iron | 0.16 | 7.3 | 0.07 |
| Aluminum | 0.20 | 2.7 | 0.53 |
| Copper | 0.09 | 9.0 | 0.94 |
| Magnesium | 0.25 | 1.7 | 0.38 |

### LIQUID METAL CONSTANTS

| Metal | Melting point, °C | Heat of fusion, cal/g | Specific heat, cal/(g)(°C) |
|---|---|---|---|
| Iron | 1540 | 65 | 0.18 |
| Aluminum | 660 | 95 | 0.26 |
| Copper | 1083 | 51 | 0.12 |
| Magnesium | 650 | 89 | 0.32 |

### SOLIDIFICATION SHRINKAGE

| Metal | Shrinkage, % |
|---|---|
| Iron | 4.0 |
| Aluminum | 6.6 |
| Copper | 4.9 |
| Magnesium | 4.2 |

# INDEX

Atomized powders, 16
  heat flow in, 166
  processing and properties of, 352

Cast irons:
  eutectic solidification in, 187–188
  heat treatment of, 338–341
  structure of, 182–185
    ductile, 186–187, 338–339
    gray, 184
    malleable, 335
    white, 183
Casting, heat flow in, 4–16
Casting processes:
  atomization, 16, 166
  die casting, 14
  fluidity in, 219–224

Casting processes:
  gating in, 215–218
  heat flow in, 4–16
  investment casting, 6, 7
  lost wax casting, 7
  permanent mold casting, 14
  risering in, 229–234, 239–244
  sand casting, 6, 164
  splat cooling, 16
Cells, 66–73
  elongated, 67
  faceted, 73
  hexagonal, 67
  two-phase, 126, 127
Cellular-dendritic transition, 75, 76
Cellular solidification:
  cell spacing, 83–84
  solid-state diffusion in, 85, 86

Cellular solidification:
  solute redistribution in, 76–78
Chemical potential:
  change on solidification, 280–283
  of elements in binary alloys, 267–268
Chvorinov's rule, 11
Coarsening:
  critical time for, 153
  of dendrites, 148–154, 172
Columnar structures:
  formation of, 134–141
  magnetic properties of, 347
  mechanical properties of, 344–347
Composites, directionally solidified:
  constitutional supercooling analysis, 107–109, 123, 126, 130
  convection in, 128–130
  perturbation analysis for interface stability, 112
  properties of, 347–348
  solute redistribution in, 107–109, 115, 116, 128–130
  ternary alloys, 121–126
Constitutional supercooling theory:
  binary alloys, 58–64
  composites, two-phase, 107, 108
  ternary alloys, 87–90
Convection in bulk liquids:
  in casting processes, 19, 21, 224–229
  in crystal growth, 41–44, 53–54
  effect on grain size, 154–155, 228–229
  effect of magnetic field, 54, 227
  influence on superheat, 228
Crystal growing, heat flow in, 3–5
Crystal growing processes:
  Bridgeman method, 1, 50
  crystal pulling, 3, 32
  normal freezing (normal solidification) 1, 32
  Verneuil method, 3
  zone melting (zone solidification), 2, 32
Curvature of liquid-solid interface:
  effect on partition ratio, 273
  at grain boundary grooves, 286
  undercooling due to, 94, 152, 169, 266, 273, 274

Dendrite arm spacing, 146–154
  effect on mechanical properties, 342–343
  influence of grain size, 157
  mechanism of establishment: primary arms, 147, 148
    secondary arms, 148–154
Dendrite directions:
  effect of convection, 158
  in faceted materials, 156–158
  influence of overlapping diffusion fields, 158
  in nonfaceted materials, 70, 137, 159
  preferred orientation (textures), 158, 160
  surface orientations, 158, 160
Dendrite remelting, 148, 152, 154, 172
Dendrites:
  morphology, 73, 74, 135–141, 322
  primary arms, 74, 137–141
  secondary arms, 74, 137–141
  two-phase, 127

Equilibrium:
  lever rule, 34
  at liquid-solid interface, 31, 277, 279, 283
  shapes of phases, 284–286
  solidification, 33
Eutectic, interdendritic, 110, 142
  composition of, 110
Eutectic solidification:
  in castings and ingots, 180–182
    Al-Si alloy, 181
    cast irons, 187–188
    influence of $G/R$, 180
  faceted-faceted, 106
  faceted-nonfaceted, 105, 106
  interface stability (*see* Composites, directionally solidified)
  lamellar: "extremum" growth of, 100
    faults, 103–104
    interface shape in, 98
    lamellar spacing in, 102–104
    solute redistribution in, 96–97
    terminations, 103
    undercooling in, 94–95

Eutectic solidification:
  rod, 104
  ternary alloys, 123
Eutectics:
  crystallography, 114
  grains, 114
  influence on properties, 344
  ternary, 123
Evaporation coefficient, 49, 50
Exudation, 22

Facets, 51, 52, 319–324
Fluid flow, interdendritic, 234–238
  (*See also* Convection in bulk liquids;
    Fluidity; Gating)
Fluidity, 219–224
  of alloys, 221–224
  analysis of, 219–220
  definition, 219
Fraction solid (in castings and ingots),
    160–166
  distribution during solidification,
    164–166
  effect of solid diffusion, 161
  relation to temperature, 161
Free energy:
  change on solidification, 264
  molar, 267, 268
  partial molar, 267
  and phase diagram construction,
    268–270

Gas, dissolved: removal, 206–208
    by inert gas flush, 208
    by vacuum, 207
  solubility, 202–206
Gas porosity, 208–210
  condition for formation, 209
  in rimming steels, 210
Gating, 215–217
  gating ratio, 216–217
  heat loss in, 217–218
  sprue design, 215–216
Grain multiplication, 154, 172
  (*See also* Dendrite remelting)

Grain refinement:
  by addition of inoculant, 298–300
  by convection, 154–155, 228–229
  by vibration, 300
Grain size:
  effect on mechanical properties, 341
  effect on workability, 351
Grain structure of castings and ingots:
  chill, 124, 155
  columnar, 134–141
  effect of convection on, 154–155,
    228–229
  equiaxed, 134, 154–156
  faceted dendrites, 157–160
  nondendritic grains, 156
Growth kinetics:
  continuous, 305–307
  lateral growth, 308
  relation to growth morphologies,
    319–324
  screw dislocation, 312–318
  twin plane, 318–319
  two-dimensional nucleation, 309–312

Heat of fusion, 264
Heat diffusivity, 10
Homogenization, 328–331
  of low alloy steel, 331
  simple model for, 329–330
Hot tearing, 253–256

Inclusion solidification:
  clustering, 197
  coalescence, 196
  collisions in, 194–196
  flotation of, 194–196
  primary, 193–200
  pushing of, 198–200
  secondary, 200–202
  trapping, 200
Inclusions:
  alteration by heat treatment, 334
  effect on mechanical properties, 344
  effect of working on, 350

Ingot casting, heat flow in: multidimensional heat flow, 24
   unidirectional solidification of alloys, 21–23
   unidirectional solidification of pure metals, 16–20
Ingot casting process, 25, 26, 167
Interface, liquid-solid: character of, 301–304
   composition at, 31, 277, 279–283
   temperature at, 279–283
Interface resistance (to heat flow), 13–16, 20–22
Interface stability (see Composites, directionally solidified)

Local solidification time, 22, 145, 154
Local solute redistribution equation:
   in cellular solidification, 81
   in dendritic solidification, 142, 145, 189, 246, 247
   (See also Scheil equation)

Macrosegregation:
   analysis of, 246–249
   channel segregates, 244, 249–250
   in continuous casting, 252
   description of, 244–245
   exudation, 252
   inverse segregation, 247–248
Malleable iron, 335–338
   heat treatment, 338–341
   kinetics of graphitization, 336–338
   types, 335
Metastable phases:
   phase diagrams for, 277
   phase equilibrium, 275–278
   in single component materials, 264–265
Microsegregation:
   effect on workability, 351
   formation of (see Solute redistribution, in dendritic growth)
   index of residual, 328
   removal by homogenization, 328–331

Monotectic solidification:
   in castings and ingots, 191–192
   of Fe–O–S alloys, 200–201
   plane front, 117
Mushy zone, 22, 26, 163

Nonequilibrium lever rule, 34–36
   (See also Scheil equation)
Nucleation:
   of gas porosity, 207
   heterogeneous: analysis of, 295–297
      effect of vibration on, 300
      grain refinement, 298–300
   homogeneous, 290–295
      analysis of, 290–292
      experiments on, 293–294
   two-dimensional, 309–312

Particle entrapment, 117–119
Particle pushing, 117–119
Partition ratio:
   average, 126
   effect of interface curvature, 273
   effect of pressure, 274
   effective, 41
   equilibrium, 32
   relation to thermodynamic quantities, 272
Peritectic solidification:
   in castings and ingots, 177–180
   plane front, 117, 178
Perturbations to growth, 53, 54
Plane front solidification, 31
Porosity:
   alteration by heat treatment, 334
   character of, 237–239, 241
   effect of working, 349–350
   gas caused, 208–210, 237–239
   shrinkage caused, 235–239
Pressure, effect of: on liquidus slope, 275
   on melting point, 172, 265–266, 274–275
   on nucleation, 300
   on partition ratio, 274

INDEX 363

Risering:
  alloys, 239–243
    cast irons, 242–243
    feeding distance, 241–242
    mass feeding, 239
  pure materials, 229–234
    analysis for, 229–230
    atmospheric pressure, 232
    improving efficiency of, 231–232
    placement, 232

Scheil equation:
  as applied to dendritic solidification, 142, 145
  as applied to ternary alloys, 189
  in crystal growth, 34–35
  as modified by diffusion in the solid, 51
  as modified by interdendritic fluid flow, 246, 247
  as modified by volatilization, 50
Seed crystal, 1, 53
Segregation, inverse, 22
Single crystal, perfection of, 52, 53
Solute boundary layers:
  with convection, 41, 45
  with no convection, 36, 87
  without interfacial equilibrium, 279–280
Solute redistribution:
  in cellular solidification, 76–78
  in crystal growth of polyphase alloys:
    with convection, 127–130
    with no convection, 107–109
    nonsteady state, 115–117
    ternary alloys, 120–122
  in dendritic growth: effect of coarsening, 146
  in peritectic reaction, 177–178
  Scheil equation for, 142
  solid diffusion, 144–145
  of ternary alloys, 188–191
    single phase, 189
    solidification path, 189
    two-phase, 189–191

Solute redistribution:
  in single crystal growth: banding, 38, 43, 52
  with convection, 40–44
  with no convection, 36–40
  in Czochralski growth (crystal pulling), 44–46
  facet effect, 50, 51
  final transient, 38
  of initial transient, 37, 43
  solid diffusion, 50
  with volatile constituents, 49–50
  in zone melting, 46–48
  thermodynamic requirements, 279–283
Solute trapping, 281–282
Solution treatment, 331–334
  effect on inclusions, 334
  effect on porosity, 334
  simple model for, 331–332
Supercooling:
  bulk, 167
  constitutional, 58–64, 87–90
  thermal, 66, 167
  (*See also* Undercooling)
Superheat, 19, 21
  effect of convection on, 228
Surface energy, liquid-solid: anisotropy of, 286–287
  contribution to free energy change on solidification, 266

Texture, 137
Thermal diffusivity, 9, 12
Thermal stresses, 54

Undercooled melts, solidification of, 167–172
  dendrite structure, 172
  dendrite tip temperature, 169–172
  growth velocity, 167
  metastable phases, 264–265
  nucleation in, 172

Undercooling, 52
    of liquid-solid interface: due to curvature, 94, 152, 169, 266, 273–274
    due to interface kinetics, 52, 168–172, 305–324
    due to solute buildup, 95
    (*See also* Supercooling)

Viscosity:
    effect of fluidity, 222

Viscosity:
    of liquid-solid mixtures, 256–258

Working, effect on cast structures, 349–352

Zone leveling, 48
Zone refining (zone purification), 31, 46–48